MOVING TOWARD NONANIMAL APPROACHES IN MEDICAL RESEARCH AND TESTING

MOVING TOWARD NONANIMAL APPROACHES IN MEDICAL RESEARCH AND TESTING

Edited by

Catharine E. Krebs
Physicians Committee for Responsible Medicine, Washington, DC, United States

Gail A. Van Norman
Department of Anesthesiology and Pain Medicine, University of Washington, Seattle, WA, United States

ELSEVIER

ACADEMIC PRESS
An imprint of Elsevier

Academic Press is an imprint of Elsevier
125 London Wall, London EC2Y 5AS, United Kingdom
50 Hampshire Street, 5th Floor, Cambridge, MA 02139, United States

ISBN: 978-0-443-26514-3

For information on all Academic Press publications visit our website at https://www.elsevier.com/books-and-journals

Publisher: Andre Wolff
Acquisitions Editor: Wendi Baker
Editorial Project Manager: Prachi Mittal
Production Project Manager: Fahmida Sultana
Cover Designer: Mark Rogers

Typeset by Integra Software Services

Catharine E. Krebs and Gail A. Van Norman dedicate this book with deep gratitude to Bertie Krebs, who taught Catharine how to care, how to love, and how to grieve, and Wilma Jane Van Norman, who loved all creatures.

Contents

Part II

Advances in nonanimal methods for research and testing

6 Microphysiological systems

Konrad Schmidt, Alessandro Bentivogli, and Sarah Hedtrich

7 The application of generative adversarial networks (GANs) as new approach methods (NAMs) toward nonanimal testing

Xi Chen, Ting Li, Weida Tong, and Minjun Chen

8 3D bioprinting as a promising alternative to animal models: Technologies and applications

Ahmed S.M. Ali, Dongwei Wu, and Jens Kurreck

9 The role of artificial intelligence in overcoming animal testing

Thomas Hartung, Thomas Luechtefeld, and Alexandra Maertens

10 Biobanks: Yielding extraordinary returns for human health and disease

Victoria Corbit, Azra Hasan, Thomas J. Bell, and Melissa W. VonDran

11 Emphasizing ancestry in human-based biomedical research

Erika Moore, Alison Veintimilla, Nana Owusu-Boaitey, and Connie J. Mulligan

Part III

Achieving a practical shift toward nonanimal-based research and testing

12 Overcoming inertia against nonanimal research and testing in the United States

Catharine E. Krebs, Emily Anderson, Janine McCarthy, Eryn Slankster-Schmierer, Jarrod Bailey, and Elizabeth Baker

13 Overcoming inertia against nonanimal approaches around the world

Mohammad Abdulkader Akbarsha, Lauren Dalat de Sousa Coelho, Abhay Deshpande, Yogendra Kumar Gupta, Yoko Hirabayashi, Hajime Kojima, Francesca Pistollato, Marcella Miranda Siqueira Furtuoso Rodrigues, Jordana Andrade Santos, and Marize Campos Valadares

14 Perspectives from pharmaceutical industry regarding nonanimal-based research and testing

Sandra P. Smieszek, Christina Tyner, and Mihael H. Polymeropoulos

15 Overcoming animal methods bias and its impacts on nonanimal research assessments

Emily R. Trunnell and Catharine E. Krebs

16 Unlocking funding for human-relevant biomedical research: Navigating the landscape of nonanimal approaches

Angela N. Hvitved and Francesca Pistollato

17 Building confidence in new approach methodologies through reporting, validation, and benchmarking

Kristifor Sunderic

18 Phasing out animal tests: An analysis of the successful transition from rabbits to human-relevant methods in Europe

Justine C. Watkins, Thomas Hartung, Ingrid Visseren-Hamakers, Dilyana Filipova, Howard White, Luísa Ferreira Bastos, Love Hansell, and Merel Ritskes-Hoitinga

19 Safety within REACH? a transition analysis for the European Commission roadmap to animal-free safety assessment of industrial chemicals

Love Hansell, Luísa Ferreira Bastos, Kelsey Beltz, Merel Ritskes-Hoitinga, Thomas Hartung, Julia Pochat, Anne Van Veen, I.J. Visseren-Hamakers, Justine Watkins and Howard White

Contributors

Mohammad Abdulkader Akbarsha Jamal Mohamed College (Autonomous); Mahatma Gandhi-Doerenkamp Center; Indian Society for Alternatives to Animal Experiments, Bharathidasan University, Tiruchirappalli, India

Ahmed S.M. Ali Applied Biochemistry, Berlin University of Technology, Berlin, Germany

Emily Anderson Physicians Committee for Responsible Medicine, Washington, DC, United States

Jarrod Bailey Physicians Committee for Responsible Medicine, Washington, DC, United States

Elizabeth Baker Physicians Committee for Responsible Medicine, Washington, DC, United States

Michael Balls University of Nottingham, Queen's Medical Center, Nottingham, United Kingdom

Luísa Ferreira Bastos Eurogroup for Animals, Brussels, Belgium

Thomas J. Bell National Disease Research Interchange (NDRI), Philadelphia, PA, United States

Kelsey Beltz The Good Lobby, Brussels, Belgium

Alessandro Bentivogli Berlin Institute of Health at Charité—Universitätsmedizin Berlin, Berlin, Germany

Minjun Chen National Center for Toxicological Research, U.S. Food and Drug Administration, Jefferson, AR, United States

Xi Chen National Center for Toxicological Research, U.S. Food and Drug Administration, Jefferson, AR, United States

Lauren Dalat de Sousa Coelho Federal University of Goiás, Goiânia, Brazil

Victoria Corbit National Disease Research Interchange (NDRI), Philadelphia, PA, United States

Abhay Deshpande Indian Society for Alternatives to Animal Experiments, Bharathidasan University, Tiruchirappalli; Jai Research Foundation, Vapi, Gujarat, India

Arianna Ferrari Center for Innovation Systems & Policy, AIT Austrian Institute of Technology GmbH, Vienna, Austria

Dilyana Filipova Doctors Against Animal Experiments, Cologne, Germany

Yogendra Kumar Gupta Indian Society for Alternatives to Animal Experiments, Bharathidasan University, Tiruchirappalli; National Poison Information Centre, All India Institute of Medical Sciences, New Delhi; National Scientific Coordinator, Pharmacovigilance Programme of India (PVPI), Indian Pharmacopoiea Commission, Ghasiabad, India

Love Hansell Institute for Management Research, Radboud University, Nijmegen, Netherlands

Thomas Hartung Johns Hopkins University, Baltimore, MD; Center for Alternatives to Animal Testing (CAAT), Bloomberg School of Public Health, Johns Hopkins University; Johns Hopkins University; Doerenkamp-Zbinden-Chair for Evidence-Based Toxicology, Baltimore, MD, United States; CAAT-Europe, University of Konstanz, Konstanz, Germany

Azra Hasan National Disease Research Interchange (NDRI), Philadelphia, PA, United States

Sarah Hedtrich School of Biomedical Engineering, The University of British Columbia; Centre for Blood Research & Life Sciences Institute, University of British Columbia, Life Sciences Centre, Vancouver, BC, Canada

Kathrin Herrmann Center for Alternatives to Animal Testing (CAAT), Johns Hopkins

Bloomberg School of Public Health, Baltimore, MD, United States

Yoko Hirabayashi Federal University of Goiás, Goiânia, Brazil; Center for Biological Safety and Research, National Institute of Health Sciences, Kawasaki, Japan

Angela N. Hvitved Alternatives Research & Development Foundation (ARDF), Jenkintown, PA, United States

Hajime Kojima Federal University of Goiás, Goiânia, Brazil; Center for Biological Safety and Research, National Institute of Health Sciences, Kawasaki; Sanyo-Onoda City University, Sanyo-Onoda, Japan

Catharine E. Krebs Physicians Committee for Responsible Medicine, Washington, DC, United States

Jens Kurreck Applied Biochemistry, Berlin University of Technology, Berlin, Germany

Ting Li National Center for Toxicological Research, U.S. Food and Drug Administration, Jefferson, AR, United States

Thomas Luechtefeld Center for Alternatives to Animal Testing (CAAT), Bloomberg School of Public Health, Johns Hopkins University, Baltimore; Insilica LLC and ToxTrack Inc., Bethesda, MD, United States

Alexandra Maertens Center for Alternatives to Animal Testing (CAAT), Bloomberg School of Public Health, Johns Hopkins University, Baltimore, MD, United States

Janine McCarthy Physicians Committee for Responsible Medicine, Washington, DC, United States

Erika Moore Fischell Department of Bioengineering, University of Maryland, College Park, MD, United States

Connie J. Mulligan Department of Anthropology, Genetics Institute, University of Florida, Gainesville, FL, United States

Nana Owusu-Boaitey Fischell Department of Bioengineering, University of Maryland, College Park, MD, United States

Pandora Pound Safer Medicines Trust, Kingsbridge, United Kingdom

Nirva Kapasi Patel Brooks McCormick Jr. Animal Law and Policy Program, Harvard Law School, Cambridge, MA, United States

Francesca Pistollato Research and Toxicology, Humane World for Animals, Brussels, Belgium

Julia Pochat Eurogroup for Animals, Brussels, Belgium

Mihael H. Polymeropoulos Vanda Pharmaceuticals, Inc., Washington, DC, United States

Merel Ritskes-Hoitinga Faculty of Veterinary Medicine, Institute for Risk Assessment Sciences (IRAS), Utrecht University, Utrecht, Netherlands; Department of Clinical Medicine, Aarhus University, Aarhus, Denmark

Marcella Miranda Siqueira Furtuoso Rodrigues Federal University of Goiás, Goiânia, Brazil

Jordana Andrade Santos Federal University of Goiás, Goiânia, Brazil

Konrad Schmidt Berlin Institute of Health at Charité—Universitätsmedizin Berlin, Berlin, Germany

Eryn Slankster-Schmierer Physicians Committee for Responsible Medicine, Washington, DC, United States

Sandra P. Smieszek Vanda Pharmaceuticals, Inc., Washington, DC, United States

Kristifor Sunderic National Center for Advancing Translational Sciences, National Institutes of Health, Rockville, MD, United States

Weida Tong National Center for Toxicological Research, U.S. Food and Drug Administration, Jefferson, AR, United States

Emily R. Trunnell People for the Ethical Treatment of Animals, Washington, D.C., United States

Christina Tyner Vanda Pharmaceuticals, Inc., Washington, DC, United States

Marize Campos Valadares Federal University of Goiás, Goiânia, Brazil

Anne Van Veen Faculty of Veterinary Medicine, Institute for Risk Assessment Sciences (IRAS), Utrecht University, Utrecht, Netherlands

Alison Veintimilla Fischell Department of Bioengineering, University of Maryland, College Park, MD, United States

I.J. Visseren-Hamakers Institute for Management Research, Radboud University, Nijmegen, Netherlands
Ingrid Visseren-Hamakers Radboud University, Nijmegen, Netherlands
Melissa W. VonDran National Disease Research Interchange (NDRI), Philadelphia, PA, United States
Justine Watkins Faculty of Veterinary Medicine, Institute for Risk Assessment Sciences (IRAS), Utrecht University, Utrecht, Netherlands
Justine C. Watkins Utrecht University, Utrecht, Netherlands
Howard White Research and Evaluation Centre, London, United Kingdom
Dongwei Wu Applied Biochemistry, Berlin University of Technology, Berlin, Germany

About the editors

Catharine E. Krebs, PhD, is a medical research program manager at the Physicians Committee for Responsible Medicine, where she engages with US policymakers to advance human-centered research and reduce animal use. She has authored numerous congressional statements and public comments to federal agencies. She is a Fellow of the Oxford Centre for Animal Ethics and leads the Coalition to Illuminate and Address Animal Methods Bias. An invited speaker at national and international meetings, she regularly presents on research policy and advancing nonanimal scientific approaches. She earned her PhD in human genetics from the University of California, Los Angeles.

Gail A. Van Norman is a Professor Emeritus of the Department of Anesthesiology and Pain Medicine at the University of Washington in Seattle. She is an academic clinician/ethicist who has spent over 35 years in clinical practice and published a number of articles regarding the ethics and background of animal research, as well as regulatory issues in pharmaceutical research and the failure of animal studies to provide effective scientific translation to human medicine. Her reviews of animal and nonanimal research published in *JACC Basic to Translational Science* have been among the most highly read, downloaded, and cited articles throughout the journal's history. She has been an invited speaker at national and international meetings regarding the issues of animal research and translational science failures.

Preface

"Ask the experiments why they experiment on animals, and the answer is: 'Because the animals are like us.' Ask the experimenters why it is morally okay to experiment on animals, and the answer is: 'Because the animals are not like us.' Animal experimentation rests on a logical contradiction."

Philosopher Charles Magel

Throughout our search for understanding and treating human disease, our relationship with nonhuman animals has been fraught with both moral and scientific controversy. Animal experiments have formed the bedrock of efforts to model human disease biology, develop therapeutically efficacious treatments for human disease, and predict human toxicity of chemicals and drugs. But it is now clear that the deeply rooted premise upon which animal experimentation is based—that animal-based methods accurately predict human biology, toxicity, and clinical effects—is a false one. Actually, animals are not sufficiently "like us" after all. Greater than 90% of drugs fail clinical trials after passing preclinical animal testing, at great cost in dollars invested, in time lost on failed experiments, and in the lives and suffering of millions of animals and humans.

Although experimenting on animals is probably as old as science itself, reliance on animal experiments for approval of human medical therapeutics is relatively recent: in the US, animal experiments were not required before marketing human drugs until after passage of the Federal Food, Drug and Cosmetics Act in 1938. And it was only after the trials at Nuremberg following World War II in the mid-1940s that research standards came to require that human experiments be supported by results from animal experiments.

Specific concerns about the welfare of animals used in medical experimentation were raised over 150 years ago but were primarily based in moral arguments about whether humans have obligations to prevent animal suffering. Few questioned the scientific logic of using animals in medical experiments. Over the last 20 years, however, scientists have been forced to examine an inescapable reality: that animal experiments simply do not translate well to humans. During that same period, nonanimal research methods, including new approach methodologies (NAMs), underwent eye opening growth and development, and are today proving in many cases to be less time consuming, less expensive, and more accurate in modeling human biology and predicting results for human therapeutics than animal experiments.

While science has been sprinting ahead, legislative, regulatory, and funding changes permitting and promoting the use of effective nonanimal methods over animal experimentation have been less nimble—in many cases clinging to outdated reliance on animal studies, and (ironically) requiring the often less reliable experiments in animals to validate the emerging methods. Only recently have legislative and regulatory developments begun to break through obstructions to the adoption of nonanimal methods, and these policy advancements have been supported by government officials on both sides of the aisle. For instance, conservative libertarian Senator Rand Paul and liberal democrat Senator Cory Booker came together to lead the FDA Modernization Act 2.0, which incorporates nonanimal methods into statutory requirements for drug approvals. And following years of effort under several presidential administrations, the National Institutes of Health (NIH) has moved to prioritize human-based research technologies while reducing animal use, including through the creation of a new office to coordinate agency-wide efforts and funds for a new Standardized Organoid Modeling Center.

In this complex and rapidly evolving landscape, it is hard to know where one should start in order to gain basic knowledge of the ethical, scientific, historical, regulatory, and funding aspects of the shift to NAMs and other nonanimal approaches. A number of books focus on NAMs for the researcher, or else on ethical controversies of animal research, but none has recently provided an overview that is suitable for someone who is beginning to explore the multiple aspects of the shift towards nonanimal methods and who needs a more complete picture. It is our intention that this text will provide a starting point for almost anyone who is beginning their journey on this exciting road in science and medicine—whether they be early-career researchers, academic scholars, journalists, advocates for reduction of animal use in medical research, or regulators and legislators involved in the acceptance and regulation of NAMs.

To this end, we have brought together a group of internationally renowned ethicists, legal experts, biomedical researchers and NAMs innovators to address aspects of transitioning toward nonanimal methods in three parts:

Part I: The Need to Move Away from Animal-Based Research and Testing is comprised of five chapters that address why it is both ethically and scientifically imperative that we move away from animal studies in the modeling of human diseases, the development of human therapeutics, and the safety testing of chemicals and other products. These draw on the experience and knowledge of renowned authors Ariana Ferrari (Austrian Institute of Technology), Nirva Patel (the Brooks McCormick Jr. Animal and Law Policy Program at Harvard Law School); Pandora Pound (Safer Medicines Trust); Kathrin Hermann (Johns Hopkins Center for Alternatives to Animal Testing); Jarrod Bailey (Physicians Committee for Responsible Medicine); and Michael Balls (University of Nottingham).

Part II: Advances in Nonanimal Methods for Research and Testing includes six chapters exploring perspectives from current researchers and scholars on the forefront of NAMs development, including Sarah Hedtrich and colleagues (The University of British Columbia); Minjun Chen, Weida Tong, and colleagues (US Food and Drug Administration); Jens Kurreck and colleagues (Technische Universität Berlin); Thomas Hartung (Johns Hopkins Center for Alternatives to Animal Testing) and colleagues; Melissa W. VonDran and colleagues (National Disease Research Interchange); and Erika Moore and colleagues (University of Maryland). Each chapter brings the reader a basic understanding of techniques, advantages, and challenges of research involving different NAMs, such as microphysiological systems, *in silico* modeling, bioprinting, artificial intelligence, and human tissue biobanking, as well as the importance of considering ancestry in human-based research.

Part III: Achieving a Practical Shift Toward Nonanimal-Based Research and Testing begins with six chapters examining obstacles to adoption of nonanimal methods and strategies for overcoming them. In the first, Catharine E.

Krebs and colleagues (Physicians Committee for Responsible Medicine) describe inertia against nonanimal methods in the United States. Following this, we get the perspectives of four groups on inertia against nonanimal methods in multiple international environments, including India (Mohammad A. Akbarsha, Jamal Mohamed College, and colleagues), Brazil (Marize Campos Valadares and colleagues, Federal University of Goiás), the European Union (Francesca Pistollato, Humane World for Animals), and Japan (Hajime Kojima and Yoko Hirabayashi, National Institute of Health Science).

Mihael H. Polymeropoulos (CEO) and Sandra P. Smieszek (Head of Genetics) from Vanda Pharmaceuticals detail their industry perspective on the advantages of NAMs and remaining needs, followed by Emily R. Trunnell (People for the Ethical Treatment of Animals) and Catharine E. Krebs, who discuss the bias toward animal methods as a barrier in need of addressing to reach NAMs full potential. Angela Hvitved (Alternatives Research & Development Foundation) and Francesca Pistollato present challenges and opportunities in funding of NAMs research, and Kristifor Sunderic (NIH), explores building confidence and trust in NAMs through standardization, validation, benchmarking, and reporting.

In the book's final two chapters, two research groups each present their analyses regarding the progression of acceptance of nonanimal methods in real-world settings, examining direct and indirect barriers, and discussing critical elements for successful transition to NAMs. In the first, Love Hansell (Radboud University) and colleagues examine the roadmap of the European Commission to animal-free safety assessments, discussing avenues in the short and long-term that may facilitate transition. In the second, Justine Watkins (Utrecht University) and colleagues analyze the successful transition of pyrogen testing methods from rabbits to human-relevant methods in the European Union, comparing the progress and barriers in the EU with those in the US, where progress has lagged.

We are on the cusp of pivotal change in the way animals are used in biomedical research and regulatory testing. Whether you are an early career researcher, biomedical scientist, toxicologist, cosmetic or household product developer, regulator, ethicist, science educator, advocate for replacing animal methods, lawyer, legislator, or simply interested in the directions nonanimal methods are taking us, we believe there is something for you in this book.

Acknowledgments

The editors would like to thank Catharine's team at the Physicians Committee, including Emily Anderson, Jarrod Bailey, Elizabeth Baker, Ann Lam, Janine McCarthy, and Mikalah Singer, for their input and guidance on the book; all of our wonderful authors, who contributed their time, energy, and thoughtful input to this book; and the wonderful staff at Elsevier for their patience and help through the publication process.

PART I

The need to move away from animal-based research and testing: Historical, ethical, legislative, and scientific context

CHAPTER

1

Beyond alternatives: Unpacking the ethics and politics of animal research and the need for systemic change

Arianna Ferrari

Center for Innovation Systems & Policy, AIT Austrian Institute of Technology GmbH, Vienna, Austria

Abstract

This chapter provides a critical historical and ethical analysis of animal experimentation, challenging the assumption that ongoing reforms and technological advances are sufficient for systemic change. It traces the development of animal research from antiquity to contemporary biomedical science, highlighting the entrenchment of animal use within scientific, regulatory, and socio-political structures. While the emergence of frameworks such as the 3Rs and new approach methodologies (NAMs) represents progress, they often fall short of challenging the fundamental ethical and epistemological status quo. The chapter explores how animals are increasingly engineered and instrumentalized for commercial, medical, and scientific purposes—from xenotransplantation to gene editing—raising profound ethical questions that exceed traditional welfare concerns. By contextualizing the limitations of incrementalist reform within broader political, cultural, and epistemic paradigms, the chapter argues for a radical rethinking of the role of animals in science. It calls for a shift from technocratic solutions toward a political critique capable of confronting speciesism and envisioning a genuine exit from the animal experimentation paradigm.

Keywords: Animal testing; Ethics; Alternatives; Animal rights; Genetic modification

Short history of animal experimentation and its critique

From ancient times to World War II

The history of animal experimentation is not just a story of scientific advancement but also one of ethical reflection and debate. From its early beginnings in ancient Greece to its central role in 19th-century biomedical research, animal testing has played a pivotal role in medical science. However, alongside this scientific progress, there has always been a critical

https://doi.org/10.1016/B978-0-443-26514-3.00014-X

ethical discussion regarding the treatment of animals, and this issue continues to be relevant today as societies strive to balance the benefits of animal testing with the moral considerations of animal welfare and rights.

The history of animal testing is closely intertwined with the development of medicine and the scientific understanding of the human body. In the *Corpus Hippocraticum*, the first medical manual of the Western world, early references to animal experimentation can be found, particularly through Galen, who used the vivisection of monkeys, pigs, and goats to gain critical insights into the functions of the brain, spinal cord, nerves, and muscles. These early experiments were groundbreaking in understanding human physiology but were also part of an intense scientific discourse about the ethics and methods of medical practice.

A significant debate occurred in ancient Alexandria between the Dogmatists and the Empiricists regarding whether human vivisection was necessary to understand the internal organs. The Dogmatists supported this practice, while the Empiricists opposed it, arguing that the different structures of living and dead tissues made human vivisection unreliable. This debate set the stage for centuries of discussions on the use of vivisection and animal testing in medicine.[1] In the medieval period, Galen's authority prevented further questioning of his anatomical theories, leading to their unquestioned application to human beings. However, during the Renaissance, dissection of human corpses became more widespread, although it often faced criticism. In the 16th century, the belief that human experimentation was cruel and a violation of divine law led to a greater emphasis on animal experimentation. The distinction between human anatomy and zoological dissection (zootomy) became clearer during this time, with proponents claiming that only through animal vivisection could explanations for physiological mechanisms be found. Philosophers such as Francis Bacon and Albrecht von Haller stressed the importance of vivisection in gaining physiological knowledge.[1]

René Descartes' influence on the view of animals as mechanical beings was central in the 17th century. Descartes argued that animals were no different from machines and lacked consciousness, a view that justified animal experimentation. His mechanistic theory was widely accepted and influenced the practice of vivisection during this period. Descartes' ideas also shaped scientific views well into the 18th century. Philosophers like Michel de Montaigne, who acknowledged the cognitive abilities of animals, challenged Descartes' views, but his perspective remained dominant for a long time.[2]

The 18th century saw a growing sensitivity to the rights of animals and the emergence of the first movements against animal testing. Philosophical discussions about Descartes' dualism, combined with the rise of animal rights movements, led to critical debates on the moral justification of animal testing. During this time, both naturalists and philosophers began questioning the ethical implications of vivisection, with growing concern about the treatment of animals.[1] This period also witnessed the development of modern political and human rights concepts, such as those found in the American Declaration of Independence and the French Revolution, which influenced how animals were viewed and treated.

In the 19th century, medical science underwent a profound transformation. Vitalism, which posited that life processes were governed by a life force distinct from physical and chemical laws, gradually lost its dominance. Instead, an experimental culture emerged that viewed animal testing as an essential tool for advancing medical knowledge. This shift was largely due to advances in bacteriology, cellular pathology, and experimental medicine.

Charles Darwin's theory of evolution, presented in *The Origin of Species* (1859), further stimulated interest in comparative anatomy and physiology, reinforcing the use of animal experimentation in medical research. Claude Bernard, a key figure in 19th-century physiology, made animal testing central to his experimental medicine. He argued that all medical hypotheses should be tested in laboratories, primarily using animals. This view led to a significant increase in animal experimentation. The development of anesthesia, which allowed for painless procedures on animals, further expanded the scope of animal testing.

However, this period also saw growing resistance to the cruelty of such experiments, with the rise of the anti-vivisection movement and criticism of unnecessary animal suffering.[1] Some scientists, including Marshall Hall, developed ethical guidelines for animal experiments, published in scientific journals like *The Lancet*, such as ensuring that experiments were only conducted when necessary, minimizing animal suffering, and avoiding unnecessary repetition of experiments.[3] Despite these ethical frameworks, the necessity and justification of animal testing remained contentious, with supporters arguing that animal testing was crucial for medical progress, while opponents believed that the scientific benefits did not justify the cruelty involved.[4] This ongoing debate about the morality of animal testing during the 19th century was framed in the broader context of changing views on animals and their treatment. As animals were seen as more similar to humans, practices like eating meat or hunting were increasingly questioned. This shift in attitudes also fueled the opposition to animal experimentation, especially among vegetarians and those advocating for animal rights. In the 19th century, the British Anti-Vivisection Society, founded in 1875, was a key organization opposing the use of animals in scientific experiments, while the Society for the Prevention of Cruelty to Animals (SPCA) and the American SPCA, founded in 1824 and 1866, respectively, campaigned against cruelty in hunting and vivisection. These societies helped raise awareness about animal welfare and contributed to the development of modern animal rights movements.

The debate over animal testing continued into the early 20th century, particularly in Britain and Germany, where new animal welfare organizations were formed and numerous polemical essays were published.[3] After World War I, the influence of anti-vivisection societies waned, with these groups becoming more moderate and engaging in dialog with the scientific community. Ironically, in 1909, Walter Cannon, a leading advocate of animal research, established the first code of ethics for animal experimentation in the United States, motivated by a desire to address animal welfare concerns and counter antivivisectionist criticism. While the code was adopted by many institutions, its enforcement was minimal, and the controversy surrounding animal experimentation, especially regarding the use of animals from shelters, escalated in the mid-20th century.[4]

The pressure to rely on animal experiments due to the atrocities of experimenting on humans during World War II

Following the atrocities committed against humans in the name of medical science during the World War II, there was a profound moral and ethical awakening within the scientific and medical communities. The horrific experiments conducted by Nazi doctors, most notably at concentration camps like Auschwitz and Dachau, as well as the infamous Unit 731 experiments in Imperial Japan, revealed the depths of cruelty that could be justified under the guise of scientific progress. These events catalyzed an urgent global response, leading to the establishment of ethical frameworks, international conventions, and legal regulations to ensure that scientific research respects human dignity and fundamental rights.

The Nuremberg Code, drafted in 1946, became the first internationally recognized set of ethical principles for human experimentation. The Declaration of Helsinki, established in 1964 by the World Medical Association, constitutes a further development of the Nuremberg Code and outlines 10 ethical principles for human experimentation, widely considered a foundational document in human research ethics. While it is not legally binding under international law, its authority stems from its incorporation into and influence on national and regional laws and regulations. The ten key principles outlined in the Code emphasized voluntary consent as an absolute requirement for participation in research; the necessity of scientific validity and social benefit in research; a requirement to avoid unnecessary suffering and injury; and the obligation to terminate research if it poses significant risk to participants. It became a cornerstone for guiding ethical human research, reinforcing the idea that human experimentation should be carried out with strict oversight and that alternatives, including animal studies, should be pursued whenever possible. As a result of these ethical shifts, animal testing was positioned as a necessary alternative to direct human experimentation in many areas of biomedical research, especially where human testing was deemed too risky or unethical. While this shift was critical in protecting human subjects, it also entrenched the use of animals as proxies in medical and scientific testing, contributing to the continued reliance on animal models in research, despite ongoing ethical and scientific debates about their validity and relevance.

The Declaration of Helsinki has undergone multiple revisions, reflecting the evolving complexities of medical research ethics, particularly in areas like genetic research, placebo-controlled trials, and research in developing countries. The Belmont Report, published by the United States National Commission for the Protection of Human Subjects of Biomedical and Behavioral Research in 1978, was another major milestone in research ethics. Motivated by ethical violations such as the Tuskegee Syphilis Study (1932–1972), it outlined three core ethical principles: (1) Respect for Persons—recognizing autonomy and ensuring informed consent; (2) Beneficence—maximizing benefits while minimizing harm; and (3) Justice—ensuring equitable selection of research participants. The Belmont Report influenced global research ethics, shaping laws such as the Common Rule in the United States and ethics guidelines in international research regulations.

In addition to these foundational documents, many countries enacted laws and created oversight bodies to ensure ethical compliance in research. Some of the major regulatory bodies include *Institutional Review Boards* in the United States, the *European Medicines Agency (EMA)* guidelines for human trials, the *Council for International Organizations of Medical Sciences* guidelines, and the *UNESCO Universal Declaration on Bioethics and Human Rights* (2005). For a deeper discussion on the legislative history of animal research, including these critical developments, see Chapter 2.

The regulation of animal research and the 3Rs

This ethical awakening in human research also spilled over into discussions about animal research ethics. As scientific communities acknowledged the moral imperatives of treating human subjects ethically, concerns about the ethical treatment of animals in research gained traction. Furthermore, the interest in experimental methods that do not use animals or use them in a more ethical manner gained momentum.

According to the work of the medical historian Parascandola,[4] the 3Rs movement traces its origins to the Universities Federation for Animal Welfare (UFAW), which was founded in 1926 by Charles Wesley Hume, who emphasized science-driven, sympathetic animal welfare. In 1947, UFAW published the first guide on laboratory animal care, and two years later, Hume highlighted the importance of reducing animal numbers and refining experiments. In 1954, William Russell and Rex Burch, hired by UFAW, conducted research on humane experimental techniques, leading to the first recorded use of the term "alternative" in reference to humane treatment and the eventual formalization of the "Three Rs" in a 1957 symposium. The 3Rs framework (Replacement, Reduction, Refinement) was then introduced by Russell and Burch with their seminal work *The Principles of Humane Experimental Technique* in 1959, reflecting an effort to apply ethical considerations to non-human subjects, though it remains less robustly enforced compared to human research ethics. In his historical reconstruction, Müller has argued that the 3Rs were originally based on an animal-centered ethic but were strategically limited to gain researchers' cooperation, focusing on improving animal welfare without challenging the legitimacy of animal research or promoting alternatives to animal testing.[5]

The 3Rs framework gained recognition in English-speaking countries only by the late 1960s and early 1970s. In the United States, the Laboratory Animal Welfare Act was passed in 1966, followed by the Animal Welfare Act in 1970, though it excluded mice, rats, and birds from protection—an exclusion that remains in place today. In 1969, Dorothy Hegarty founded the Fund for the Replacement of Animals in Medical Experiments (FRAME) in the UK to promote alternatives like tissue culture and computer simulations, bridging animal welfare groups and the scientific community (in 2024, FRAME's name was changed to *Replacing Animal Research*). The 1970s also saw the Humane Society of the United States intensify its focus on animal experimentation, establishing the Institute for the Study of Animal Problems in 1976 and launching campaigns against animal testing. Despite limited progress in animal testing during the next two decades, the 1980s saw significant developments, including the 1986 European Convention on the Protection of Vertebrate Animals and a US Congress report on alternative methods.

The late 1960s and early 1970s also saw a critical reassessment of the role and goals of medicine, with philosophers like Hans Jonas and Ivan Illich criticizing reductionist medical models. This period witnessed the rise of bioethics and animal ethics, with animal testing becoming a key topic in these fields. The "Oxford Group," a network of philosophers and activists at the University of Oxford, played a key role in developing the philosophical foundation for animal ethics and the debate on rights, with Richard Ryder's 1975 book *Victims of Science* and Peter Singer's *Animal Liberation* being pivotal in this movement. Influenced by these works, Henry Spira led a successful campaign in 1976 to stop inhumane animal experiments at the American Museum of Natural History and later focused on combating cruel cosmetic testing practices, such as the Draize test. Although Peter Singer has consistently critiqued animal experimentation, he supports a risk-benefit approach and does not categorically condemn all experiments.[6] In contrast, Tom Regan, who belongs to the first phase of animal ethics and followed a deontological approach, has maintained a more stringent stance on animal ethics from the outset, unequivocally condemning animal experiments.[7] He argued that research itself does not present a direct conflict because the very act of subjecting animals to laboratory experiments constitutes a violation of their inherent rights, whereas the rights of unhealthy humans have not been similarly violated.

On the experimental side, important discoveries in methods were highlighted in David H. Smyth's 1978 book *Alternatives to Animal Experiments*, which provided an in-depth exploration of alternative research methods and the challenges surrounding their adoption, aiming to raise awareness within the scientific community. While Smyth explores the potential of these techniques to replace animals in certain instances, he also gives considerable focus to their limitations, highlighting the challenges they face. James Gowans, secretary of the Medical Research Council, in the book's foreword, acknowledged both the benefits of animal research in medicine and the increasing public demand for stricter controls and alternative techniques, noting a divide between scientists who were already working on alternatives and those who believed legislation was necessary to encourage further development.[4]

NAMs and the pressure to end animal experiments today

Along with the development of alternative methods, in 2007, the US National Academy of Sciences introduced the term "new approach methodologies" (NAMs) to refer to in vitro, in chemico, or computational methods that, when used alone or together, enhance chemical safety assessments and help reduce animal use.[8] In silico methods use computer simulations and mathematical models to predict outcomes, such as the effects or distribution of specific substances within an organism. *In chemico methods* focus on studying chemical reactions, particularly to analyze the interaction behavior of substances. In vitro methods involve the use of cell cultures, where cells are grown in an artificial environment, such as a Petri dish, outside the organism. Ex vivo methods, on the other hand, utilize tissue samples obtained directly from a living organism to examine the effects of chemicals outside the organism, while animal studies are classified as in vivo methods. This NAM-based approach was developed in the field of systemic toxicity, which has made significant progress over the past decade[9] and was not intended to replace animal toxicity tests entirely, but rather to transform toxicological safety assessment by focusing on exposure and mechanistic data. Nonanimal approaches, such as the use of human liver chips, have demonstrated impressive predictive accuracy in toxicity testing, with one study achieving over 80% specificity and 100% sensitivity in predicting hepatotoxicity.[10]

Since then, significant progress has been made in applying NAMs, driving the shift toward a new testing paradigm rooted in relevant biological principles. Regulatory animal experiments are legally required as part of the medicine and vaccine development process and are frequently cited by companies and research institutions to justify the use of animals in nonregulatory research. However, NAMs and other modern, sophisticated, and more efficient technologies aim to challenge the entrenched belief that animal testing is the gold standard. This shift has also given rise to the concept of Next Generation Risk Assessment (NGRA), an exposure-driven, hypothesis-based approach that integrates in silico, in chemico, and in vitro methods (and not animal experiments), with NGRA serving as the ultimate goal. For a deeper discussion on NGRA, see Chapter 18.

NAMs are becoming increasingly important in the entire biomedical research enterprise, and the COVID-19 pandemic played a key role in accelerating their adoption. NAMs were crucial not only for understanding disease progression but also for repurposing existing drugs to address urgent health needs. The US Food and Drug Administration (FDA) accepted efficacy data from organ-on-a-chip models for clinical trial submissions, marking a significant milestone in the acceptance of nonanimal approaches in drug development.[11] Furthermore, as Ritskes-Hoitinga and her team have demonstrated, the pandemic significantly altered vaccine development processes, particularly during the approval of the Pfizer/BioNTech BNT162b2 vaccine (Comirnaty) by the European Medicines Agency (EMA).[12] The EMA granted conditional marketing authorization for the Pfizer/BioNTech vaccine in just 70 days, utilizing emergency procedures that included a rolling review of data as it became available. During the exploratory phase of SARS-CoV-2 research, animal studies were minimized, with human-specific data being allowed to replace animal data when sufficiently robust. This shift, driven by advances like genome sequencing and in silico models, may set a new precedent for vaccine development and medical research, suggesting that fewer animal studies could be necessary even in nonpandemic situations.[12]

However, it has also become clear that issues related to the efficacy and translatability of results are not the only considerations that can influence the move away from animal experiments. As Swaters and her group, in a paper on the history of regulatory testing, have pointed out, animal testing became entrenched as the gold standard due to past public health disasters, such as the sulfanilamide elixir tragedy and the thalidomide crisis, leading to stringent regulations that mandated extensive animal trials.[13] Despite increasing scientific evidence questioning the translatability of animal models to human biology, regulatory inertia and risk aversion have sustained the reliance on animal experiments. In many cases, pharmaceutical companies face legal and bureaucratic barriers when attempting to bypass redundant animal tests, reinforcing the epistemic rigidity of regulatory frameworks.[13]

Shifting toward human-relevant methods in research necessitates not only scientific advancements but *also a profound paradigm shift*—one that impacts both regulatory frameworks and all experimental practices. Some scholars have sought to expand this framework to incorporate considerations of the entire lifecycle of animals used in experiments, from breeding to maintenance, and have proposed more nuanced tools to assess the social benefits of research alongside the values driving it. For example, Brink and Lewis introduced an expanded ethical model for animal research, the 12 Rs framework.[14] This model integrates additional ethical principles into three overarching domains: Animal Welfare Rs, Social Value Rs, and Scientific Integrity Rs. These domains intersect to ensure rigorous ethical oversight while incorporating harm-benefit analysis, regulatory compliance, and cultural considerations. The aim is to harmonize global ethical standards for animal research, balancing scientific validity with social accountability.

Similarly, Beauchamp and DeGrazia's 2020 "Principles of Animal Research" introduced a set of six guiding principles, including the "Principles of Social Benefit."[15] These emphasize that animal research should only be conducted when no viable alternative exists, the anticipated benefits to human society outweigh the potential harms, and the research provides significant value justifying the harm to animals. The "Principles of Animal Welfare" focus on ensuring that any harm to animals is essential for achieving scientific goals, that their basic needs—such as food, housing, and medical care—are met, and that there are limits to the harm inflicted on them. While these principles enhance the 3Rs by demanding a more explicit consideration of the benefits of animal research and setting limits on experimentation, they fall short in offering detailed guidance on how to practically apply these concepts. Notably, critical questions—such as how to assess the balance between animal harm and human benefit in complex cases—remain unanswered. DeGrazia and Beauchamp presuppose that research institutions and review boards will rigorously apply these principles, but they fail to provide a concrete strategy for ensuring compliance. As a result, their framework does not sufficiently challenge the ethical status quo or provide a substantial mechanism for reform in the ethical culture of animal research. By implicitly assuming the continuation of animal research, the framework remains vulnerable to critique from those who argue that, given the advancements in alternative methodologies, ethically justifiable animal research is increasingly untenable. Nevertheless, the framework operates on a logic of minimizing animal suffering rather than offering a systemic critique of the animal experimentation paradigm.

To date, the strategies to exit animal experiments are very few, though a notable example can be found in the Netherlands. The political will in the Netherlands to transition away from animal experimentation is evident in the government's commitment to becoming a global leader in animal-free innovation, initially targeting 2025 as the year to significantly reduce animal testing. However, the ambition faces several challenges, including the continued legal requirements for animal use in preclinical safety and efficacy testing, as well as the scientific community's reliance on animal models for certain types of research. Despite these challenges, the Dutch government and various funding bodies, such as ZonMw, the Dutch Research Council (NWO), and the Ministry of Health, Welfare and Sport have launched several initiatives to accelerate the development and implementation of human-based research models. Key initiatives include the Human Measurement Models Program, which funds the creation of human-centric alternatives for disease research, and the Animal-Free Models Program, which focuses on legislative acceptance and the widespread adoption of animal-free methods. These efforts, supported by public-private partnerships, are part of a broader strategy to shift research culture, reduce animal suffering, and improve the effectiveness of medical research, though the path to fully realizing a transition to animal-free research remains complex and requires both scientific innovation and cultural change.

Of course, it is easier to follow the incrementalist or reformist approach, and surely advances in policies regarding the replacement of animals in toxicology and regulatory experiments are a huge step in this direction. However, this incremental approach does not seem to be sufficient to change the socioeconomic and cultural paradigm that relies on animal experiments, because it drives us to look for (new) areas in which the use of animals is intrinsic to the very goals set.

In the broader debate, NAMs are often seen as "advanced" alternatives to animal testing. However, it is crucial to distinguish between replacing animal experiments in regulatory toxicology—which aims to assess the health risks of food, feed, chemicals, and consumer products—and other scientific research areas focused on improving human health, curing diseases, or enhancing animal traits for commercial or entertainment purposes. The goal of safety assessment in regulatory toxicology is more narrowly defined and focused, whereas the objectives of advancing health, altering properties, or enhancing the capabilities of living beings are much broader. Even within biomedical research, there is a difference between conducting trials to assess the toxicity of a treatment (such as a drug or therapy) in preclinical studies and creating models of disease. While issues of translatability remain in the creation of animal disease models, it is more epistemologically challenging to argue for a direct one-to-one replacement in this context. The questions posed within biomedicine often require the use of animal models (in vivo) for their resolution or are intrinsically linked to the need for animals to obtain certain products—such as drugs extracted from their bodily fluids or cells, tissues, and organs for xenotransplantation—or to test devices like brain-machine interfaces. Thus, while NAMs face challenges in gaining recognition and relevance within institutional and scientific contexts still deeply entrenched in the animal research paradigm, there remains a broad array of scientific questions and problems that inherently rely on the use of animals.

As long as animal experiments are generally permissible and the decision is made solely based on the efficiency of one methodology over another, some ethical questions will remain unaddressed. Indeed, the maximum efficacy would be achieved through the use of human beings in investigating human diseases, but this is not permissible for ethical reasons that lie at the core of all regulations and legislation developed after WW2.

Modifying animals to serve as products through xenotransplantation, gene pharming, and gene editing

The ongoing debate regarding the legitimacy of animal experiments, alongside the efforts of the 3Rs framework to regulate and assess these practices, is often grounded in the presupposition that animal experimentation serves as a tool to acquire knowledge. This knowledge is expected to eventually lead to medical advancements, such as cures for diseases in both humans and animals. In addition, it aims to protect human, animal, and environmental health, and overall enhance well-being. However, this view has become increasingly complicated and difficult to defend in all cases, as advancements in animal breeding, genetic engineering, and genome editing have led to an expansion in the ways animals are utilized in scientific endeavors. Animals are no longer simply subjects of research in a traditional sense; they are also actively modified to fulfill specific, often commercially driven, purposes.

In fields like xenotransplantation, animals—particularly pigs—are genetically modified to provide cells, tissues, or organs suitable for human transplants. In gene-pharming, animals are altered so that their bodily fluids, such as blood, plasma, or milk, possess properties that can be exploited to produce pharmaceuticals. In agriculture, genetic modifications are made to increase animals' productivity, whether it is for meat, milk, or other resources.[16,17] Moreover, genetic engineering and genome-editing are also used to enhance the performance or characteristics of animals for sport and companionship, particularly in dogs and horses.[17] These practices represent a significant departure from traditional animal experimentation because they focus on modifying animals to serve specific functions, which raises critical ethical and scientific questions.

In 2020, the FDA approved a genetically modified pig, GalSafe, for both human consumption and potential therapeutic applications, such as xenotransplantation. Developed by Revivicor, the pig has been modified to remove the alpha-Gal antigen, which causes allergic reactions in some individuals with Alpha-Gal Syndrome, making it safe for those affected to consume pork without triggering allergies.[18] Additionally, in 2021, experimental xenotransplantations using GalSafe pig organs were conducted in brain-dead human patients, aiming to reduce the risk of organ rejection. Meanwhile, in June 2020, genetically edited cattle by Acceligen were classified as non-GMO by Argentine authorities, including double-edited Celtic/SLICK Holstein cattle for hornlessness and improved heat tolerance, and SLICK-edited Red Angus cattle. In 2021, Acceligen also produced MSTN knockout Nelore cattle in Brazil, which have increased muscle fiber count for leaner, tenderer beef, and were similarly classified as non-GMO by Brazilian authorities, though the commercialization status of these products remains unclear. The positive evaluation of Acceligen's PRLR-SLICK knockout cattle by North American authorities, including the FDA, was based on data from only four genetically edited calves, two of which were successfully modified and studied[19]—a number that would be unthinkable in human drug approval processes. This reflects a fundamentally different risk assessment framework for animals compared to humans.

In January 2022, the first gene-edited pig heart was transplanted into a human patient with severe heart failure who was not eligible for allotransplantation. The donor pig had 10 modified genes, including knockouts to remove xenogeneic antigens and genes to reduce inflammation, prevent coagulation, and downregulate complement activation. While the patient was supported by the pig heart for 7 weeks, the graft eventually failed after 60 days.[19,20] Following this, in June and July 2022, two heart transplants were conducted at New York University using the same genetically modified pigs as donors on brain-dead human beings. The transplanted hearts remained functional for 72 h without any signs of rejection. On September 20, 2023, a Baltimore team performed a second pig-to-human heart transplant in a 58-year-old patient who was ineligible for an allogeneic transplant due to severe peripheral vascular disease and internal bleeding complications. The patient passed away 40 days after the transplant, likely due to early signs of rejection. Since its inception, xenotransplantation has posed broad ethical and social issues in the human domain, together with the more obvious problems for animal rights. Key questions include criteria for patient selection, the problem of informed consent, and the risk of long-term health threats such as infection, despite advancements in genetic modification to reduce immune rejection. The inclusion of economically disadvantaged or vulnerable populations in trials, along with the uncertainty of the procedure's benefits, poses serious questions about the fairness and integrity of these studies, highlighting the need for more stringent ethical safeguards and better evidence of safety before widespread application. Additionally, broader social policy issues such as fair resource distribution and consideration of public attitudes must be addressed when developing policies.[21]

With the development of gene-editing tools, the idea of fundamentally changing some animal properties to fit medical or commercial purposes for human beings has intensified. Gene-editing tools can make targeted cuts in DNA, which are then repaired by natural cellular mechanisms, allowing specific genetic modifications to be introduced and permitting the introduction of multiple genetic alterations in an animal's genome simultaneously. Some of the most well-known SDNs include zinc finger nucleases, transcription activator-like effector nucleases, and the CRISPR/Cas9 system. Whereas traditional genetic engineering methods required several years to breed, for example, transgenic pigs, CRISPR now allows for the same modifications to be made in just a few months because it enables direct genetic modifications in early-stage embryos, or zygotes, to achieve genetic changes in a single generation without the need for fully matured animals.[17]

The belief that genome editing of animals is simpler, cheaper, more precise, and therefore revolutionary should be approached with caution. There is still a lack of comparative studies between classical genetic engineering methods and genome editing for each specific area.[22] In addition, genome editing of animals often requires the incorporation of methods from "classical" genetic engineering, which is associated with inefficiencies and welfare problems for the animals. Definitive, quantitative assessments of CRISPR's efficiency are hard to come by, as estimates vary significantly and are influenced by various factors, such as the target site and the specific CRISPR molecule used. The category "efficient" is normatively shaped and depends on the importance placed on the use of animal organisms and the risk of unintended effects. Finally, even if a technique is simpler and cheaper for altering animals, it does not automatically mean that the number of animals used will be reduced.

As was evident in the 1980s, when genetic engineering was heavily promoted as a solution to enhance the 3Rs but instead led to a significant increase in animal experimentation,[23] a similar trend appears to be unfolding with the advent of gene-editing tools. There are several unintended effects of gene-editing tools that can affect the welfare of the animals created, and these effects are generally distinguished between *on-target* and *off-target effects*. On-target efficiency refers to the ability of genome-editing tools to precisely target the desired location in the genome and induce the intended modification. This efficiency depends on factors such as species, cell type, and the specific modification, with estimates for CRISPR efficiency ranging from 2% to 100%.[22] Off-target effects refer to unintended genomic alterations that occur at locations distant from the targeted region, arising from factors such as sequence similarity, suboptimal specificity of the editing tool, and genomic context. These effects include unintended mutations, chromosomal rearrangements, insertion of foreign DNA, epigenetic changes, and toxic effects such as apoptosis in sensitive cells. Detection of off-target effects requires the complete sequencing and analysis of the organism's genome, but no comprehensive testing method exists to detect all possible off-target mutations. Despite these advances, further research is needed, especially concerning the role of repair mechanisms, which can often introduce undesirable effects, such as off-target mutations that compromise animal health by disrupting tumor-suppressor genes.[24]

Mosaicism also represents a problem for gene-edited animals, in which genetically distinct cell populations exist within an organism, with some cells carrying the edited genes and others remaining unchanged. This phenomenon typically occurs when genetic modifications are made at early developmental stages, such as in embryos or during early cell division, and can arise due to varying editing efficiency, incomplete cell division and differentiation, or genomic barriers that hinder gene integration. Mosaicism is a well-known issue in the so-called genome-edited large animals—such as monkeys, pigs (for xenotransplantation), goats, sheep, and cows[25]—as well as in aquaculture fish species.[26] This leads to the creation of modified individuals that do not always exhibit the desired properties.

Moreover, for certain applications involving large mammal species (animal species larger than rodents, which are the predominant animals used in experiments), gene-editing technologies are often used in combination with more traditional genetic engineering methods, such as microinjection and somatic cell nuclear transfer (SCNT). The welfare problems connected with SCNT cloning for animals include oversize organs and abnormal growth patterns, as well as an increased likelihood of stillbirths and premature deaths. Cloned animals often experience higher rates of genetic defects, immune system dysfunction, and compromised fertility. Multiple gene edits increase the risk of unintended off-target effects and require highly precise design and control.[27] The efficiency of genetic modifications in these large animals is still relatively low due to factors such as the opacity of the zygote, difficulties in integrating the transgene at the desired genomic location, mosaicism, lack of stable embryonic stem cell lines, longer reproductive cycles, and higher maintenance costs compared to mice.[28]

In the commercial use of genetically modified animals, their welfare is often compromised in pursuit of profit. A case in point is the 2021 approval in Japan for the commercial sale of genome-edited fish—red sea bream and tiger pufferfish—developed by the Regional Fish Institute in collaboration with Japanese universities.[29,30] While the red sea bream, modified with CRISPR to knock out the myostatin gene, exhibited larger size and improved feed efficiency, homozygous fish had lower survival rates; similarly, the tiger pufferfish, edited for increased appetite and faster growth through the knockout of leptin receptor genes, raised concerns about potential long-term health risks.

From an ethical standpoint, it is crucial to consider not only the welfare of transgenic or gene-edited animals but also the animals involved in their creation. Depending on the methods used, multiple generations of animals may be affected, as it is not always possible to directly modify only the zygotes. Furthermore, depending on the applications, other animals are also involved. For example, in xenotransplantation, the suffering extends beyond the donor pigs—whose cells, tissues, or organs are harvested—to include primates used in testing these materials.[31]

Genome-editing tools, particularly CRISPR-Cas9, have been used to create genetically modified dogs to restore pathogenic mutations in purebred dogs or to generate inbred disease models for research.[32] Moreover, SCNT has been used in dogs for commercial purposes for a couple of decades, especially in the US and Asian markets. Genome-edited dogs can be created with CRISPR-Cas9 and SCNT, though the efficiency remains low at about 3.7%.[32–34] This means that there are many miscarriages or that the dogs born are not clones. In horses, genome-editing is being explored to enhance breeding, particularly for sports applications, although much research remains to be done before approaching commercialization because the efficiency also remains very low.[35,36]

Conclusions

In experimental research on animals, individuals are not only bred and used to comply with regulatory requirements for the safety of products, drugs, devices, and medical tools; there are, indeed, many uses in which animals are viewed as raw materials to be modified, augmented, or exploited for human benefit. Whether it is the harvesting of organs for transplantation, the production of pharmaceuticals, or the enhancement of agricultural outputs, these practices expand the role of animals from objects of study to tools designed to meet human-driven objectives.

This expansion raises significant ethical concerns as the traditional interpretation of the 3Rs framework, no matter if approached from either an anthropocentric or non-anthropocentric perspective, proves inadequate in addressing the complexities inherent in these new applications of animal modification. As long as practices like xenotransplantation—which, also from a human ethics perspective, raise significant issues such as exacerbating healthcare inequalities and posing public health risks—are promoted as viable solutions to the shortage of tissues and organs, it will remain impossible to genuinely advocate for an exit strategy from animal experimentation. The same applies to the broader spectrum of animal exploitation, whether for companionship, entertainment (e.g., in sports), food, or as bioreactors for pharmaceutical production. The ongoing discourse continues to be confined within the limits of welfarism, focusing on reforms that may improve the conditions of animals bred and used for scientific purposes, but without addressing the fundamental problem: the very use of animals in these contexts.

Within the context of using genetic tools to improve commercial traits or to foster xenotransplantation, the guiding principle of "replacement" is virtually impossible to apply. In such practices, the animal is crucial to the desired result. Xenotransplantation, by definition, requires animal use—there are no nonanimal "alternatives" if the specific practice continues to be accepted. In principle, such practices must be given up entirely to replace the use of animals. The frameworks underpinning such practices are guided by the anthropocentric spirit of the Nuremberg Code and the Declaration of Helsinki, reinforcing a hierarchical relationship between humans and other species, where animals are seen as raw material to fulfill human purposes. This perspective is deeply rooted in speciesism, a worldview that privileges human interests over those of other species, thus failing to engage critically with the ethical and philosophical foundations of animal rights and protection.

Therefore the true barrier to ending animal research is not merely scientific or ethical but fundamentally political. The institutionalization of animal research reflects deeply entrenched societal values that prioritize human interests over the welfare of animals.[37] These values justify the exploitation and modification of animals to serve human needs, embedding animal experimentation as an unquestioned norm. Challenging this system requires more than just scientific and ethical advancements—it demands a political movement that confronts and dismantles the entire culture of animal research.

A radical political critique of animal experimentation demands a shift in both societal attitudes and legislative frameworks. Legislative changes, redefined funding priorities, and shifts in professional norms are essential components of a larger political movement aimed at achieving real and lasting change in the way animal research is conducted. This movement must ensure that ethical considerations are not merely an afterthought but are fully integrated into the decision-making process at all levels of research and regulation. Only by challenging the very foundation of animal use in science, industry, and entertainment can we begin to build a just society where animals are no longer seen as tools for human benefit, but as beings whose intrinsic value must be respected.

References

[1] Maehle A-H. Kritik Und Verteidigung Des Tierversuchs: Die Anfänge der Diskussion Im 17. Und 18. Jahrhundert. Stuttgart: Steiner; 1992.

[2] Guerrini A. Experimenting with Humans and Animals: From Galen to Animal Rights, 2nd ed. Baltimore, Md: Johns Hopkins University Press; 2003.

[3] Monamy V. Animal Experimentation: A Guide to Issues; [A Balanced Treatment Of: Ethics, Law and Regulation, History, Opposition and Alternatives]. New York: Cambridge University Press; 2000.

[4] Parascandola J. A history of the development of alternatives to animals in research and testing. In: New Directions in the Human-Animal Bond Series. West Lafayette, Indiana: Purdue University Press; 2024.
[5] Müller ND. Beyond Anthropocentrism: the Moral and Strategic Philosophy behind Russell and Burch's 3Rs in Animal Experimentation. Sci Eng Ethics. 2024;30(5):44. https://doi.org/10.1007/s11948-024-00504-1.
[6] Singer P. Animal Liberation Now, Bodley Head, UK; 2023.
[7] Regan T. The Case for Animal Rights. University of California Press; 1983.
[8] Sewell F, Alexander-White C, Brescia S, et al. New Approach Methodologies (NAMs): identifying and Overcoming Hurdles to Accelerated Adoption. Toxicol Res. 2024;13(2):tfae044. https://doi.org/10.1093/toxres/tfae044.
[9] Krewski D, Acosta D, Andersen M, et al. Toxicity Testing in the 21st Century: a Vision and a Strategy. J Toxicol Environ Health B. 2010;13(2–4):51–138. https://doi.org/10.1080/10937404.2010.483176.
[10] Marshall LJ, Bailey J, Cassotta M, Herrmann K, Pistollato F. Poor Translatability of Biomedical Research Using Animals—a Narrative Review. Altern Lab Anim. 2023;51(2):102–135. https://doi.org/10.1177/02611929231157756.
[11] Rumsey JW, Lorance C, Jackson M, et al. Classical Complement Pathway Inhibition in a 'Human-On-A-Chip' Model of Autoimmune Demyelinating Neuropathies. Adv Ther. 2022;5(6):2200030. https://doi.org/10.1002/adtp.202200030.
[12] Ritskes-Hoitinga M, Barella Y, Kleinhout-Vliek T. The Promises of Speeding Up: changes in Requirements for Animal Studies and Alternatives during COVID-19 Vaccine Approval–A Case Study. Animals. 2022;12(13):1735. https://doi.org/10.3390/ani12131735.
[13] Swaters D, Van Veen A, Van Meurs W, Turner JE, Ritskes-Hoitinga M. A History of Regulatory Animal Testing: what Can We Learn? Altern Lab Anim. 2022;50(5):322–329. https://doi.org/10.1177/02611929221118001.
[14] Brink CB, Lewis DI. The 12 Rs Framework as a Comprehensive, Unifying Construct for Principles Guiding Animal Research Ethics. Animals. 2023;13(7):1128. https://doi.org/10.3390/ani13071128.
[15] Beauchamp TL, DeGrazia D. Principles of Animal Research Ethics. New York, NY: Oxford university press; 2020.
[16] Gim G-M, Uhm K-H, Kwon D-H, et al. Germline Transmission of MSTN Knockout Cattle via CRISPR-Cas9. Theriogenology. 2022;192(October):22–27. https://doi.org/10.1016/j.theriogenology.2022.08.021.
[17] Gim G-M, Jang G. Outlook on Genome Editing Application to Cattle. J Vet Sci. 2024;25(1):e10. https://doi.org/10.4142/jvs.23133.
[18] Bailey J. Genetic modification of animals: scientific and ethical issues. In: Kathrin Herrmann, Jane K, eds., Animal Experimentation: Working Towards a Paradigm Change. Boston: Brill; 2019, pp. 443–479.
[19] Griffith BP, Goerlich CE, Singh AK, et al. Genetically Modified Porcine-to-Human Cardiac Xenotransplantation. N Engl J Med. 2022;387(1):35–44. https://doi.org/10.1056/NEJMoa2201422.
[20] Platt JL, Cascalho M. The Future of Transplantation. N Engl J Med. 2022;387(1):77–78. https://doi.org/10.1056/NEJMe2207105.
[21] Krebs CE, McCarhey J, Sullivan K, et al. Considering the risks and costs of solid organ xenotransplantation. Adv Biol. 2025;9(4):2400453. https://doi.org/10.1002/adbi.202400453.
[22] Camenzind S. Xenotransplantation: neue gentechnische Möglichkeiten - neue ethische Fragen? In: Willemsen A, ed. Beiträge Zur Ethik Und Biotechnologie, Vol 16, Bern: Eidgenössische Ethikkommission für die Biotechnologie im Ausserhumanbereich; 2023:59. Available at: https://www.ekah.admin.ch/inhalte/ekah-dateien/dokumentation/stellungnahmen/stellungnahmen_2020-2023/EKAH-Bericht_Xenotransplantation._Neue_Moeglichkeiten__neue_Fragen_November_2023_verabschiedet.pdf. Accessed 11 August, 2026.
[23] Ferrari A. Genetically Modified Laboratory Animals in the Name of the 3Rs? ALTEX. 2006;23(4):294–307.
[24] Ishii T. Genome-Edited Livestock: ethics and Social Acceptance. Anim Front. 2017;7(2):24–32. https://doi.org/10.2527/af.2017.0115.
[25] Mehravar M, Shirazi A, Nazari M, Banan M. Mosaicism in CRISPR/Cas9-Mediated Genome Editing. Dev Biol. 2019;445(2):156–162. https://doi.org/10.1016/j.ydbio.2018.10.008.
[26] Blix TB, Dalmo RA, Wargelius A, Myhr AI. Genome Editing on Finfish: current Status and Implications for Sustainability. Rev Aquac. 2021;13(4):2344–2363. https://doi.org/10.1111/raq.12571.
[27] Ren J, Hai T, Chen Y, et al. Improve Meat Production and Virus Resistance by Simultaneously Editing Multiple Genes in Livestock Using Cas12iMax. Sci China Life Sci. 2024;67(3):555–564. https://doi.org/10.1007/s11427-023-2407-0.
[28] Eenennaam V, Alison L. 120 The Future of Genome Editing in Food Animal Species. J Anim Sci. 2020;98(Supplement_3):48–49. https://doi.org/10.1093/jas/skaa054.089.

[29] Ohama M, Washio Y, Kishimoto K, Kinoshita M, Kato K. Growth Performance of Myostatin Knockout Red Sea Bream Pagrus Major Juveniles Produced by Genome Editing with CRISPR/Cas9. Aquaculture. 2020;529(December):735672. https://doi.org/10.1016/j.aquaculture.2020.735672.

[30] Matsuo M, Tachikawa M. Implications and Lessons From the Introduction of Genome-Edited Food Products in Japan. Front Genome Ed. 2022;4(June):899154. https://doi.org/10.3389/fgeed.2022.899154.

[31] Camenzind S. Xenotransplantation in the Age of Genome Editing: results From the Expert Report for the Federal Ethics Committee on Nonhuman Biotechnology With a Special Focus on Animal Ethics. Xenotransplantation. 2024;31(6):e70008. https://doi.org/10.1111/xen.70008.

[32] Feng C, Wang X, Shi H, et al. Generation of ApoE Deficient Dogs via Combination of Embryo Injection of CRISPR/Cas9 with Somatic Cell Nuclear Transfer. J Genet Genomics. 2018;45(1):47–50. https://doi.org/10.1016/j.jgg.2017.11.003.

[33] Kim M-J, Park J-E, Hyun-Ju O, et al. Preservation through Cloning of Superior Canine Scent Detection Ability for Cancer Screening. J Vet Clin. 2015;32(4):352–355. https://doi.org/10.17555/jvc.2015.08.32.4.352.

[34] Choi YB, Kim GA, Oh HJ, et al. Cloning of the Short-Tailed Gyeongju Donggyeong Dog via SCNT: conserving Phenotypic Inheritance. J Vet Med Sci. 2015;78(2):329–331. https://doi.org/10.1292/jvms.15-0107.

[35] Bennett EP, Petersen BL, Johansen IE, et al. INDEL Detection, the 'Achilles Heel' of Precise Genome Editing: a Survey of Methods for Accurate Profiling of Gene Editing Induced Indels. Nucleic Acids Res. 2020;48(21):11958–11981. https://doi.org/10.1093/nar/gkaa975.

[36] Tozaki T, Ohnuma A, Kikuchi M, et al. Short Insertion and Deletion Discoveries via Whole-Genome Sequencing of 101 Thoroughbred Racehorses. Genes. 2023;14(3):638. https://doi.org/10.3390/genes14030638.

[37] Ferrari A. Contesting animal experiments through ethics and epistemology: in defense of a political critique of animal experimentation. In: Herrmann K, Jayne K, eds. Animal Experimentation: Working Towards a Paradigm Change. BRILL; 2019, pp. 194–206. https://doi.org/10.1163/9789004391192_008.

CHAPTER

2

Legislative history of animal-based research and implications for the shift to nonanimal methods

Nirva Kapasi Patel

Brooks McCormick Jr. Animal Law and Policy Program, Harvard Law School, Cambridge, MA, United States

Abstract

Animal research in the United States has witnessed an increasing emphasis on nonanimal models. The existing legislations have been critiqued for insufficient species coverage, lack of rigorous enforcement, and inadequate transparency. Ethical concerns, coupled with scientific progress, have accelerated the development of alternative research methods. Internationally, legislative frameworks are evolving to support nonanimal testing, with milestones such as the European Union's Directive 2010/63/EU, which prioritizes alternative methods. Countries like Japan and India have also enacted guidelines and bans to limit animal testing. These initiatives reflect the principles of the 3Rs—replacement, reduction, and refinement—and promote innovative technologies like in vitro testing and organ-on-chip systems. Despite advancements, significant challenges remain in achieving widespread adoption of nonanimal research methodologies.

Keywords: Animal research; Nonanimal models; Alternative methods to animal testing; Animal testing; Clinical research; Animal science; Research activity; Drug

Overview of the history of animal research

The historical use of animals in biomedical research can be traced as far back as the 4th and 3rd centuries BCE. Aristotle (384–322 BCE) and Erasistratus (304–258 BCE) were among the first to perform experiments on nonhuman animals. Galen (129–216 AD), a prominent physician in ancient Rome, introduced the experimental method in medical investigation. Avenzoar, an Arabic physician in 12th-century Moorish Spain, tested surgical procedures on animals before applying them to human patients.

Despite ethical and moral concerns, the pursuit of medical advancements historically took precedence over considerations for animal welfare. Notably, Claude Bernard (1813–1878), a

Moving Toward Nonanimal Approaches in Medical Research and Testing
https://doi.org/10.1016/B978-0-443-26514-3.00015-1

French physiologist known as the founder of experimental physiology, highlighted and advocated for the critical role of animal testing in biomedical research. His controversial methods, which included subjecting animals to extreme conditions and procedures, underscored the era's limited concern for animal suffering. Nevertheless, his work significantly influenced the establishment of animal testing as a cornerstone of scientific inquiry, while sparking the antivivisection movement of the 19th century.[1]

The 20th century witnessed significant reforms in the regulation of animal research, which were driven by growing ethical concerns and scientific scrutiny. Legal and regulatory frameworks were developed to ensure the humane treatment of animals used in research. However, contemporary scientific literature increasingly questioned the reliability of animal models due to issues such as laboratory environment effects, discrepancies between animal and human diseases, and species-specific physiological and genetic differences.[2]

The legislative history of animal-based research reflects a complex interplay between scientific advancement and ethical considerations. While historical practices often neglected animal welfare, modern regulatory frameworks strive to balance the necessity of animal research with ethical obligations to minimize suffering and seek alternatives. Key ethical documents, such as the Nuremberg Code and the Helsinki Declaration, along with initiatives such as the Basel Declaration, continue to shape the evolving landscape of animal research, promoting responsible and humane scientific practices.

Public advocacy has played a crucial role in helping to shape animal research legislation and enforcement in the United States. Citizens and advocacy organizations can influence animal research policies through several key channels, particularly through congressional action, administrative processes, and legal mechanisms. At the congressional level, constituents can contact their representatives to support strengthening the Animal Welfare Act (AWA) and increasing funding for enforcement, while organizations can provide expert testimony at congressional hearings on animal research oversight. Through administrative channels, the public can engage by submitting comments during federal rulemaking processes when agencies like the United States Department of Agriculture propose new regulations, filing Freedom of Information Act (FOIA) requests to obtain inspection reports and other documents to increase transparency. Animal welfare organizations can bring lawsuits to challenge inadequate enforcement of existing regulations, file cases under the Administrative Procedure Act to contest agency policies or decisions, and pursue strategic litigation to establish legal precedents that strengthen protections. These advocacy efforts have historically driven major advances in laboratory animal welfare, from the original passage of the AWA to the strengthening of enforcement mechanisms, helping to ensure accountability and advance reforms through the democratic process.

Emerging guidelines and principles that influence animal research legislation

World War II and its aftermath represent a turning point in the history of research involving human subjects. The Nuremberg Doctor Trials, formally known as the "Doctors' Trial,"[3] were a series of prosecutions held from December 9, 1946, to August 10, 1947, and focused on the inhumane medical experiments conducted by Nazi doctors during World War II. Defendants were accused of conducting brutal and often lethal medical experiments on prisoners of war, concentration camp inmates, and other nonconsenting individuals.

Experiments included freezing, infecting with malaria, mustard gas exposure, nerve regeneration, poisoning, and saltwater experiments. These trials were pivotal in establishing principles of medical ethics and human rights and ultimately led to the Nuremberg Code, which set standards for ethical conduct in medical research.

The Nuremberg Code of 1948 was a direct result of the trial, serving as a response to unethical human experimentation. It was the first international document to advocate for the following principles: (1) participants in medical research must do so voluntarily and give informed consent without coercion, deceit, or manipulation; (2) the experiment should not be random or unnecessary in nature but should yield fruitful results for the benefit of society; (3) the experiment should be designed and based on the results of animal experimentation and a knowledge of the natural history of the disease, such that the anticipated results will justify the performance of the experiment; (4) the experiment should be conducted to avoid all unnecessary physical and mental suffering and injury; (5) no experiment should be conducted where there is reason to believe that death or disabling injury will occur, except where the physicians also serve as subjects; (6) the degree of risk of the experiment should never exceed that determined by the humanitarian importance of the problem to be solved by the experiment; (7) proper preparations and adequate facilities should protect the experimental subject against even remote possibilities of injury, disability, or death; (8) the experiment should be conducted by scientifically qualified persons, and the highest degree of skill and care should be required through all stages of the experiment; (9) the human subject should, at any time, be able to end participation in the experiment if they have reached a physical or mental state where continuation is impossible; and (10) the scientist in charge must be prepared to terminate the experiment at any stage if they have probable cause to believe that the continued experiment is likely to result in injury, disability, or death to the experimental subject.[4,5] Ethical standards in research were significantly influenced by the Nuremberg Code. Although animals cannot provide informed consent, the principles underscore the importance of minimizing harm, yielding fruitful results for the benefit of society, and ensuring that research is well informed by existing scientific literature.

Twenty years later, in 1964, the Helsinki Declaration was adopted by the World Medical Association (WMA),[6] outlining more directly ethical guidelines for medical research involving human subjects. Amended many times and endorsed by numerous countries, the declaration is a cornerstone document setting ethical principles for conducting medical research involving human subjects. It emphasizes the fundamental principle of respect for all human subjects, recognizing their autonomy and the necessity of obtaining informed consent.[7] It mandates that research must be scientifically justified and methodologically sound, with potential benefits outweighing the risks. Researchers are required to minimize harm and maximize benefits, ensuring the welfare of participants takes precedence over scientific and societal interests. All research proposals must undergo independent ethical review and approval, ensuring an impartial evaluation of the study's design, informed consent process, and risk-benefit analysis. Informed consent is a crucial element, requiring that participants are fully informed about the study's aims, methods, potential risks, and benefits, and that consent is obtained freely without coercion. Special protections are afforded to vulnerable populations. The declaration also emphasizes the importance of protecting participants' privacy and the confidentiality of their data. Transparency in research is encouraged through the registration of clinical trials and dissemination of results, including both positive and negative findings. Furthermore, researchers must consider posttrial provisions, ensuring participants have access to beneficial interventions identified during the study. These principles collectively ensure that medical research is conducted ethically, balancing scientific advancement with the rights and welfare of human subjects.[7]

Several key declarations and guidelines have been established to help govern ethical medical research involving human subjects and, in some cases, animals. Some of the most prominent ones are listed in Table 1.

Both the Basel Declaration and the CIOMS Guidelines contribute to the ethical framework for animal research by promoting principles that prioritize animal welfare, minimize harm and suffering, and ensure that research involving animals is conducted with integrity and respect for ethical considerations. The principles from both the Nuremberg Code and the Declaration of Helsinki were incorporated into US federal regulations in the 1980s and applied to most government agencies, including the National Institutes of Health (NIH), which is a major source of funding for human and animal experimentation (45 CFR § 46.101).[15]

The "Three Rs" (Replacement, Reduction, and Refinement) serve as a foundational framework for ethical animal research. "Replacement" advocates for finding alternatives to using

TABLE 1 Declarations and guidelines established to ensure ethical medical research involving human and/or animal subjects.

Declaration or guideline (year introduced)	Description
Declaration of Geneva (1948)[8]	Adopted by the WMA as a physician's pledge, it sets out ethical principles for the medical profession, emphasizing respect for human life and patient autonomy
Proposed International Ethical Guidelines for Biomedical Research Involving Human Subjects (1982)[9]	Developed by the Council for International Organizations of Medical Sciences (CIOMS) in collaboration with the World Health Organization, these guidelines provide detailed ethical standards for biomedical research, particularly in low-resource settings
Belmont Report (1979)[10]	Issued by the US National Commission for the Protection of Human Subjects of Biomedical and Behavioral Research, it outlines ethical principles and guidelines for research involving human subjects, including respect for persons, beneficence, and justice
Good clinical practice guidelines (1996)[11]	Developed by the International Council for Harmonization of Technical Requirements for Pharmaceuticals for Human Use, these guidelines provide a unified standard for designing, conducting, recording, and reporting trials that involve the participation of human subjects
Guidelines (2016)[12]	These updated guidelines address various ethical issues, including research in low-resource settings, biobanking, and research with vulnerable populations
Universal Declaration on Bioethics and Human Rights (2005)[13]	Adopted by the United Nations Educational, Scientific, and Cultural Organization, it sets out principles for bioethics, including respect for human dignity, human rights, and fundamental freedoms
Basel Declaration (2010)[14]	Focused on animal research, it promotes the ethical treatment of animals in scientific research and emphasizes the importance of the "Three Rs" (Replacement, Reduction, and Refinement)

animals in research whenever possible, including nonanimal methods such as computer simulations, cell cultures, and mathematical models. The "Reduction" principle aims to minimize the number of animals used in research studies. Researchers are encouraged to design experiments that require the fewest possible animals to achieve meaningful scientific results. This involves careful experimental design, statistical analysis, and data sharing to ensure that studies are efficient and scientifically rigorous. "Refinement" focuses on improving the welfare and well-being of animals used in research. This principle involves enhancing experimental procedures, housing conditions, and care practices to minimize pain, suffering, and distress experienced by animals. The Three Rs help to uphold the ethical standards of animal welfare while facilitating scientific progress. As such, the Three Rs serve as essential guidelines for ensuring that animal research is conducted in a humane, responsible, and scientifically valid manner. For a more detailed discussion about the Three Rs and animal research ethics, please see Chapter 1.

Collectively, these declarations and principles form a comprehensive framework for ensuring ethical conduct in medical and biomedical research, safeguarding the rights and well-being of experimental subjects.

A note on the varying legal definitions of animal

The definition of "animal" varies significantly across different legal and regulatory frameworks, impacting the scope and effectiveness of animal research protections. In the United States, two primary regulatory frameworks employ notably different definitions. The AWA defines animals narrowly, specifically excluding rats, mice, and birds bred for research—species that comprise approximately 95% of laboratory animals. This significant exclusion stems from a 2002 legislative amendment and substantially limits the AWA's protective reach in practice. In contrast, the Public Health Service (PHS) Policy on Humane Care and Use of Laboratory Animals (PHS Policy) adopts a broader definition, encompassing all vertebrate animals, including rats, mice, and birds, thereby providing more comprehensive oversight for research conducted with federal funding. Internationally, definitions show considerable variety, often reflecting different cultural, scientific, and ethical perspectives. The European Union's Directive 2010/63/EU[16,17] provides one of the most comprehensive definitions, protecting all live vertebrates, including independently feeding larval forms, mammalian fetuses in the last trimester of development, and cephalopods like octopi—marking the first major legal protection of invertebrates in research. The UK's Animals (Scientific Procedures) Act of 1986[18] similarly protects all living vertebrates and cephalopods, while also extending protection to certain fetal and embryonic forms. Japan's Act on Welfare and Management of Animals[19] covers vertebrates used in research but excludes most invertebrates, though it includes specific provisions for proper methods even with excluded species. These differences have profound practical implications. They determine which species receive legal protections, influence institutional oversight requirements, shape research design decisions, and affect cross-border research collaborations. When research spans multiple jurisdictions, investigators must navigate these varying definitions and comply with the most stringent applicable standards. The evolution of these definitions also reflects growing scientific understanding of animal cognition and sentience, as evidenced by the inclusion of cephalopods in several regulatory frameworks following evidence of their cognitive sophistication.

The US regulatory framework

To ensure the safety of consumer products achieved through scientific research, it is critical to provide regulatory oversight and ethical considerations to safeguard the welfare of animals in research while upholding scientific integrity. In the United States, several federal agencies, including the Food and Drug Administration (FDA), the NIH, the Environmental Protection Agency, and the United States Department of Agriculture (USDA), enforce regulations and guidelines governing animal research. This section analyzes the regulatory requirements established by these agencies, focusing on the AWA, the PHS Policy, Institutional Animal Care and Use Committees (IACUCs), the Guide for the Care and Use of Laboratory Animals (the Guide), and the FDA Modernization Act.

The Animal Welfare Act

Signed into law by President Lyndon B. Johnson on August 24, 1966, the Animal Welfare Act of 1966 regulates the transport, sale, and handling of dogs, cats, guinea pigs, nonhuman primates, hamsters, and rabbits used for research or other purposes.[20] The Congressional policy justification to regulate animals and activities under the AWA rests on the power of the US Congress to regulate interstate or foreign commerce.[21] The AWA is overseen and upheld by the USDA, granting the power to the Secretary of Agriculture to enforce this federal legislation.[22]

Furthermore, the AWA applies to research institutions that receive federal funding or are involved in interstate commerce.[23] Research facilities under the jurisdiction of the AWA include all schools (except elementary or secondary schools), institutions, organizations, or persons that use or intend to use live animals in research, tests, or experiments and that (1) purchase or transport live animals in commerce or (2) receive funds under a grant, award, loan, or contract from a department, agency, or instrumentality of the United States for the purpose of carrying out research, tests, or experiments.[23]

There are over 14,000 dealers, exhibitors, research facilities, carriers, and intermediate handlers with a license or registration under the AWA.[24] The Animal and Plant Health Inspection Service (APHIS) requires each research facility to submit an annual report indicating the number of animals used, the species, surgeries conducted, and pain medications and drugs administered.[25] The AWA sets forth minimum standards for the humane care and treatment of animals in research facilities, focusing on proper housing, feeding, veterinary care, and handling of research animals.[25]

The 1985 Amendment established the Animal Welfare Information Center (AWIC) to provide researchers with a database of alternatives to painful animal experiments.[25] The AWIC, founded by the National Agricultural Library in cooperation with the National Library of Medicine, provides information on preventing unintended duplication of animal experimentation and improved methods of animal experimentation, including methods that could reduce or replace animal use and those that minimize pain and distress to animals.[25] The 1985 Amendment also required each research facility in the United States to establish an IACUC to review all experimental protocols involving live, warm-blooded animals.[25]

The AWA establishes regulations to ensure the humane treatment and care of animals used in research, exhibition, transport, and commerce, known as the Animal Welfare Regulations (AWR).[26] Key aspects of the AWR involve obtaining licenses or registrations from the USDA for regulated activities involving animals, establishing minimum standards of care, authorizing the USDA to conduct inspections of regulated facilities, maintaining records documenting activities involving animals, providing transportation provisions, and establishing IACUCs to oversee animal research protocols.[26] Overall, the AWR under the AWA establishes standards and requirements to promote the humane treatment and care of animals in various settings, including research, exhibition, transportation, and commerce.[26]

The PHS Policy

Issued by the Office of Laboratory Animal Welfare of the NIH, the PHS Policy[27] sets forth guidelines for institutions receiving PHS funding—from the NIH, Centers for Disease Control and Prevention, FDA, and others—as well as other entities that have a Memorandum of Understanding with the NIH, including the National Science Foundation,[28] the National Aeronautics and Space Administration,[29] and the Department of Veterans Affairs.[30] The PHS Policy is a set of principles, guidelines, and regulations aimed at ensuring the humane treatment of animals used in scientific research, testing, and education. Specific requirements pertain to the establishment of IACUCs, appropriate housing and husbandry practices, and measures to minimize pain and distress.[27]

Any institution receiving PHS funding and conducting animal research must periodically obtain an Animal Welfare Assurance from the NIH Office of Laboratory Animal Welfare, which, among other details, requires reporting the gross square footage of each animal facility, the species housed therein, and the average daily inventory, by species, of animals in each facility. Institutions are also required to report noncompliance and take corrective action to address noncompliant animal care and use practices.[27] One notable distinction between the PHS and AWA is that the PHS applies to all vertebrate animals used in PHS-funded research, while such nonhuman animals are not specifically covered under the AWA. This regulatory gap means that experiments using invertebrates often face minimal welfare requirements or review, potentially leading to unnecessary suffering and missed opportunities to refine experimental methods.[31] As the use of invertebrates is increasing in neuroscience, behavior studies, and environmental research, this lack of protection may impede the development of best practices that could enhance both animal welfare and scientific validity. The few jurisdictions that do protect certain invertebrates, notably the EU's inclusion of cephalopods, demonstrate that extending such protections is both feasible and scientifically valuable.

The PHS Policy is overseen by the Department of Health and Human Services and was developed under the Health Research Extension Act of 1985.[27] The PHS endorses the "U.S. Government Principles for the Utilization and Care of Vertebrate Animals Used in Testing, Research, and Training," which stipulate ten principles, including that animal experiments be designed and performed considering their relevance to human or animal health, the advancement of knowledge, or the good of society.[27] The PHS Policy also requires that grant applicants describe the rationale for involving animals, which is reviewed by NIH scientific review groups in determining the appropriateness of the proposed animal use.[32]

Institutional Animal Care and Use Committees

As mentioned, both the AWA and PHS Policy establish IACUC requirements to oversee the welfare of research animals.[25,27] IACUCs are self-regulating bodies required to create their own regulations within the scope of federal guidelines and regulations and must consist of at least five appointed members, including a veterinarian, a practicing scientist experienced in animal research, a person whose primary concerns lie in a nonscientific area, and a member of the public.

IACUC oversight includes reviewing institutional animal programs, reporting compliance and deficiencies, and reviewing, approving, requiring modifications to, or withholding approvals of covered animal activities. In these latter protocol reviews, IACUCs are required to determine that animal activities avoid or minimize discomfort, distress, and pain; appropriately specify euthanasia methods; and appropriately specify living conditions.[27] Inspections are required by both the PHS Policy and the AWA.[25,27] PHS Policy requires IACUCs to inspect all the institution's animal facilities at least once every six months.[27] APHIS and USDA conduct at least one unannounced inspection annually of each research facility covered by the AWA.[25] These inspections must review the condition of the animals, veterinary care, IACUC records, research protocols, and the adequacy of the institution's facilities. Any deficiencies noted during the inspections must be documented, and the institution must develop and implement a plan to correct them. Failure to correct deficiencies may result in the suspension of PHS funding or the suspension or revocation of the institution's AWA registration.[27]

Criticism of IACUCs often centers around issues of self-regulation and the rarity of protocol disapprovals. Because IACUCs are typically composed of members within the same institution conducting the research, concerns arise about potential conflicts of interest and pressures to prioritize research goals over animal welfare.[33–35] This self-regulation model can create an environment where critical evaluation of research is compromised. Furthermore, the infrequency with which IACUCs outright disapprove research protocols—with some studies suggesting rejection rates of <1%[36]—raises questions about the depth and effectiveness of their oversight. While low disapproval rates could be attributed to researchers proactively addressing IACUC concerns, they also suggest a potential lack of rigorous scrutiny.[37]

The Guide for the Care and Use of Laboratory Animals (The Guide)

The Guide provides detailed recommendations for the humane care and use of laboratory animals.[38] The Guide was created in 1963 and is periodically updated to reflect the latest standards in animal welfare. Although in its Eighth Edition, it has not been updated since 2011. Institutions that receive PHS funding or are accredited by the Association for Assessment and Accreditation of Laboratory Animal Care International (AAALAC) must comply with the recommendations of the Guide. The Guide covers various aspects of animal care, including housing, environment, diet, veterinary care, and the qualifications and training of personnel involved in animal care and use, though it only pertains to vertebrate animals.

The FDA Modernization Act 2.0

Signed into law by President Joseph R. Biden Jr. on December 29, 2022, the FDA Modernization Act 2.0 of 2022 allows alternatives to animal testing in investigational new drug applications submitted to the FDA.[39] This modernization effort does not eliminate animal testing; rather, it allows for greater flexibility in testing methods and emphasizes the use of nonanimal alternatives when appropriate. New testing methods may include computer modeling, cell-based assays, organ-on-a-chip technology, and other innovative approaches that do not involve animals. This change is part of a broader effort to reduce reliance on animal testing while still ensuring the safety and efficacy of new drugs and treatments.[40] While promising progress has been made, the implementation of the FDA Modernization Act 2.0 still remains incomplete. The FDA Modernization Act 3.0 is a bipartisan effort to mandate the agency to update its regulations that still require animal testing in order to fulfill the aim of the earlier legislation.[41]

The regulatory outlook

The regulatory framework governing animal research in the United States, while intended to ensure the humane treatment and care of research animals, faces criticisms regarding its effectiveness and potential to perpetuate reliance on animal models. The AWA and PHS Policy, along with IACUC oversight and the Guide, create a system for monitoring animal conditions, but some argue that these regulations are insufficient. Enforcement is inconsistent, and the emphasis on "humane treatment" within the existing research paradigm may not adequately address the fundamental ethical questions surrounding animal use. Furthermore, the system is often criticized for prioritizing the continuation of animal research over actively promoting and incentivizing the development and implementation of nonanimal alternatives. While the FDA Modernization Act represents a step toward reducing reliance on animal testing, critics argue that progress is slow and that systemic resistance within the research community hinders a more rapid transition to nonanimal approaches. The regulatory structure, therefore, while providing a framework for oversight, may inadvertently work to uphold the existing research system and its continued reliance on animal models, rather than aggressively driving a shift toward more ethical and potentially more effective nonanimal methods.

EU policy and procedures regarding authorization of animal research

The European Union (EU) has implemented strict laws and procedures overseeing the approval of animal research to protect the welfare of the participating animals and further scientific understanding. A number of Directives, or legal acts, have been passed by the EU that outline objectives for its member states to meet. The Directives are meant to be a guide for EU member states when they draft their own legislation, not legally binding acts.[16] In 1986, the EU issued its first Directive (Directive 86/609/EEC)[42] to protect animals in science. Another legal instrument also passed in 1986 was the European Convention for the Protection of Vertebrate Animals used for Experimental and Other Scientific Purposes,[43] published by the Council of Europe. Directive 86/609/EEC remained the standard until 2002, when Europe reviewed the existing framework because of scientific and technological advancements. Following the UK's departure from the EU, concerns emerged about the UK losing access to important regulations, including those that promoted the 3Rs in chemical testing.[44] It remains to be seen whether the UK will fall behind the EU or whether the country will use this as an opportunity to diverge from EU policy to advance protections for animals used in experiments.

Researchers in the EU are required to submit comprehensive project proposals that delineate the research objectives, methodologies, anticipated results, and an evaluation of potential harm to animals, along with strategies to mitigate suffering. The ethical assessment of research proposals constitutes a pivotal aspect of the authorization process. Each project undergoes rigorous examination by animal welfare bodies or ethics committees, consisting of specialists in animal welfare, scientific disciplines, and ethical considerations.

Despite the EU's robust regulatory framework, the ongoing challenge of reconciling animal welfare with scientific progress persists.[45] Maintaining this balance requires continual improvement of authorization protocols, advocacy for alternative approaches, and efforts to enhance transparency and public participation. Furthermore, technological advancements such as advanced computational models and organ-on-a-chip systems offer promising alternatives to conventional animal testing, presenting opportunities to reduce reliance on animal experimentation.[46]

International legislative context and comparisons

Key organizations shaping international standards for animal research include the World Health Organization, the World Organization for Animal Health, and the Council of Europe's European Convention for the Protection of Vertebrate Animals Used for Experimental and Other Scientific Purposes.[45,47]

The global scientific community appears to share a consensus regarding the responsibility held by laboratories for the welfare of animals involved in research endeavors. The guidelines regulating the use of laboratory animals are influenced by societal norms, cultural customs, and historical precedents, all overseen by legal and cultural institutions responsible for their creation and enforcement. Globally, the Three R principles are frequently embraced.

Latin America

In Latin America, much of the laws and regulations regarding animal research are driven by economic factors, with Brazil, Mexico, and Uruguay having the most specific legislation relating to animal care and use. Other Latin American countries have little or no legal framework or governmental infrastructure.[47]

Brazil

The history of animal research laws and regulations in Brazil is marked by significant developments aimed at improving animal welfare. Early efforts to protect animals began with Decree No. 24,645 in 1934, which provided basic protections against animal cruelty. This decree was later replaced by Decree No. 9662 in 2019, which continued to emphasize the state's role in safeguarding animals.[48]

In 1967, the Fauna Protection Act (Law No. 5.197)[49] was introduced, imposing strict penalties for crimes against wildlife. However, the Environmental Crimes Law (Law No. 9.605)[50] of 1998 replaced this act, offering a more lenient framework by considering many environmental crimes as lesser offenses, often substituting imprisonment with community service or fines.[51] This law includes specific provisions for crimes against wild fauna, such as unauthorized hunting or capturing, with penalties ranging from fines to imprisonment.[51]

There has been a growing movement toward stronger animal protection laws. For example, the Penal Code was amended in 2012 to increase penalties for animal cruelty, and Senate Bill 631/2015 seeks to provide clearer definitions of abuse and harsher penalties.[48] In addition, Brazil has taken steps to regulate the use of animals in scientific research, reflecting a broader trend of improving animal welfare standards in various sectors.

Overall, Brazil's approach to animal welfare has evolved from basic protections to more comprehensive regulations, driven by both legislative changes and public advocacy.[48,51]

Mexico

The history of animal research laws and regulations in Mexico reflects a progressive alignment with international standards for animal welfare and ethical scientific practice. In 1988, the General Law of Ecological Balance and Environmental Protection laid the groundwork for regulating animal research. It underscored the importance of humane treatment and the prevention of cruelty to animals.[52] Subsequently, the Federal Law of Animal Health, enacted in 2007, introduced specific provisions for the care and handling of animals in research, including the necessity of adequate nutrition, hydration, and veterinary care.[53]

A significant milestone was the establishment of the Official Mexican Standard NOM-062-ZOO-1999, which outlines the technical specifications for the production, care, and humane treatment of laboratory animals. This standard mandates that research institutions form IACUCs to oversee and ensure the ethical conduct of animal research.[54] Moreover, the regulatory framework requires that any research involving animals must be approved by the National Commission for the Protection Against Sanitary Risks, ensuring that all protocols adhere to stringent ethical guidelines designed to minimize animal suffering.[55]

At the local level, various states have enacted their own animal protection laws to address specific welfare concerns within their jurisdictions, complementing federal regulations and providing a more comprehensive protection network for animals used in research. State laws in Aguascalientes and Baja California, among others, set forth detailed anticruelty provisions and responsibilities for animal caretakers.[54]

These developments indicate Mexico's ongoing commitment to enhancing the welfare of animals in research, ensuring ethical practices, and integrating with global standards for the humane treatment of animals in scientific contexts.

Uruguay

Uruguay has developed and enforced specific regulations for laboratory animal research, despite its relatively modest scientific community. The country's initial animal-related legislation dates back to 1918, when it focused mainly on outlawing activities such as bullfighting. However, significant progress toward regulating animal research was made in 2000 with the introduction of the "University of the Republic (UdelaR) Regulations: Use of Animals in Testing, Research and Teaching."[47] Inspired by Brazil's regulatory model, Uruguay subsequently enacted additional regulations in 2009 and 2014, establishing comprehensive frameworks for overseeing animal-based research. These regulations cover various aspects, including the establishment of the National Commission for Animal Experimentation (CNEA), the composition and authority of IACUCs, guidelines for animal breeding and use, and penalties for noncompliance. Moreover, the CNEA plays a crucial role in promoting animal welfare and implementing the Three Rs in research practices.[47]

The National Registry of Institutions oversees the registration of animal research and breeding centers, regulates the use and transport of animals for scientific purposes, and manages the Personnel Accreditation National System, which ensures training and licensing for individuals involved in animal research. The CNEA, chaired by the Ministry of Education and Culture, comprises representatives from various government ministries, educational institutions, research agencies, industry associations, veterinary societies, and animal welfare organizations. Since its inception, the CNEA has made significant progress in developing operational rules, establishing subcommittees, creating online resources, and implementing the National Control System for monitoring research facilities and experimental procedures. Looking ahead, the CNEA aims to enhance the National Control System, expand the National Register of Experimental Procedures, and promote the development of alternative methods and the harmonization of animal research policies within the region.[47]

The Pacific Rim

Various countries in the Pacific Rim, such as Singapore, Thailand, Indonesia, and Malaysia, have diverse levels of infrastructure and regulatory frameworks for animal research, influenced by economic development, culture, and religion. Notably, Singapore has a robust research environment, leading to the formulation and adoption of the "Guidelines on the Care and Use of Animals for Scientific Purposes" by the National Advisory Committee for Laboratory Animal Research in 2004. These guidelines delineate the roles and responsibilities of institutions, scientists, and animal care personnel in animal research endeavors, with a focus on implementing the principles of the Three Rs.[56]

India

In India, the Prevention of Cruelty to Animals Act of 1960 established the foundation for regulating animal research and welfare.[57] This legislation led to the creation of the Committee for the Purpose of Control and Supervision of Experiments on Animals (CPCSEA), which is responsible for overseeing the ethical treatment of animals used in research. The CPCSEA's guidelines, known as the "Guidelines for Laboratory Animal Facility," cover comprehensive aspects of animal care, including veterinary care, quarantine measures, procurement, husbandry, and personnel training.[58]

Ethical considerations in Indian scientific experimentation with animals are guided by five core principles: the advancement of knowledge, minimizing animal pain and suffering, moral responsibility toward research animals, appropriate euthanasia practices, and ensuring suitable living conditions. In addition, the Prevention of Cruelty to Animals Act of 1960 emphasizes the prevention of unnecessary pain and suffering and mandates the establishment of animal welfare committees to monitor compliance with these principles. The draft amendments to the law increase penalties for violations and reinforce the importance of humane treatment of animals.[59]

Australia

In Australia, protective legislation for laboratory animals is implemented at the state level, leading to variations in regulatory frameworks across different territories. Each state operates under its own set of regulations, but all are anchored in the Australian Code for the Care and Use of Animals for Scientific Purposes. Initially introduced in 1969 and last amended in 2004, this code emphasizes the Three Rs to minimize animal suffering and promote ethical research practices.[60] Despite the advantages of decentralized oversight, there have been calls for a centralized regulatory body to govern animal research at the national level in Australia, as recommended by Marston et al.[61] These calls stem from the desire for more consistent enforcement of ethical standards and streamlined regulatory processes across all states and territories.[61]

China, Japan, and Korea

Regulations concerning animal research vary widely in China, Japan, and Korea, reflecting different approaches to animal welfare and research oversight. In China, animal welfare legislation was first introduced in 1983, with the Ministry of Health initially responsible for issuing regulations. The Ministry of Science and Technology administers the "Statute on the Administration of Laboratory Animals" (Order No. 2), which comprehensively addresses experimentation and laboratory animal care. Local governments further reinforce this legislation through their own laws, creating a multilayered regulatory system that includes central, regional, and institutional oversight levels.[60]

Japan's regulatory framework is more decentralized, governed by the Law for the Humane Treatment and Management of Animals (Law No. 105, 1973). In Japan, institutions are largely self-regulated, and while the Ministry of Education mandates the establishment of IACUCs, these committees are guided by ministerial guidelines rather than stringent legislative measures. These guidelines cover humane endpoints, pain management, euthanasia, and safety precautions, but there are no formal government inspections or mandatory reporting obligations.[62]

South Korea also follows a detailed regulatory framework similar to China, with specific laws and guidelines governing the use of animals in research. The regulatory environment in Korea includes mandatory requirements for licensing and registration of research facilities, as well as the establishment of IACUCs to oversee animal welfare.[63]

The differences among these countries highlight the varied approaches to balancing scientific advancement with animal welfare, with each country integrating elements of the Three Rs to varying extents in its respective regulatory practices.

Animal research accountability and transparency

Transparency in publicly funded animal research is critical for several reasons. Ethically, the public has the right to know how their taxpayer dollars are being spent and whether research aligns with societal values, particularly concerning animal welfare.[64] Scientifically, transparency is essential when validating research findings and facilitating reproducibility.[65] Without adequate transparency, the public, who essentially provides taxpayer dollars for research, cannot hold institutions accountable. Institutions are susceptible to fund misuse and unethical treatment of animals.[66]

Several factors contribute to the lack of transparency in publicly funded animal research. Research institutions often operate with a high degree of secrecy, citing intellectual property concerns and competitive advantages. While these are legitimate considerations, they often come at the expense of public accountability and transparency. Existing regulations, such as the AWA in the United States, are frequently criticized for their limitations.[67] The AWA, for instance, requires USDA-registered and federal research facilities to submit an annual report documenting their use of animals, including the number of animals held and used in research and a classification of pain and distress.[68] However, the AWA does not cover all species used in research, and enforcement of the regulations can be inconsistent. Reporting requirements are often minimal and lack detail, focusing on numbers and broad categories rather than the specifics necessary for meaningful public scrutiny and evaluation of the ethical and scientific validity of animal research. The PHS Policy covers more species, but its reporting requirements are even more minimal than the AWA—in terms of numbers, it merely requires institutions to report an average daily inventory of animals once per Assurance Period, approximately every five years, and these numbers are not publicly available. Thus the number of mice and rats used each year in US laboratories—by far the most commonly used species—is nearly impossible to obtain via publicly available information.[69]

The lack of transparency in publicly funded animal research has significant consequences. Without transparency, there is no way to ensure that ethical standards for the treatment of animals are being upheld.[60] This can lead to animal suffering and unethical research practices going unnoticed and unaddressed. Public trust in scientific research and institutions is crucial. Lack of transparency can lead to suspicion, mistrust, and opposition to animal research.[64] This mistrust can extend to other areas of scientific inquiry, undermining the broader scientific enterprise. Transparency is a cornerstone of scientific integrity. Without it, the reproducibility and validity of research findings may be compromised,[65] which can lead to wasted resources and efforts, as well as a delay in scientific and medical advancements. In addition, lack of transparency can obscure misuse of funds and prevent proper oversight and accountability.

Improving transparency in publicly funded animal research requires coordinated efforts from government agencies, research institutions, and the scientific community. Regulatory bodies should enforce detailed and standardized reporting of all animal research, encompassing the number of animals used, species, research objectives, outcomes, and welfare considerations.[67] These reports must be publicly accessible and easily understandable. Establishing open-access databases to register and update all publicly funded animal research projects can significantly enhance transparency.[64] These databases should include comprehensive details about research methodologies, funding, and findings.

Although the NIH RePORTER is a publicly available database for NIH-funded grants, its user interface is not particularly user-friendly, posing challenges in accessing specific information. For instance, a search for "primate" may not yield results if the grant lists "primates" in the plural form. Frequent audits and inspections conducted by independent entities can guarantee compliance with ethical standards and transparency requirements. Making the subsequent audit reports public can contribute to fostering accountability.[67]

AAALAC International claims to ensure high standards for animal care and use in research. However, concerns about the effectiveness and transparency of their oversight persist. Despite AAALAC's rigorous accreditation process, its voluntary nature leads to inconsistencies in animal care standards across institutions.[69] In addition, the lack of public access to AAALAC inspection reports and accreditation findings limits transparency and accountability, making it difficult for stakeholders to assess the true conditions and ethical standards of accredited facilities.[64,67]

Engaging the public through education and outreach about the importance, processes, and ethical considerations of animal research can help build trust and support for necessary research activities.[64] Thus transparent communication about both successes and challenges in animal research is crucial, and shifting the scientific culture toward one that values and rewards transparency is essential. This includes encouraging the publication of negative results, detailed methodology reporting, and comprehensive data sharing.

Transparency in publicly funded animal research is not merely a bureaucratic necessity but a moral and scientific imperative. It ensures the ethical treatment of animals, fosters public trust, upholds scientific integrity, and guarantees the responsible use of public funds. While significant barriers exist, concerted efforts by all stakeholders can lead to greater transparency and accountability, ultimately benefiting science, society, and animal welfare. As we move forward, it is essential to recognize that transparency is not an obstacle to scientific progress but a foundation on which it is built.

Conclusion

The evolution of animal research legislation represents a complex interplay between scientific advancement and ethical considerations. While global regulatory frameworks have progressed from initial limited protections to more comprehensive oversight mechanisms, significant questions remain about the adequacy of existing safeguards. The institutional self-regulation model, coupled with a research culture that often prioritizes productivity and profit, raises fundamental concerns about the effectiveness of existing protective measures and their implementation across research facilities. The existing legislative paradigm in many countries, while establishing baseline standards for animal welfare in research, inherently perpetuates a system that views animals as resources rather than sentient beings deserving of inherent respect. This philosophical tension manifests in practical challenges regarding enforcement, species coverage, and transparency.

The premise of balancing animal welfare against research needs has created a regulatory environment that, despite its intentions, may inadvertently impede the development of innovative nonanimal methodologies that could advance both ethical and scientific objectives. The path forward requires a fundamental shift that extends beyond merely strengthening existing regulations. A comprehensive approach must encompass active investment in and validation of nonanimal methodologies. Moreover, enhanced transparency and public accountability mechanisms are essential to ensure compliance with ethical standards and foster public trust in scientific research. This transition necessitates creating regulatory pathways that explicitly prioritize the adoption of alternative methods, supported by robust funding initiatives and educational programs to facilitate the implementation of nonanimal research paradigms.

AI disclosure

During the preparation of this work, the author used Grammarly to access basic spellcheck features of the Grammarly app. After using this tool/service, the author reviewed and edited the content as needed and takes full responsibility for the content of the publication.

References

[1] Rupke NA (Ed.). Vivisection in Historical Perspective. Croom Helm; 1987.

[2] Knight A. The Costs and Benefits of Animal Experiments. Palgrave Macmillan; 2011.

[3] The Nuremberg Code. In Trials of War Criminals Before the Nuremberg Military Tribunals under Control Council Law No. 10. Vol 2. Washington, DC: U.S. Government Printing Office; 1949. pp. 181–182.

[4] Shuster E. Fifty years later: the significance of the Nuremberg Code. N Engl J Med. 1997;337(20):1436–1440.

[5] Nuremberg Code. Trials of War Criminals Before the Nuremberg Military Tribunals Under Control Council Law No. 10. Vol 2. U.S. Government Printing Office; 1947. pp. 181–182.

[6] World Medical Association. Declaration of Helsinki. Available at: https://www.wma.net/wp-content/uploads/2018/07/DoH-Jun1964.pdf. Accessed Aug 11, 2026.

[7] World Medical Association. WMA Declaration of Helsinki—ethical principles for medical research involving human subjects. Available at: https://www.wma.net/wp-content/uploads/2024/10/DoH-Oct2013.pdf. Accessed Aug 11, 2026.

[8] World Medical Association Declaration of Geneva, Adopted October 1949. Available at: https://www.wma.net/wp-content/uploads/2024/10/DoH-Oct2013.pdf. Accessed Aug 11, 2026.

[9] Council for International Organizations of Medical Sciences. Geneva, Switzerland. Proposed International Ethical Guidelines for Biomedical Research Involving Human Subjects; 1982. Available at: https://cioms.ch/publications/product/proposed-international-guidelines-for-biomedical-research-involving-human-subjects/. Accessed Aug 11, 2026.

[10] Department of Health, Education and Welfare The National Commission for the Protection of Human Subjects of Biomedical and Behavioral Research. The Belmont Report: ethical principles and guidelines for the protection of human subjects of research. https://www.hhs.gov/ohrp/sites/default/files/the-belmont-report-508c_FINAL.pdf; 1979. Accessed Oct 16, 2024.

[11] International Conference on Harmonization of Technical Requirements for Registration of Pharmaceuticals for Human Use. Guideline for Good Clinical PracticeE6(R1). 1966. Available at: https://acrpnet.org/wp-content/uploads/dlm_uploads/2016/11/ICH-E6-R1-Guideline-For-Good-Clinical-Practice.pdf. Accessed Aug 11, 2026.

[12] Council for International Organizations of Medical Sciences (CIOMS). International Ethical Guidelines for Health-Related Research Involving Humans. CIOMS; 2016. Available at: https://cioms.ch/publications/product/international-ethical-guidelines-for-health-related-research-involving-humans/. Accessed Aug 11, 2026.

[13] UNESCO. Universal declaration on bioethics and human rights. 2005. Available at: https://www.unesco.org/en/legal-affairs/universal-declaration-bioethics-and-human-rights?hub=387. Accessed Aug 11, 2026.

[14] Animal Research Tomorrow. The Basel Declaration. Adopted Nov. 2010;29. Available at: https://animalresearchtomorrow.org/en/basel-declaration. Accessed Aug 11, 2026.

[15] U.S. Department of Health and Human Services. Code of Federal Regulations, Title 45, § 46.101; 2024. Available at: https://www.ecfr.gov/current/title-45/subtitle-A/subchapter-A/part-46. Accessed Aug 11, 2026

[16] European Union. Directive 2010/63/EU of the European Parliament and of the Council of 22 September 2010 on the protection of animals used for scientific purposes. Off J Eur Union. 2010;276:33–79.

[17] Smith JA, Andrews PLR, Hawkins P, et al. Cephalopod research and EU Directive 2010/63/EU: requirements, impacts and ethical review. J Exp Mar Biol Ecol. 2013;447:31–45.

[18] Legislation.gov.uk. Animals (Scientific Procedures) Act 1986. https://www.legislation.gov.uk/ukpga/1986/14/contents. Accessed March 11, 2025.

[19] Japanese Law Translation. Act on Welfare and Management of Animals. Act No.105 of October 1, 1973. https://www.japaneselawtranslation.go.jp/en/laws/view/3798/en; 1973. Accessed March 4, 2025.

[20] Cornell Law School Legal Information Institute. U.S. Code of Federal Regulations. 7 U.S.C. § 2131 et seq. https://www.law.cornell.edu/uscode/text/7/chapter-54. Accessed Oct 16, 2024.

[21] Cornell Law School Legal Information Institute. U.S. Code of Federal Regulation, Title 7, Chapter 54, Section 2131. https://www.law.cornell.edu/uscode/text/7/2131. Accessed March 11, 2025.

[22] Cornell Law School Legal Information Institute. U.S. Code of Federal Regulation, Title 7, Chapter 54, Section 2146. https://www.law.cornell.edu/uscode/text/7/2146. Accessed Oct 16, 2024.

[23] Cornell Law School Legal Information Institute. U.S. Code of Federal Regulation, Title 7, Chapter 54, Section 2132. https://www.law.cornell.edu/uscode/text/7/2132. Accessed Oct 16, 2024.

[24] U.S. Department of Agriculture. Animal and Plant Health Inspection Service: list of persons licensed or registered under the Animal Welfare Act (AWA); 2024. https://aphis.my.site.com/PublicSearchTool/s/. Accessed 16 October 2024.

[25] Cornell Law School Legal Information Institute. U.S. Code of Federal Regulation, Title 7, Chapter 54, Section 2143. https://www.law.cornell.edu/uscode/text/7/2143. Accessed Oct 15, 2024.

[26] Cornell Law School Legal Information Institute. U.S. Code of Federal Regulation, Title 7, Chapter 54. https://www.law.cornell.edu/uscode/text/7/chapter-54. Accessed Oct 16, 2024.

[27] National Institutes of Health. PHS Policy on Humane Care and Use of Laboratory Animals. Revised 2015. https://olaw.nih.gov/policies-laws/phs-policy.htm. 2015. Accessed March 3, 2025.

[28] National Institutes of Health. Office of Laboratory Animal Welfare: MOU Between NIH and NSF; 2020. https://grants.nih.gov/policy-and-compliance/policy-topics/animal-welfare/laws-regulations/regulatory-references#mou-between-nih-and-nsf. Accessed Mar 3, 2025.

[29] National Institutes of Health. Office of Laboratory Animal Welfare: MOU between NIH and NASA. 2024. https://grants.nih.gov/policy-and-compliance/policy-topics/animal-welfare/laws-regulations/regulatory-references#mou-between-nih-and-nasa. Accessed Mar 3, 2025.

[30] National Institutes of Health. Office of Laboratory Animal Welfare. MOU between NIH and VA. https://grants.nih.gov/policy-and-compliance/policy-topics/animal-welfare/laws-regulations/regulatory-references#mou-between-nih-and-va; 2023. Accessed Mar 3, 2025.

[31] Passantino A, Elwood RW, Coluccio P. Why protect decapod crustaceans used as models in biomedical research and in ecotoxicology? Ethical and legislative considerations. Animals. 2021;11(1):73. https://www.mdpi.com/2076-2615/11/1/73. Accessed Mar 3, 2025.

[32] National Institutes of Health. Office of Laboratory Animal Welfare: vertebrate animals section. https://olaw.nih.gov/guidance/vertebrate-animal-section.htm. Accessed Mar 3, 2025.

[33] Carbone L. What Animals Want: Expertise and Advocacy in Laboratory Animal Welfare Policy, 1st ed., Oxford University Press; 2004.

[34] Hansen LA. Institution animal care and use committees need greater ethical diversity. J Med Ethics. 2013;39(3):188–190.

[35] Silverman J, Lidz CW, Clayfield JC, et al. Decision making and the IACUC: part 1—protocol information discussed at full-committee reviews. J Am Assoc Lab Anim Sci. 2015;54(4):389–398.

[36] Dresser R. Developing standards in animal research review. J Law Med Ethics. 1989;17(2):158–172.

[37] Plous S, Herzog H. Reliability of protocol reviews for animal research. Science. 2001;293(5529):608–609.

[38] National Academies of Sciences. Engineering, and Medicine. Guide for the Care and Use of Laboratory Animals, 8th ed., The National Academies Press; 2011.

[39] S.5002—117th Congress (2021-2022) FDA Modernization Act 2.0. Passed 9/29/2022. https://www.congress.gov/bill/117th-congress/senate-bill/5002. Accessed Mar 3, 2025.

[40] H.R. 7667—Food and Drug Amendments of 2022. Food and Drug Omnibus Reform Act of 2022, Title VII, Subtitle A, Section 201. https://www.congress.gov/bill/117th-congress/house-bill/7667/text#toc-H48536F4F27354D158D49E24CD1F77B6D. 2022. Accessed March 3, 2025.

[41] Booker, Schmitt, Paul, King, Braun, Whitehouse, Kennedy, Luján, Blumenthal introduce the FDA Modernization Act 3.0; 2024. https://www.booker.senate.gov/news/press/booker-schmitt-paul-king-braun-whitehouse-kennedy-lujan-blumenthal-introduce-the-fda-modernization-act-30. Accessed March 3, 2025.

[42] Council of the European Communities. Directive 86/609/EEC of 24 November 1986 on the protection of animals used for experimental and other scientific purposes. Off J Eur Communities. 1986. https://eurlex.europa.eu/legal-content/EN/LSU/?uri=oj:JOL_1986_358_R_0001_01.

[43] Council of Europe. European Convention for the protection of vertebrate animals used for experimental and other scientific purposes (ETS 123). https://www.coe.int/en/web/conventions/full-list?module=treatydetail&treatynum=123. Accessed March 11, 2025.

[44] Dunn R. Brexit: a boon or a curse for animals used in scientific procedures? Animals. 2021;11(6). https://www.mdpi.com/2076-2615/11/6/1547. Accessed March 11, 2025. doi:10.3390/ani11061547.

[45] Gribble K. Animal Research Ethics: evolving views and practices. Anim Front. 2018;8(2):32–37.
[46] Marx U, Akabane T, Andersson TB, et al. Biology-inspired microphysiological systems to advance patient benefit and animal welfare in drug development. ALTEX. 2020;37(3):365–394.
[47] Revera E, Hernandez R, Carissimi AS, Pekow C. Laboratory animal legislation in Latin America. ILAR J. 2016;57(3):293–300.
[48] World Animal Protection. Brazil. https://api.worldanimalprotection.org/country/brazil. Accessed Oct 17, 2024.
[49] UN Environmental Program. Act No.5.197 providing for the protection of wild fauna. LEAP (Law and Environmental Assistance Program). https://leap.unep.org/en/countries/br/national-legislation/act-no-5197-providing-protection-wild-fauna; 1967. Accessed Oct 16, 2024.
[50] Animal Legal and Historical Center. Brazilian Environmental Crimes Law: law No.9.605. https://www.animallaw.info/statute/brazil-crimes-brazilian-environmental-crimes-law. Accessed Oct 16, 2024.
[51] Animal Legal & Historical Center. Overview of Brazil's legal structure for animal issues. Retrieved from https://www.animallaw.info/article/overview-brazils-legal-structure-animal-issues. Accessed March 11. 2025.
[52] Animal Legal & Historical Center. Mexico. Available at: https://www.animallaw.info/site/search-results?googlesearch=LGEEPA. Accessed Oct 16, 2024.
[53] Animal Legal & Historical Center. Mexico. Available at: https://www.animallaw.info/intro/mexico. Accessed Oct 17, 2024.
[54] World Animal Protection. Mexico. Available at: https://api.worldanimalprotection.org/country/mexico. Accessed Oct 16, 2024.
[55] National Institute of Allergy and Infectious Diseases. Clinical Research Regulation for Mexico. ClinRegs; 2023. https://clinregs.niaid.nih.gov. Accessed March 11, 2025.
[56] Retnam L, Chatikavanij P, Kunjara P, et al. Laws, regulations, guidelines and standards for animal care and use for scientific purposes in the countries of Singapore, Thailand, Indonesia, Malaysia and India. ILAR J. 2016;57(3):312–323.
[57] Animal Welfare Board of India. Prevention of Cruelty to Animals Act, 1960. Available at: https://www.globalanimallaw.org/downloads/database/national/india/the_prevention_of_cruelty_to_animals_act_1960.pdf. 1960 Accessed Oct 16, 2024.
[58] Committee for the Purpose of Control and Supervision of Experiments on Animals (CCSEA). Acts, rules, and guidelines; 2024. https://ccsea.gov.in/Content/54_1_ACTS,RULESANDGUIDELINES.aspx. Accessed 16 October 2024.
[59] Government of India. Draft Prevention of Cruelty to Animal Act (Amendment) Bill, 2022 for amendment of prevention of cruelty to animal act, 1960-reg. https://prsindia.org/files/parliamentry-announcement/2022-12-07/Public%20notice-Draft%20PCA%20bill-2022.pdf 2022. Accessed Oct 16, 2024.
[60] National Health and Medical Research Council. Australian Code of Practice for the Care and Use of Animals for Scientific Purposes. Commonwealth of Australia: Canberra; 2004. Retrieved from https://www.nhmrc.gov.au/about-us/publications/australian-code-care-and-use-animals-scientific-purposes. Acessed March 11, 2025.
[61] Marston HM, Bennett AJ. Ethical principles and guidelines for scientific research with animals. Adv Exp Med Biol. 2016;123:67–80.
[62] MacArthur Clark JA, Sun D. Guidelines for the ethical review of laboratory animal welfare People's Republic of China National Standard GB/T 35892-2018 [Issued 6 February 2018 Effective from 1 September 2018]. Animal Models Exp Med. 2020; https://onlinelibrary.wiley.com/doi/10.1002/ame2.12111.
[63] Animal Protection Act [of South Korea], Wholly Amended by Act No. 18853, 2022. https://law.go.kr/lsInfoP.do?lsiSeq=241931&chrClsCd=010203&urlMode=engLsInfoR&viewCls=engLsInfoR#0000. Accessed March 4, 2025.
[64] Ferdowsian HR. Beck N. Ethical and scientific considerations regarding animal testing and research. PLoS ONE. 2011;6(9):e24059. doi:10.1371/journal.pone.0024059.
[65] Ioannidis JPA. Why most published research findings are false. PLoS Med. 2005;2(8):e124. doi:10.1371/journal.pmed.0020124.
[66] Orlans FB. Ethical themes of animal experimentation. In: Evans DL, Kessel WP, eds. The Science of Animal Research: A Guide for Informal Science Educators. National Academy Press; 1993, pp. 93–105.
[67] Bayne K, Anderson LC, Bennett BT. Animal research: a review of developments, ethical issues, and standards. ILAR J. 1998;57(3):301–311.

[68] USDA Animal and Plant Health Inspection Service. U.S. Department of Agriculture. Research Facility Annual Usage Summary Report. 2024. Available at: https://www.aphis.usda.gov/awa/research-facility-report/annual-summary. Accessed March 4, 2025.

[69] Carbone L. Estimating mouse and rat use in American laboratories by extrapolation from Animal Welfare Act-regulated species. Nat Sci Rep Open. 2021. Available at: https://www.nature.com/articles/s41598-020-79961-0. Accessed March 4, 2025.

CHAPTER

3

The validity and translatability of preclinical animal research

Pandora Pound

Safer Medicines Trust, Kingsbridge, United Kingdom

Abstract

There is ample evidence that most preclinical animal research is conducted poorly and does not translate to humans. Significant efforts are underway, therefore, to improve the quality and conduct of animal studies. This chapter considers the various factors with the potential to influence the success of preclinical animal experiments, analyzing them as affecting either internal or external validity. It argues that while certain aspects of internal and external validity can be addressed in preclinical research, a key aspect of external validity cannot—and that is the differences between species. Since all aspects of validity need to be robust for an animal experiment to translate to humans, there is little point in improving other aspects of validity if species differences present an insurmountable challenge to clinical translation.

Keywords: Internal validity; External validity; Animal models; Stroke; Sepsis

Introduction

There is now widespread acknowledgment that most preclinical animal experiments are conducted poorly and fail to translate to the clinical setting. In other words, findings from animal experiments are not generally useful for developing medical treatments for humans. This acknowledgment is based on plentiful, robust evidence, as this chapter will demonstrate. Rather than abandoning the project and using approaches and technologies based on human biology, significant efforts are underway to improve the quality and conduct of animal studies in the hope that better clinical translation will ensue. But are the two linked? Will animal studies that adhere to rigorous scientific standards necessarily lead to better translation?

My short answer to this question is no, but this chapter provides the longer answer. Here, we will explore factors that impact the translation of animal study findings to humans, focusing on the use of animals as "models" of human diseases, where animals are manipulated in an attempt to reproduce a human disease or condition. For an "animal model" of stroke, for example, an animal might undergo surgery to block an artery supplying blood to

Moving Toward Nonanimal Approaches in Medical Research and Testing
https://doi.org/10.1016/B978-0-443-26514-3.00016-3

the brain in an attempt to mimic a human stroke. An experimental drug may then be administered to that animal to see whether it improves the outcome of the stroke. The idea is that the findings generated in the "animal model" will be applicable to humans.

To continue with this example, animal models of stroke have been spectacularly unsuccessful. The vast majority of drugs that were successful in animal studies turned out to either have no efficacy in humans, as with NXY-059[1] or calcium channel blockers,[2] or to be harmful to humans, as with diaspirin,[3] emlimobab,[4] selfotel,[5] and tirilazad,[6] all of which increased the risk of death in clinical trials. The failure of neuroprotective agents is infamous; of more than 1,000 candidate neuroprotective drugs tested in animals, *not one* was found to benefit humans with stroke.[7] This means that, despite decades of research and millions of animals' lives, few options exist for people admitted to the hospital with a stroke. The only drug available for acute stroke is tissue plasminogen activator, or tPA. This thrombolytic (clot-busting) drug was repurposed following its success in humans in the treatment of myocardial infarction (heart attacks), so it was not even dependent upon animal models of stroke for its development. However, tPA is far from ideal. It can only be administered to selected patients who are admitted to a center of expertise within four and a half hours of stroke onset, meaning that a mere 11%–12% of stroke patients receive tPA in England, Wales, and Northern Ireland,[8] for example. Furthermore, of those who receive it, around half will remain dependent or die.[9]

Given this appalling situation, there have been several attempts to improve the scientific rigor of animal studies of stroke, dating back as far as 1999.[10] A quarter of a century later, however, these efforts have failed to bear fruit in terms of benefits for stroke patients. The highly anticipated drug, NXY-059, was developed in adherence to the highest scientific standards of the time, yet after showing promise in animal studies, it still failed to translate to humans.[1] Why is this? And why has animal research failed to translate elsewhere?[11–32]

To answer this question, we must differentiate the many factors that have the potential to affect the success of a preclinical animal experiment. These fall into four broad categories: (1) the experiment itself, i.e., how well the experiment is designed and conducted, including how well the data are analyzed and reported; (2) the adequacy of "animal models," i.e., the extent to which they represent the human disease of interest and reproduce the clinical context in which that disease is treated; (3) the characteristics of animals used in experiments and how well these represent the human population of interest; and (4) the similarities between humans and the species of animal used. In turn, these four categories fall under two overarching groupings: those of internal validity and external validity (Table 1).

TABLE 1 Factors with the potential to affect the success of preclinical animal experiments.

Internal validity	External validity
The experiment itself, i.e., how well it is designed and conducted, including how well the data are analyzed and reported	The adequacy of "animal models", i.e., the extent to which they represent the human disease of interest and reproduce the clinical context in which that disease is treated
	The characteristics of animals used in experiments and how well they represent the human population of interest
	Similarities between humans and the species of animals used

For an animal experiment to translate successfully to humans, all these aspects need to be sound. Or, put more simply, translation will only occur if both internal and external validity are robust.[33]

Internal validity

There are many factors that may threaten the validity of an experiment—factors, in other words, that may undermine the research and make it difficult to trust its findings. When designing experiments, it is vital to take certain steps to minimize the risk of any bias so that the study's conclusions are as robust and trustworthy as possible. In the case of experiments that aim to test the effects of a drug, the only difference between the groups of animals should be the different treatments they receive. Steps to reduce the risk of bias in this context include a method for randomizing animals to treatment or control groups (random allocation), a way of concealing this allocation from those assigning animals to the groups (allocation concealment), and a way of concealing which animals are in the treatment or control groups from those assessing the outcome of experiments (blinded outcome assessment).

Unfortunately, awareness of the risk of bias is low among scientists using animals[34] and only a small proportion take the necessary steps to guard against it. Only a few—between 12%[35] and 37%[36–39]—report using random allocation to assign animals to experimental groups. Similarly, between zero[36] and 15%[38] of animal studies report using allocation concealment, and the reported use of blinded outcome assessment is also low, ranging between zero[36] and 35%.[38] Thus, in a shocking two-thirds of animal studies, scientists fail to take appropriate measures to minimize bias, meaning that their findings cannot be trusted. Studies that fail to use randomization are more likely to report positive findings than studies that do; in other words, they overstate the benefits of the experimental treatment.[37,38] Studies that are unblinded also overestimate the effect of interventions.[15,37,40] Unfortunately, these untrustworthy studies are often the precursor to human trials. If animal studies appear to show a positive effect of a drug, that drug will likely be tested in humans, yet the positive effect may have been falsely inflated due to bias. This means that clinical trial volunteers may be put at unnecessary risk.[41]

Bias can also occur if researchers fail to conduct a power calculation in advance to determine the sample size for their study. Stating the sample size up front prevents researchers from continuing to accumulate data until they find a significant result. Unfortunately, *fewer than 1%* of animal studies report performing formal power calculations in advance,[35,39] creating doubt about the scientific rigor of those studies. Furthermore, sample sizes have traditionally been relatively small within animal research. While this may seem positive from the animals' point of view, small samples lead to underpowered studies[42] that produce misleading, untrustworthy, and unusable results.[43] Ultimately, those animals' lives will have been utterly wasted.

Factors that may influence or confound a study's findings are known as confounding variables, and in the context of animal studies, these are myriad. Unless researchers take measures to control or exclude these variables, they may skew the data and affect the outcomes of research, reducing the validity of the study and its results. Confounding variables include those relating to the animal itself (e.g., its stock/strain, source, sex, age, weight, pathogen status); those relating to the facility in which it is held (e.g., diet, bedding, housing, water delivery, lighting, noise, vibration, temperature, humidity); the laboratory in which the experiments are conducted; variables relating to the animal handler, such as their sex and whether they are wearing scent or aftershave; and variables relating to the experiment, such as pain and the drugs that manage pain (e.g., surgical anesthesia and postoperative analgesics), the way drug doses are administered, and animal stress and suffering.[44–48]

Because so many variables can affect the scientific rigor of an animal study, it has proven difficult for a scientist to independently repeat another scientist's study and produce identical outcomes. When this is not possible, a study is said to lack reproducibility. The *ability* to reproduce a study, on the other hand, provides an indication of the scientific rigor of that study. Unfortunately, pharmaceutical companies that have tried to reproduce academic animal studies have found that many are not reproducible, casting doubt on their validity.[49–51]

Reporting and publishing animal research

In an ideal world, research is reported as fully as possible so that the reader has a clear idea of exactly how an experiment was conducted, how many animals were involved, and how many (if any) were lost from the study and why. All the analyses should be reported, not just those with positive results, and all studies, including those with "negative" findings, should be published. Many animal research publications, however, are incompletely reported,[52] lacking even basic information, such as the number of animals used in experiments,[26] and failing to include the loss of animals through death or exclusion.[53] Such losses (also known as attrition) need to be reported, as they can dramatically alter the results of a study. If they are not reported, the influence of the losses cannot be accounted for, and the effects of the experimental treatments will likely be overestimated.[53]

Failing to report all the analyses performed and outcomes studied, and selectively picking only those with the best results, is known as selective analysis and outcome reporting. As with all other types of bias, it leads to an overestimation of the beneficial effects of the experimental treatment, ultimately generating a body of evidence with an inflated proportion of studies that report statistically significant results. In 2013, researchers used a statistical technique to estimate whether the number of published animal studies with positive results was too good to be true.[54] They assessed 4445 animal studies for 160 potential treatments of neurological disorders and found that, whereas only 919 studies would be expected to have significant results, 1,719 studies actually reported positive results, suggesting that the authors of those studies had been selective in the analyses and outcomes they reported.

Publication bias also makes animal research appear more successful than it actually is.[55] This is a phenomenon whereby studies are more likely to be published if they present "positive" findings, for example, that an experimental drug treatment was effective. But if "negative" or inconclusive findings are rarely (or never) published, this creates a misleading illusion of success. Publication bias is a significant problem in animal research, where, in some fields, it has been estimated that the benefits of animal studies are overstated by a third[56] or more[36] as a result of negative studies not being published.

Bias can also result from funding sources in the form of conflicts of interest. To enable readers to judge whether bias has been introduced for this reason, researchers are encouraged to be transparent about any conflicts of interest related to funding. However, only a minority of animal studies have been found to include a conflict of interest declaration.[36,39] Lastly, peer review bias can also impact research conduct and outcomes and is explored further in Chapter 15.

Many problems of internal validity could, in theory, be prevented by the prospective registration of animal study protocols on a publicly accessible platform. If details about the proposed animal study were recorded in advance, including the hypothesis, proposed sample sizes, analyses to be conducted, and outcomes to be reported, this would improve transparency and might prevent scientists from retrospectively altering their hypotheses and other aspects of the study to fit their data. It could also encourage the publication of studies that do not produce statistically significant findings and those that show no treatment effects. Prospective registration could ultimately reduce the phenomenon whereby treatment effect sizes are overestimated and reduce the number of animals used in research by preventing the unnecessary duplication of animal studies. However, despite the existence of animal study registries, uptake is low, especially compared with the use of registries for clinical studies.[57]

External validity

Problems with replicating human diseases and conditions

As noted earlier, the aim of an "animal model" is to recreate a human disease or condition in an animal so that the animal can be used as a surrogate for humans. The idea is that scientists can use the animal to gain an understanding of, and test drugs for, that disease, rather than studying humans with the disease. The trouble is, humans are complex creatures; we each have a unique life history and genetic makeup, and each of us experiences unique influences on our health, including diet, environment, geography, and the quality of our relationships and social networks. Although we may suffer from broadly similar conditions such as stroke, cancer, Alzheimer's disease, and so on, our presentation of these conditions will be unique, again depending on a myriad of factors, including age, comorbidities, and the medicines we take. In this part of the chapter, I will use the example of sepsis to illustrate some of the problems inherent in using animals to study human diseases.

Sepsis is a life-threatening condition that arises when the body's response to infection results in harm to its own organs and tissues. The immune response to sepsis involves the interplay of thousands of genes and depends on the pathogen that caused it, the site of the infection, and the health of the person. Differences in patients, the source of infection, and its route of entry all mean that there is huge variation in the way sepsis progresses and in the ways patients respond to treatment. Like cancer, there is no single trajectory and no typical patient. Worldwide, around 11 million people die each year from sepsis, and in the UK, five people die of sepsis every hour. It is a leading cause of inpatient death because there is no specific treatment, despite decades of investment in animal models of sepsis.

Scientists have found it difficult to reproduce the complex characteristics of sepsis in animals[58]—indeed, it has been described as a "uniquely human disease."[59] Nevertheless, two main animal models exist: the cecal ligation and puncture (CLP) model and the endotoxin model. The CLP model involves puncturing part of the large intestine, which results in bacteria entering the bloodstream, causing sepsis. Findings from the CLP model are only relevant to human patients in whom sepsis is due to abdominal perforations that lead to peritonitis.[60] Indeed, a review conducted by the National Advisory General Medical Sciences Council Working Group on Sepsis confirmed that the CLP model does not reflect the way in which most humans develop sepsis[59] and only captures characteristics of short-term infection. The endotoxin model involves injecting a toxin to trigger inflammation and, ultimately, sepsis. The problem with the endotoxin model is that mammals differ hugely in their response to endotoxins. Mice—the animal most commonly used to model sepsis—are between 1000 and 100,000 times more resistant to endotoxins than humans.[23] This is probably because mice have lived for millions of years in environments teeming with microbes and have thus evolved a better strategy for dealing with infection without inducing the overwhelming inflammation seen in humans.[61]

Scientists acknowledge that animal models are unable to mimic the multifaceted ways in which sepsis manifests and progresses in humans. The lack of correspondence between the animal model and the human disease is not unique to sepsis; there are issues with animal models of all human diseases, especially those that are complex and chronic.[62] The lack of correspondence is significant, however, because it means that the research is unlikely to have clinical impact. In addition, contributing to the lack of translation is the fact that experimental drugs may be administered to animals in ways that are not clinically relevant. For instance, some experimental sepsis drugs are administered to animals *prior* to sepsis being induced,[63,64] a situation that will never translate to the clinic because, in humans, the drugs will obviously only be administered after the development of symptoms.[58] Furthermore, some argue that if animals are to faithfully represent a human sepsis patient, they should be given the same supportive care and treatment that sepsis patients receive, such as antibiotics, fluid resuscitation, mechanical ventilation, and/or dialysis. The challenge is that incorporating additional interventions into animal studies is costly, complex in terms of experimental design, and highly likely to increase the harms endured by animals.

Animal samples do not represent human populations

In January 2024, Chinese scientists reported having successfully cloned the first rhesus monkey, claiming that cloned monkeys "could speed up drug testing, as genetically identical animals give like-for-like results, providing greater certainty in trials."[65] However, the animals used in biomedical research are already a homogeneous population; they tend to come from highly inbred strains, meaning that they are already genetically very similar, and they are generally young, healthy, and of similar weight. They are also fed a standardized diet and housed in highly controlled, pathogen-free environments. While controlling all these variables might seem logical from an experimental point of view because it increases the internal validity of the study, it is not so desirable if the findings are intended to have applicability to diverse patient populations—in other words, if the study aims to have *external* validity.

For this reason, some have suggested using more genetically diverse animal samples or housing animals in less highly controlled environments. This could lead to "uncontrolled heterogeneity,"[58] though, making it difficult to interpret study findings unless the heterogeneity was introduced systematically. Systematic introduction of heterogeneity, however, would be costly because it would require larger samples of animals (to demonstrate meaningful differences in the effect of the experimental treatment) and, therefore, changes in infrastructure. In the field of sepsis, some have suggested using animals of different ages to reflect the sepsis population, which includes the very young to the very old (the mice used in sepsis models are typically aged two or three months).[60] Another suggestion is to incorporate animals with comorbidities[59,60] because many older sepsis patients have comorbidities. From the point of view of increasing external validity, this would seem to make sense, since it would better reflect the clinical experience. Certainly, in the field of stroke, many potential treatments were found to be less effective in the presence of hypertension,[66] meaning that animal studies of stroke that use animals *without* hypertension overestimate the effects of the experimental intervention.[40] However, incorporating comorbidities into animal models will, of course, increase the severity of harms experienced by animals, raising significant ethical questions. It also raises scientific concerns: the pain and stress endured by animals—and the medications used to control these—are all confounding variables which, unless controlled for, will compromise internal validity.

Attempts to improve external validity might increase animal harm, compromise internal validity, and be very costly. Ignoring external validity, on the other hand, compromises clinical translation. However, all these difficulties and dilemmas pale into insignificance in the face of a far greater challenge to the animal model paradigm: species differences.

Species differences and the problem of homology

In his radical and groundbreaking book, *On the Origin of Species*, Charles Darwin proposed that all species are descended over time from common ancestors, with each gradually transforming as those possessing a variation enabling them to thrive lived to pass on their improved traits to the next generation. We noted earlier, for example, that mice have evolved a better strategy than humans for dealing with infection because of the environments they have inhabited. But, of course, Darwin's theory of evolution fundamentally undermines the practice of conducting preclinical research on animals.[67] As Shanks and LaFollette have pointed out, two species may appear on the surface to be very similar, but because they have faced different evolutionary pressures, they have found different ways to achieve similar ends. So, even if two species have similar biological properties, it does not mean that their underlying mechanisms are similar. Furthermore, a change in one place in an organism often has ripple effects elsewhere in the organism. So, while some species may be biologically similar in many respects, they are also likely to be significantly different biologically.[68] Each species is unique, distinct in important ways from other species—special, in other words.[69]

Yet the premise of preclinical animal research relies solely on the similarities that species share as a result of their common ancestry. This principle, known as homology, is clearly demonstrated on the website of Understanding Animal Research (UAR), the organization that represents the UK animal research industry: "Obviously there are differences between animals and people. But under the skin, the biology of humans and other animals, particularly mammals, is remarkably similar. We have the same organs, controlled by the same nerves and hormones, as many other species."[70] The trouble is, species are both similar *and* different, and UAR's perspective completely ignores the importance of the uniqueness of each species as a result of natural selection. Although there may be many superficial similarities, there will also be important differences due to each species having adapted to distinct ecological niches. Furthermore, while the genetic similarities between humans and animals are often highlighted by scientists using animals, as though these were some sort of clincher, it is the way in which genes are *expressed* that counts—in other words, the way in which the information in genes is converted into functions.

Scientists are increasingly highlighting important differences between species that need to be acknowledged,[23,67,71–76] noting that in complex living systems, even minor differences can cause significant disparities in biological processes and outcomes.[77] Yet, within animal research, these differences are either ignored, downplayed, or treated as confounding variables[76] rather than as the fundamental challenge they represent to the entire enterprise. Where animal researchers *do* admit to differences, they conceptualize these as quantitative, focusing on differences in size, weight, or metabolic rate and attempting to accommodate them using scaling techniques. This sort of thinking enables them to believe that one animal can stand in for another, but this is not possible: each species is *qualitatively* different.

I noted above that animal models of stroke have been spectacularly unsuccessful in terms of generating clinical benefits for humans with stroke. A few years ago, a colleague and I explored what animal researchers in the field thought about this failure by analyzing 80 opinion papers on the topic. The authors of these papers agreed that translational stroke research was in crisis; indeed, in over a quarter of the papers, the authors even acknowledged that this translational failure was due to species differences. However, when it came to proposing solutions to the crisis, the vast majority continued to suggest tweaking animal models and introducing measures to improve internal validity. Only one recommended abandoning animal models and using human biology-based methods instead.[78]

One attempt to address the problem of species differences in the field of animal research more generally has been the development of transgenic animals, particularly transgenic mice. However, inserting human genetic material into mice does not make them more like humans, and it has not improved clinical translation.[73] A basic principle of biology is that the function of genes depends on the biochemical context; thus a human gene will not function in the same way in mice as it does in humans.[62] Consequently, while the explosion of studies using transgenic mice has resulted in a significant increase in the number of animals used in experiments, there has been no corresponding improvement in clinical translation.

In most respects, a lack of internal and/or external validity can be addressed using appropriate measures, as Table 2 shows. However, the problem of species differences *cannot* be addressed. Tweaking other aspects of validity or creating transgenic animals has not solved—and will not solve—the problem of species differences or lead to better clinical translation.

Lack of clinical translation is a significant problem. Only 5% of drugs that have passed animal testing go on to be approved for use in patients.[79] Similarly, drugs that appear safe in animals can go on to be toxic in humans.[75] Consequently, as long as scientists persist in relying on animal experiments preclinically, patients will continue to suffer from an ongoing lack of safe and effective treatments for even our most common diseases, as well as from the adverse effects of medicines that were found to be safe in animals. Patients will also be at risk if they participate in drug trials that proceed on the basis of misleading animal studies.[80] Indirectly, they will suffer from the loss of potentially beneficial treatments, either because these were discarded after being found unsafe in animals or because animal research consumes funding that could be more productively directed toward research based on humans.[62]

TABLE 2 Potential responses to the lack of internal and external validity in preclinical animal research.

Internal validity		External validity	
Problem	**Response**	**Problem**	**Response**
Risk of bias	Improve experimental design	Animal models do not represent the human condition or clinical context	Attempt to improve the model and its similarity to the clinical context
Confounding variables	Improve experimental design	Animal samples do not represent human populations of interest	Attempt to include animals with characteristics that are more representative of the human populations of interest
Selective reporting/ publishing	Publish the protocol in advance; report all analyses and outcomes as outlined in the advance protocol; report all details of the animal study, including attrition; publish "negative" or inconclusive studies	Species differences	Use transgenic models (but these have failed to improve clinical translation); attempt to tweak internal validity and aspects of external validity (which will not address species differences); continue to ignore species differences; abandon the paradigm

Conclusion

Fully acknowledging the problem of species differences in preclinical animal research entails accepting that animals cannot be used as valid models for human diseases. Unfortunately, because there is still reluctance to recognize this, the problem of species differences is simply ignored. Yet this aspect of external validity trumps all other aspects of validity, both internal and external. There is absolutely no point in improving experimental design, encouraging full reporting practices, or improving animal models or the characteristics of animals included in study samples if the research is going to fail anyway due to species differences. For an animal study to be valid, *all* aspects of internal and external validity must be robust. As I have argued elsewhere,[33] attempts to improve the quality of animal studies will only bring about modest improvements.[81] Even if we were to try to perfect the design, conduct, and reporting of animal studies over the next few centuries, they would still fail to translate reliably because of species differences. Clearly, this would be a phenomenal waste of time, money, and resources. It is far better, then, to abandon the animal model paradigm and revitalize biomedical research using approaches based on *human* biology. Human-based approaches (discussed at length in Part II of this book) are not only directly relevant and therefore scientifically superior, but they are also quicker, cheaper,[82] and, of course, far more ethical.

References

[1] Shuaib A, Lees KR, Lyden P, Grotta J, et al. NXY-059 for the treatment of acute ischemic stroke. *N Engl J Med*. 2007;357(6):562–571. doi:10.1056/NEJMoa070240.

[2] Horn J, Limburg M. Calcium antagonists for ischemic stroke. A systematic review. *Stroke*. 2001;32(2):570–576. doi:10.1161/01.str.32.2.570.

[3] Saxena R, Wijnhoud AD, Carton H, et al. Controlled safety study of a hemoglobin-based oxygen carrier, DCLHb, in acute ischemic stroke. *Stroke*. 1999;30(5):993–996. doi:10.1161/01.STR.30.5.993.

[4] Enlimomab Acute Stroke Investigators. Use of anti-ICAM-1 therapy in ischemic stroke: results of the Enlimomab Acute Stroke Trial. *Neurology*. 2001;57(8):1428–1434. doi:10.1212/WNL.57.8.1428.

[5] Davis SM, Lees KR, Albers GW, Diener HC, Markabi S, et al. Selfotel in acute ischemic stroke: possible neurotoxic effects of an NMDA antagonist. *Stroke*. 2000;31(2):347–354. doi:10.1161/01.STR.31.2.347.

[6] Bath PM, Iddenden R, Bath FJ, Orgogozo JM. Tirilazad International Steering Committee. Tirilazad for acute ischaemic stroke. *Cochrane Database Syst Rev*. 2001;4:CD002087. doi:10.1002/14651858.CD002087.

[7] O'Collins VE, Macleod MR, Donnan GA, Horky LL, van der Worp BH, et al. 1,026 experimental treatments in acute stroke. *Ann Neurol*. 2006;59(3):467–477. doi:10.1002/ana.20741.

[8] Intercollegiate Stroke Working Party. SSNAP (Sentinel Stroke National Audit Programme) annual public report; 2018. Available at: https://www.hqip.org.uk/wp-content/uploads/2019/06/apr2017mar2018-ssnap-annualreport-final.pdf Accessed 10 August 2026.

[9] Howells DW, Sena ES, Macleod MR. Bringing rigour to translational medicine. *Nat Rev Neurol*. 2014;10(1):37–43. doi:10.1038/nrneurol.2013.232.

[10] Stroke Therapy Academic Industry Roundtable (STAIR). Recommendations for standards regarding preclinical neuroprotective and restorative drug development. *Stroke*. 1999;30(12):2752–2758. doi:10.1161/01.str.30.12.2752.

[11] Raza A. *The First Cell and the Human Costs of Pursuing Cancer to the Last*. Basic Books; 2019.

[12] Cummings JL, Morstorf T, Zhong K. Alzheimer's disease drug-development pipeline: few candidates, frequent failures. *Alzheimers Res Ther*. 2014;6(4):37. doi:10.1186/alzrt269.

[13] Pound P. *Animal Models and the Search for Drug Treatments for Traumatic Brain Injury*. Springer Science and Business Media LLC; 2020, pp. 287–302. doi:10.1007/978-3-030-31011-0_17.

[14] Malfait AM, Little CB. On the predictive utility of animal models of osteoarthritis. *Arthritis Res Ther*. 2015;17(1):225. doi:10.1186/s13075-015-0747-6.

[15] Vesterinen HM, Sena ES, French-Constant C, et al. Improving the translational hit of experimental treatments in multiple sclerosis. *Mult Scler*. 2010;16(9):1044–1055. doi:10.1177/1352458510379612.

[16] Parker JL, Kohler JC. The success rate of new drug development in clinical trials: crohn's disease. *J Pharm Pharm Sci*. 2010;13(2):191–197. doi:10.18433/j39014.

[17] Perrin S. Preclinical research: make mouse studies work. *Nature*. 2014;507(7493):423–425. doi:10.1038/507423a.

[18] McNamee K, Williams R, Seed M. Animal models of rheumatoid arthritis: how informative are they? *Eur J Pharmacol*. 2015;759 278–286. doi:10.1016/j.ejphar.2015.03.047

[19] Mullane K, Williams M. Animal models of asthma: reprise or reboot? *Biochem Pharmacol*. 2014;87(1):131–139. doi:10.1016/j.bcp.2013.06.026.

[20] Sheets RL, Zhou TQ, Knezevic I. Review of efficacy trials of HIV-1/AIDS vaccines and regulatory lessons learned. A review from a regulatory perspective. *Biologicals*. 2016;44(2):73–89. doi:10.1016/j.biologicals.2015.10.004.

[21] Czéh B, Fuchs E, Wiborg O, Simon M. Animal models of major depression and their clinical implications. *Prog Neuropsychopharmacol Biol Psychiatry*. 2016;64:293–310. doi:10.1016/j.pnpbp.2015.04.004.

[22] Potashkin JA, Blume SR, Runkle NK. Limitations of animal models of Parkinson's disease. *Parkinsons Dis*. 2011:658083. doi:10.4061/2011/658083.

[23] Seok J, Shaw Warren H, Cuenca AG, et al. Genomic responses in mouse models poorly mimic human inflammatory diseases. *Proc Natl Acad Sci U S A*. 2013;110(9):3507–3512. doi:10.1073/pnas.1222878110.

[24] Roep BO, Atkinson M, Von Herrath M. Satisfaction (not) guaranteed: re-evaluating the use of animal models of type 1 diabetes. *Nat Rev Immunol*. 2004;4(12):989–997. doi:10.1038/nri1502.

[25] Thomas D, Chancellor D, Micklus A, LaFever S, Hay M. *Clinical Development Success Rates and Contributing Factors 2011-2020*. Biotechnology Innovation Organization; 2021. Available at: https://go.bio.org/rs/490-EHZ-999/images/ClinicalDevelopmentSuccessRates2011_2020.pdf?_gl=1*1op1opr*_gcl_au*MzQ5OTE3NTQ1LjE3NDEyMDgwOTE.&_ga=2.95489244.1125650682.1741208091-1682639095.1741208091. Accessed March 5, 2025.

[26] Pound P, Nicol CJ. Retrospective harm benefit analysis of pre-clinical animal research for six treatment interventions. *PLOS ONE*. 2018;13(3). doi:10.1371/journal.pone.0193758.

[27] Perel P, Roberts I, Sena E, et al. Comparison of treatment effects between animal experiments and clinical trials: systematic review. *BMJ*. 2007;334(7586). doi:10.1136/bmj.39048.407928.be.

[28] Hackam DG, Redelmeier DA. Translation of research evidence from animals to humans. *JAMA*. 2006;296(14):1731–1732. doi:10.1001/jama.296.14.1731.

[29] Marshall LJ, Triunfol M, Seidle T. Patient-derived xenograft vs. organoids: a preliminary analysis of cancer research output, funding and human health impact in 2014–2019. *Animals*. 2020;10(10). doi:10.3390/ani10101923.

[30] Contopoulos-Ioannidis DG, Ntzani EE, Ioannidis JPA. Translation of highly promising basic science research into clinical applications. *Am J Med*. 2003;114(6):477–484. doi:10.1016/S0002-9343(03)00013-5.

[31] Bailey J, Balls M. Clinical impact of high-profile animal-based research reported in the UK national press. *BMJ Open Sci*. 2020;4(1). doi:10.1136/bmjos-2019-100039.

[32] Guthrie S, Kirtley A, Garrod B, et al. A "DECISIVE" approach to research funding. lessons from three retrosight studies. *Rand Health Q*. 2016;6(1):6.

[33] Pound P, Ritskes-Hoitinga M. Is it possible to overcome issues of external validity in preclinical animal research? Why most animal models are bound to fail. *J Transl Med*. 2018;16(1):304. doi:10.1186/s12967-018-1678-1.

[34] Reichlin TS, Vogt L, Würbel H. The researchers' view of scientific rigor—survey on the conduct and reporting of in vivo research. *PLOS ONE*. 2016;11(12). doi:10.1371/journal.pone.0165999.

[35] Kilkenny C, Parsons N, Kadyszewski E, et al. Survey of the quality of experimental design, statistical analysis and reporting of research using animals. *PLoS ONE*. 2009;4(11):e7824. doi:10.1371/journal.pone.0007824.

[36] Henderson VC, Demko N, Hakala A, et al. A meta-analysis of threats to valid clinical inference in preclinical research of sunitinib. *eLife*. 2015;4:e08351. doi:10.7554/eLife.08351.

[37] Bebarta V, Luyten D, Heard K. Emergency medicine animal research: does use of randomization and blinding affect the results? *Acad Emerg Med*. 2003;10(6):684–687. doi:10.1111/j.1553-2712.2003.tb00056.x.

[38] Hirst JA, Howick J, Aronson J, et al. The need for randomization in animal trials: an overview of systematic reviews. *PLoS ONE*. 2014;9(6). doi:10.1371/journal.pone.0098856.

[39] Macleod MR, McLean AL, Kyriakopoulou A, et al. Risk of bias in reports of in vivo research: a focus for improvement. *PLoS Biol*. 2015;13(10). doi:10.1371/journal.pbio.1002273.

[40] Crossley NA, Sena D, Goehler J, et al. Empirical evidence of bias in the design of experimental stroke studies. *Stroke*. 2008;39(3):929–934. doi:10.1161/strokeaha.107.498725.

[41] Yarborough M, Bredenoord A, D'Abramo F, et al. The bench is closer to the bedside than we think: uncovering the ethical ties between preclinical researchers in translational neuroscience and patients in clinical trials. *PLoS Biol*. 2018;16(6). doi:10.1371/journal.pbio.2006343.

[42] Sena ES, Currie GL, McCann SK, Macleaod MR, Howells DW. Systematic reviews and meta-analysis of preclinical studies: why perform them and how to appraise them critically. *J Cereb Blood Flow Metab*. 2014;34(5):737–742. doi:10.1038/jcbfm.2014.28.

[43] Button KS, Ioannidis JPA, Mokrysz C, et al. Power failure: why small sample size undermines the reliability of neuroscience. *Nat Rev Neurosci*. 2013;14(5):365–376. doi:10.1038/nrn3475.

[44] Everitt JI. The future of preclinical animal models in pharmaceutical discovery and development: a need to bring in cerebro to the in vivo discussions. *Toxicol Pathol*. 2015;43(1):70–77. doi:10.1177/0192623314555162.

[45] Carbone L, Austin J, Gao C-Q. Pain and laboratory animals: publication practices for better data reproducibility and better animal welfare. *PLoS One*. 2016;11(5):e0155001. doi:10.1371/journal.pone.0155001.

[46] Friese C. Realizing potential in translational medicine. *Curr Anthropol*. 2013;54(S7):S129–S138. doi:10.1086/670805.

[47] Animal Care and Use Program, University of Michigan. Remember their senses: sound, light, and smells can have big effects on our animals. 2017. Available at: https://animalcare.umich.edu/news/announcements/remember-their-senses-sound-light-and-smells-can-have-big-effects-on-our-animals/. Accessed 10 August 2026.

[48] Jaric I, Voelkl B, Amrein I, et al. Using mice from different breeding sites fails to improve replicability of results from single-laboratory studies. *Lab Anim*. 2024;53(1):18–22. doi:10.1038/s41684-023-01307-w.

[49] Prinz F, Schlange T, Asadullah K. Believe it or not: how much can we rely on published data on potential drug targets? *Nat Rev Drug Discov*. 2011;10(9):712. doi:10.1038/nrd3439-c1.

[50] Begley CG, Ellis LM. Raise standards for preclinical cancer research. *Nature*. 2012;483(7391):531–533. doi:10.1038/483531a.

[51] Mobley A, et al. A survey on data reproducibility in cancer research provides insights into our limited ability to translate findings from the laboratory to the clinic. *PLoS ONE*. 2013;8(5):e63221. doi:10.1371/journal.pone.0063221.

[52] Kilkenny C, Browne WJ, Cuthill IC, Emerson M, Altman DG. Improving bioscience research reporting: the ARRIVE guidelines for reporting animal research. *PLoS Biol.* 2010;8(6). doi:10.1371/journal.pbio.1000412.
[53] Holman C, Piper SK, Grittner U, et al. Where have all the rodents gone? the effects of attrition in experimental research on cancer and stroke. *PLoS Biol.* 2016;14(1). doi:10.1371/journal.pbio.1002331.
[54] Tsilidis KK, Panagiotou OR, Sena ES, et al. Evaluation of excess significance bias in animal studies of neurological diseases. *PLoS Biol.* 2013;11(7):e1001609. doi:10.1371/journal.pbio.1001609.
[55] Mueller KF, et al. Dissemination bias in systematic reviews of animal research: a systematic review. *PLoS One* 2014;9(12):e116016. doi:10.1371/journal.pone.0116016.
[56] Sena ES, vander Worp HB, Bath PMW, Howells DW, Macleaod MR. Publication bias in reports of animal stroke studies leads to major overstatement of efficacy. *PLoS Biol.* 2010;8(3). doi:10.1371/journal.pbio.1000344.
[57] Jeffers MS, MacLellan A, Avey MT, et al. A call to implement preclinical study registration in animal ethics review. *PLoS Biol.* 2023;21(10). doi:10.1371/journal.pbio.3002293.
[58] Nandi M, Jackson SK, Macrae D, Shankar-Hari M, et al. Rethinking animal models of sepsis - Working towards improved clinical translation whilst integrating the 3Rs. *Clin Sci.* 2020;134(13):1715–1734.
[59] Ayala A, Coopersmith C, Kraft M, et al. NAGMSC working group on sepsis final report; 2019. Available at: https://www.nigms.nih.gov/sites/nigms/files/migrated/nagmsc-working-group-sepsis-report.pdf. Accessed 10 August 2026.
[60] Dejager L, Pinheiro I, Dejonckheere E, Libert C. Cecal ligation and puncture: the gold standard model for polymicrobial sepsis? *Trends Microbiol.* 2011;19(4):198–208. doi:10.1016/j.tim.2011.01.001.
[61] Warren HS. *Why do we use mice to study human diseases?*. TEDMED; 2013. Available at: https://www.youtube.com/watch?v=_V0LlyV1WWQ. Accessed 10 August 2026.
[62] Pound P. *Rat Trap: The Capture of Medicine by Animal Research - and How to Break Free*. Troubador Publishing; 2023.
[63] Reynolds K, Novosad B, Hoffhines A, et al. Pretreatment with troglitazone decreases lethality during endotoxemia in mice. *J Endotoxin Res.* 2002;8(4):307–314. doi:10.1179/096805102125000515.
[64] Parida S, Singh TU, Thangamalai R, et al. Daidzein pretreatment improves survival in mouse model of sepsis. *J Surg Res.* 2015;197(2):363–373. doi:10.1016/j.jss.2015.03.059.
[65] Ghosh P. *Cloned rhesus monkey created to speed medical research. BBC News*; 2024. Available at: https://www.bbc.com/news/science-environment-67987633. Accessed Nov 13, 2024.
[66] Howells DW, MacLeod MR. Evidence-based translational medicine. *Stroke.* 2013;44(5):1466–1471. doi:10.1161/STROKEAHA.113.000469.
[67] Perlman RL. Mouse models of human disease: an evolutionary perspective. In: *Evolution, Medicine, and Public Health.* 2016;2016(1):170–176. doi:10.1093/emph/eow014.
[68] Shanks N, Lafollette H. *Brute Science: Dilemmas of Animal Experimentation.* Routledge; 1997.
[69] Preuss T. NIH Neuroscience Series Seminar: what animal models do to us; 2016. Available at: https://www.videocast.nih.gov/watch=17982. Accessed March 5, 2025.
[70] Understanding Animal Research, the UK. 2024. https://www.understandinganimalresearch.org.uk/using-animals-in-scientific-research/faqs.Accessed March 5, 2025.
[71] Wall RJ, Shani M. Are animal models as good as we think? *Theriogenology.* 2008;69(1):2–9. doi:10.1016/j.theriogenology.2007.09.030.
[72] Lynch VJ. Use with caution: developmental systems divergence and potential pitfalls of animal models. *Yale J Biol Med.* 2009;82(2):53–66.
[73] Geerts H. Of mice and men: bridging the translational disconnect in cns drug discovery. *CNS Drugs.* 2009;23(11):915–926. doi:10.2165/11310890-000000000-00000.
[74] Greek R, Hansen LA. Questions regarding the predictive value of one evolved complex adaptive system for a second: exemplified by the SOD1 mouse. *Prog Biophys Mol Biol.* 2013;113(2):231–253. doi:10.1016/j.pbiomolbio.2013.06.002.
[75] Bailey J, Thew M, Balls M. An analysis of the use of animal models in predicting human toxicology and drug safety. *ATLA Altern to Lab Anim.* 2014;42(3):181–199. doi:10.1177/026119291404200306.
[76] Preuss TM. Who's afraid of Homo sapiens? *J Biomed Discov Collab.* 2006;1:17. doi:10.1186/1747-5333-1-17.
[77] Bailey J, Taylor K. Non-human primates in neuroscience research: the case against its scientific necessity. *ATLA Altern Lab Anim.* 2016;44(1):43–69. doi:10.1177/026119291604400101.
[78] Pound P, Ram R. Are researchers moving away from animal models as a result of poor clinical translation in the field of stroke? An analysis of opinion papers. *BMJ Open Sci.* 2020;4(1):e100041. doi:10.1136/bmjos-2019-100041.

[79] Ineichen BV, Furrer E, Grüninger SL, et al. Analysis of animal to-human translation shows that only 5% of animal-tested therapeutic interventions obtain regulatory approval for human applications. *PLoS Biol.* 2024;22(6):e3002667.
[80] Pound P, Ritskes-Hoitinga M. Can prospective systematic reviews of animal studies improve clinical translation? *J Transl Med*. 2020;18(1):15.
[81] Kousholt BS, Præstegaard KF, Stone JC, et al. Reporting quality in preclinical animal experimental research in 2009 and 2018: a nationwide systematic investigation. *PLoS One*. 2022;17(11):e0257962. doi:10.1371/journal.pone.0275962.
[82] Franzen N, van Harten WH, Retél VP, et al. Impact of organ-on-a-chip technology on pharmaceutical R&D costs. *Drug Discov Today*. 2019;24(9):1720–1724. doi:10.1016/j.drudis.2019.06.003.

CHAPTER

4

Systemic failures and suffering in animal-based research: Why a human-relevant paradigm is urgently needed

Kathrin Herrmann

Center for Alternatives to Animal Testing (CAAT), Johns Hopkins Bloomberg School of Public Health, Baltimore, MD, United States

Abstract

Animal experimentation is often portrayed as scientifically indispensable and ethically well policed; however, contemporary evidence indicates otherwise. This chapter identifies six systemic weaknesses within the current animal-research paradigm: (1) sustained high animal use and large "surplus" killings in breeding colonies; (2) routine under estimation and under treatment of pain; (3) selective transparency that omits ≈95% of US laboratory animals and keeps harm–benefit and retrospective assessments confidential; (4) public-funding bias that allocates ≤1% of health-research budgets to human-biology-based nonanimal methods (NAMs); (5) welfare and translational shortcomings driven by barren housing, chronic stress, and interspecies differences; and (6) fragmented regulatory oversight characterized by low refusal rates and limited accountability. Severe procedures now cluster in neuroscience, high-dose viral challenge, and fish ecotoxicity, while ≈90% of drug candidates that appear promising in animals fail in human trials. A decisive pivot to human-relevant science is both an ethical imperative and a strategic investment in biomedical innovation.

Keywords: Animal experimentation; Nonanimal methods; Funding bias; Transparency gap; Pain and severity assessment; Surplus breeding; Translational failure; Regulatory reform

Introduction

Animal experimentation remains one of the most contested practices in contemporary science. Proponents depict it as essential to biomedical progress, while critics point to intolerable ethical costs and persistent scientific shortfalls. The debate has been sharpened by

Moving Toward Nonanimal Approaches in Medical Research and Testing
https://doi.org/10.1016/B978-0-443-26514-3.00008-4

rapid advances in human-based technologies, including organoids, organ-on-chip systems, high-throughput omics, and in silico models—that offer mechanistic insight without the confounding effects of interspecies biology.[1–3] In this chapter, *new approach methodologies* (NAMs) are understood as *human-relevant* methods—that is, approaches that are human-derived and/or demonstrably predictive of human biology. Platforms sometimes marketed as "NAMs" but based on animal tissues, animal-trained systems, or hybrid animal–human platforms fall outside this definition. Advocates of human-biology-based replacement argue that a paradigm shift is overdue.[4–6]

This chapter contends that six systemic failure points in the current animal-research paradigm undermine both the ethical and the scientific justification for continued reliance on animals: (1) the scale of animal use and surplus killing; (2) persistent underrecognition and undertreatment of pain; (3) selective transparency and weak public accountability; (4) funding allocations that marginalize human-relevant approaches; (5) welfare shortfalls and poor translatability rooted in species differences; and (6) fragmented, often rubber-stamped regulation. Drawing on peer-reviewed literature, official statistics, regulatory reports, and economic analyses, the chapter shows how these domains converge on a single conclusion: the justification for animal use is weaker than ever.

This analysis is global in scope but focuses on jurisdictions with comparatively detailed reporting systems—the European Union (EU-27 + Norway), the United Kingdom (UK), the United States, Canada, and Australia. The chapter closes with evidence-based recommendations to accelerate the adoption of NAMs—defined here as human-derived or human-validated methods—so that biomedical research becomes both more humane and more predictive for human biology and health.

Scale of animal use and suffering

Regional trends

In the European Union, overall animal use declined by 11% between 2002 and 2022,[7] but numbers have remained largely stable since 2015. The most recent statistics reported a further decrease in 2023, with nearly 8 million animals used in scientific procedures across the EU and Norway, representing a 4.9% reduction compared with 2022. In addition, almost 1 million animals were used for the creation and maintenance of genetically altered (GA) lines. GA animals accounted for 22.3% of all reported animal uses in 2023, offsetting reductions in traditional toxicity testing.[8]

The United States publishes audited figures only for species covered by the Animal Welfare Act (AWA), thereby excluding purpose-bred mice, rats, fishes, and, until recently, all birds—groups that together constitute over 90% of laboratory populations. A regulatory change finalized in 2023 brought only nonpurpose-bred birds under the AWA for the first time;[9] this change was first reflected in the 2024 reporting cycle, when 63,396 birds were included in the national statistics.[10] Purpose-bred birds remain excluded from federal reporting, leaving total avian use unknown. In that year, 1014 research facilities reported using 775,297 AWA-covered animals in 2024.[10] Nevertheless, these figures capture only a small fraction of total animal use. Extrapolating from institutional data, Carbone[11] estimated that around 111 million rats and mice are used annually in US laboratories, with an estimated 40–45 million subjected to painful experiments. Estimates vary depending on methodology and assumptions; however, even conservative extrapolations imply tens of millions of mice and rats annually, with a substantial fraction falling into potentially painful categories.

Canadian statistics likewise indicate persistently high levels of animal use. In 2024, the Canadian Council on Animal Care (CCAC) reported 3.71 million animals used in research, teaching, and testing. Fishes (46.2%) and mice (32.4%) accounted for the majority of reported animal use, while cattle represented 10% of all animals used. GA animals dominate biomedical studies, and regulatory testing involving farm animals has increased alongside the intensification of animal agriculture.[12]

In Australia, the absence of a national reporting scheme means figures must be assembled from state reports, surveys, and freedom-of-information (FOI) requests. Available data indicate that approximately 6 million animals are used each year, a figure that has plateaued rather than declined.[13] Australia is unusual in counting free-living animals observed in ecological studies, which inflates totals but still demonstrates the continuing reliance on vertebrate models.

In Brazil, by far the biggest user in Latin America, regulatory authorities reported that approximately 2.5 million animals were used in research and teaching activities in 2021, with more than 11.3 million animals used between 2019 and 2023 combined.[14] These figures included rodents, birds, fishes, horses, pigs, and smaller numbers of nonhuman primates and reptiles. While the Brazilian reporting system captures a broader range of species than many international reporting systems, important limitations remain. While the Brazilian reporting system captures a broader range of species than many international reporting systems, the published statistics do not provide disaggregated information on genetically altered animals, surplus breeding, or severity categories.[14]

China, Japan, and South Korea have begun publishing partial statistics, but incomplete coverage of GA animals, fishes, and cephalopods leaves continental estimates uncertain.[15] Given China's scale in biomedical research, the unreported population is likely substantial. Taylor and Rego Alvarez estimated that China used approximately 20.5 million animals for scientific purposes in 2015, making it one of the world's largest users of laboratory animals.[15]

Across these regions, official statistics reveal plateauing trends in headline numbers but conceal extensive undercounts and hidden deaths. The drivers behind this persistence are explored in the next section.

Drivers of persistent use

The plateau in overall animal-use numbers coincides with a dramatic rise in genetically engineered mice for disease modeling.[8] These models require large breeding colonies and generate substantial "wastage" when offspring do not possess the desired genotype. According to the *European Commission's report on the statistics for the use of animals for scientific purposes in the EU*,[16,17] the number of animals killed as surplus—i.e., bred but not used in procedures—was estimated at 10.58 million in 2020, 10.13 million in 2021, and 9.57 million in 2022.[16,17] These figures represent animals killed without being used in experiments, including animals deemed unsuitable because of genotype, sex, age, or overbreeding.

The fact that surplus killings consistently exceed or rival the number of animals used in scientific research (e.g., 9.24 million animals used in procedures in 2022) underlines a largely hidden dimension of animal use in science. These deaths remain largely invisible in public discourse and are not scrutinized through the same harm-benefit assessments required for research procedures, despite their substantial welfare implications.

At the same time, regulatory chemical safety programs—such as Registration, Evaluation, Authorization, and Restriction of Chemicals (REACH) in the EU and the Toxic Substances Control Act in the United States—continue to rely heavily on animal tests despite validated in vitro and other nonanimal approaches. Analyses highlight structural barriers, including entrenched career incentives, established supply chains, and institutional inertia that favor familiar animal protocols over novel NAM platforms that require additional validation.[18-20] The FDA's *Potential Approaches to Drive Future Integration of New Alternative Methods to Animal Testing* roadmap likewise acknowledges that, although NAMs are advancing, animal tests remain the default in many preclinical safety domains.[21] The result is an ethically paradoxical landscape: technological capacity for replacement is advancing rapidly, yet systemic inertia continues to sustain animal demand.

Pain, distress, and severity assessment

Regulatory frameworks

Under Directive 2010/63/EU and the Animals (Scientific Procedures) Act 1986 (UK, as amended),[22] every procedure must be assigned a prospective severity classification—mild, moderate, severe, or nonrecovery—before work begins, and an actual severity classification once the procedure ends. Whenever real-world harm exceeds the authorized limit, the incident must be reported; in the United Kingdom, such a deviation automatically triggers a compliance investigation.[23,24]

Prospective underestimation of severity

Licensing files reveal a persistent optimism bias in prospective severity assessment. In a retrospective review of 506 German project-licence applications, covering 684 rodent recovery surgical procedures, 6% of procedures a severity classification lacked any severity classification, despite this being a legal requirement.[25] Among the remaining procedures, researchers classified almost 40% as "mild," even though all involved surgery under general anaesthesia. Using Directive 2010/63/EU and relevant guidance documents, Herrmann and Flecknell classified all procedures as at least moderate and 28% as severe, compared with only 7% classified as severe by the applicants themselves. Overall, 59% of researcher-provided severity classifications deviated from the authors' assessment, and in 58% of cases the applicants' classification was lower. Only two procedures were classified more severely by the applicants than by the authors.[25] The authors concluded that researchers frequently underestimated the pain, suffering, distress, and lasting harm associated with the proposed procedures, and that planned health-monitoring strategies were often inadequate, reflecting persistent deficits in pain recognition, refinement implementation, and training.[25]

The harm-benefit ledger: structural bias

Licensing decisions are meant to pass a simple utilitarian test: projected human benefit must outweigh animal harm after the latter has been "reduced to the absolute minimum."[26] In practice, however, neither side of the equation is measured with scientific rigor, and both are routinely tilted toward approval. Applications regularly downplay the real severity of procedures. Independent reviewers rescoring German rodent protocols found that one-third of procedures labeled "mild" should have been classified as "severe," including thoracotomies, craniotomies, and intraabdominal implantations performed under minimal or conditional analgesia.[25]

Underestimation extends beyond surgical pain. Monitoring plans are frequently vague or delayed, allowing distress to go undetected for hours or days; anesthesia and analgesia are often limited to a single 24-hour dose despite evidence of longer-lasting pain; preventive multimodal regimens are rare. Humane endpoints are frequently late or poorly defined, and killing methods still include techniques known to be aversive or unreliable (e.g., CO_2 exposure, cervical dislocation, and decapitation without adequate sedation).[27,28] Because severity is underestimated prospectively, the "cost" column in the harm–benefit ledger begins artificially low. Retrospective reviews—an important mechanism for detecting this bias—cover only a minority of projects and are seldom published.[29]

Just as harms are discounted, anticipated benefits are often overstated. Most interventions that show promise in animal studies ultimately fail during human clinical development,[30,31] a pattern linked to poor reproducibility, selective reporting, and weak model validity.[2,3] Oversight committees may lack the time, expertise, or data access to interrogate these claims in depth, further entrenching optimism bias.[32] Cultural attitudes can also contribute: speciesism—privileging human interests over equivalent animal interests—predicts reduced empathy for nonhuman pain,[33] and long-term exposure to animal suffering can create a "moral callus," normalizing severe practices.[34,35]

Cumulative suffering is likewise underrecognized. Although Directive 2010/63/EU requires retrospective review of severe projects, repeated procedures, chronic recovery, and barren housing are rarely captured from a lifetime perspective.[36] Multiple audits confirm the gap, including findings of 58% underestimation in German rodent surgery proposals.[25] Public summaries also remain uneven: only 15%–18% contain sufficient detail to reconstruct lifetime experiences, and reporting quality varies substantially across jurisdictions.[37] EU methodological notes further emphasize that statistics are reported per "use," not per animal, meaning that reused animals can appear multiple times and cumulative burden cannot be reconstructed from official datasets.[16,17] New approaches, such as the cumulative severity assessment -scores, have therefore been proposed to quantify lifetime burden more systematically.[38] Transparency analyses likewise show that lay summaries often highlight a single intervention while obscuring the broader lived experience of animals.[18]

Taken together, these patterns indicate a structurally biased harm–benefit assessment in which harms are systematically underestimated while anticipated benefits are overstated. Addressing these shortcomings requires independent severity audits, mandatory NAM expertise on review boards, outcome-based reporting of cumulative suffering, and a strategic shift toward human-relevant methods.

Actual severe use in the EU and the United Kingdom

According to official EU statistics, across the EU-27 plus Norway, the absolute number of severe procedures declined from 976,445 (10.9%) in 2018 to 784,048 (9.2%) in 2022. Regulatory safety testing once accounted for almost one-third of these cases,[39] but contributed only 18% in 2022[16,17] after EU-wide adoption of in vitro pyrogen assays[40] and the near-elimination of mouse-ascites antibody production.[16] Severe cases are now predominantly associated with basic and translational research—representing approximately 77% of all severe procedures—particularly in neuroscience, oncology, and infectious-disease models. The mouse ascites method nevertheless still caused about 24,000 severe procedures in 2022, almost all in France.[41] Ecotoxicology remains a hotspot, with 42% of fish toxicity studies still classified as severe in 2020.[42]

Great Britain shows an even steeper decline: severe experimental procedures accounted for approximately 8% of all work in 2014 but only 4% in 2022, and the latest Home Office data indicate a further decline to around 3% in 2023.[43,44] When breeding of GA lines is included, severe classifications represent about 2% of all procedures. In 2022, mice experienced 45,500 severe procedures, followed by zebrafish (≈4600), guinea pigs (≈1300, mostly batch-safety tests), and rats (≈1000). Acute LD_{50}-type studies remain the single most severe category: 47% of such UK tests were logged as severe in 2022.[24]

Analgesia deficiencies

Pain management continues to lag behind best practices. A systematic review of more than 2000 rodent craniotomy studies found that only approximately one quarter reported any perioperative analgesia, while fewer than one in six mentioned postoperative pain control, underscoring persistent deficiencies in both pain management and reporting.[45] A German review found 30% of almost 700 planned rodent surgeries included no postoperative analgesic plan, while genuine multimodal regimens were virtually absent.[28] Inadequate pain relief not only harms animals but also distorts immunological, endocrine, and behavioral data, compromising reproducibility.[46]

The foregoing analysis shows that severe suffering has declined in some domains but remains structurally underrecognized and undertreated. The next section examines why widely promoted mitigation strategies—refinement and partial replacement—have not delivered the expected welfare gains in practice, and how these limitations may also compromise scientific validity.

Welfare shortfalls and scientific validity

Environmental enrichment and the standardization fallacy

Barren, tightly standardized housing has not resolved the reproducibility problem: in a coordinated three-laboratory study, behavioral phenotypes differed more between sites than within a given protocol, revealing substantial lab-to-lab effects despite meticulous standardization.[47] At the same time, adding enrichment improves welfare and promotes normal species-typical behavior without inflating within-group variability or producing conflicting outcomes, countering the long-standing assumption that "richer" cages compromise precision.[48] Together, these findings expose the standardization fallacy—the assumption that increasingly strict homogenization automatically yields more reliable data.[49]

A more effective strategy may be systematic heterogenization: preplanning modest, theory-driven variation in factors such as housing, handling, timing, or apparatus—and, where feasible, distributing experiments across sites—to sample the real-world noise that otherwise blindsides replication. Conceptually and in proof-of-principle demonstrations, this structured variation produces effects that are more stable across contexts without demanding larger sample sizes, thereby improving external validity **within animal research** (i.e., generalizability across labs/batches/handlers) and between-lab reproducibility—while preserving internal validity and statistical efficiency.[49]

Chronic stress and epigenetic change

Routine stressors—handling, restraint, cage cleaning, olfactory cues from predators—activate the hypothalamic–pituitary–adrenal axis, altering cytokine profiles, tumor growth kinetics, and cognitive performance.[46] Epigenetic modifications can persist for generations, meaning that stress experienced by parent animals may confound phenotypes in offspring even before experiments begin.

Species differences and translational failure

Even with improved husbandry, the biology of nonhuman species imposes a ceiling on predictive value. A large comparative study by Seok and colleagues found that genomic responses to acute inflammatory insults—including trauma, burns, and sepsis—were highly reproducible across different mouse strains but bore little resemblance to the corresponding human responses.[50] This striking divergence underscores how even well-controlled animal models can systematically misrepresent human disease pathways.

Persistent failures in human translation reflect fundamental limitations in the ability of animal models to reliably predict human physiology, disease mechanisms, and therapeutic responses.[51,52] Sun and colleagues trace many of these failures to an optimization culture that, during preclinical development, prioritizes nanomolar potency and plasma pharmacokinetics while overlooking whether a compound actually reaches the relevant human tissue at tolerable doses.[52] They argue that this "structure–tissue-exposure gap" explains why efficacy signals collapse once compounds are tested for effectiveness in Phase II trials, and why toxicity resurfaces when drugs are tested for safety and tolerability in Phase I trials or during the larger confirmatory Phase III studies. Crucially, this gap cannot be bridged by animal testing, since tissue exposure, metabolism and selectivity profiles differ across species. Instead, it demands human-relevant methods—such as organoids, organ-on-chip systems, and in silico ADME (absorption, distribution, metabolism, and excretion) models—that can directly measure or predict human tissue distribution and selectivity.

As Pound and Ritskes-Hoitinga emphasize, fundamental species differences in immune pathways, metabolic enzymes, and receptor pharmacodynamics impose limits on the translational validity of animal models—limits that cannot be overcome by better husbandry or incremental refinements.[3] These challenges are therefore not merely methodological but biological in nature. As later chapters show, NAMs—including organoids, organ-on-chip systems, multiomics platforms, and computational disease models—offer opportunities to study human biology directly while avoiding many of the translational uncertainties associated with cross-species extrapolation.

Human costs: moral injury and secondary trauma

Animal experimentation inflicts not only suffering on nonhuman animals but also significant psychological costs on the humans tasked with harming them.[53] Ethnographic studies first highlighted this unease: Arluke documented the moral ambivalence of laboratory technicians,[54] while Chang and Hart showed how bonds formed with laboratory animals deepened distress when the same animals were later subjected to painful procedures or euthanasia.[55] Reeve et al. described this as the *caring–killing paradox*: when individuals are both caregivers and executioners, the resulting guilt, grief, and emotional exhaustion create a distinct form of occupational strain.[56]

Philosopher Bernard Rollin was among the first to call this phenomenon *moral stress*, identifying the harm done to individuals compelled to kill animals for reasons not in the animals' own interests.[35] Later, Johnson and Smajdor argued that animal research institutionalizes a "failure of recognition," systematically reducing animals to objects and thereby eroding the moral agency of the humans tasked with harming them.[57] Rohlf and Bennett described this process as *perpetration-induced traumatic stress*, noting symptoms akin to posttraumatic stress disorder, including intrusive thoughts, emotional withdrawal, and hyperarousal.[58]

Empirical data confirm the scope of these burdens. Randall et al., in a survey of 801 laboratory-animal professionals across Europe and Asia, found that compassion fatigue and secondary traumatic stress were widespread, particularly among staff who named or trained animals.[59] The systematic review by Rumpel et al. synthesized twelve studies and identified euthanasia as the most consistent and distressing stressor for laboratory-animal personnel.[60] Across contexts, killing animals produced both acute strain (grief, sadness, guilt) and chronic strain (burnout, compassion fatigue, secondary traumatic stress). Importantly, the review distinguished between *stressors* (such as euthanasia), *strain* (psychological outcomes), and *moderators* (such as social support), showing that while institutional coping programs may buffer some effects, they do not resolve the ethical conflict at the root of the problem.

Institutions, however, frequently reframe these experiences using the language of compassion fatigue. Coping strategies—such as rotating staff between tasks, providing workshops, or holding symbolic memorial rituals—are aimed at sustaining staff resilience rather than interrogating the harms done to animals.[61,62] Analyses of hidden curricula suggest that researchers are socialized through stepwise desensitization—beginning with minor procedures, then progressing to surgeries, and finally euthanasia—so that empathy is gradually suppressed and emotional detachment becomes professionalized.[63] This dynamic can be understood as a form of engaged followership—a concept from social psychology describing conformity to authority[64]—in which loyalty to laboratory norms and the perceived demands of "science" overrides individual moral instincts.

Recognizing moral injury and secondary trauma underscores that animal experimentation harms both nonhuman animals and human personnel. Transitioning toward nonanimal research methods would not only spare animals but also protect the mental health and moral integrity of the people compelled to harm them.[57]

Refinement and replacement: the implementation gap

Despite decades of research on refinement—including multimodal analgesia, non-aversive handling, enriched housing, and earlier humane endpoints—uptake remains inconsistent. Literature reviews continue to show poor reporting of analgesia, monitoring, and humane endpoints.[65-67] A multinational trainee survey ranked refinement as both urgent and feasible, yet respondents cited few concrete changes actually implemented.[68] Retrospective reviews of German animal-research licence applications further revealed substantial gaps between refinement principles and actual practice.[25,27,28] Approximately 30% of surgical procedures planned no postoperative analgesia, while another 10% proposed analgesia only "if needed," despite lacking structured pain-assessment protocols.[27] Preventive analgesia was frequently omitted when volatile anaesthesia was used, and multimodal analgesic regimens were exceedingly rare even for highly invasive surgeries such as thoracotomies.[27] Humane endpoints were often vague, implemented only at advanced stages of suffering, or omitted altogether, with nearly half of reviewed surgical procedures lacking explicit humane endpoints.[28] Researchers also frequently provided incomplete or inadequate health-monitoring strategies, and some applications lacked legally required information, including severity classifications.[25]

Together, these findings suggest that concerns about experimental standardization and potential confounding effects often continued to outweigh concerns about animal suffering, despite longstanding legal and scientific recognition of refinement obligations. Restricted access to project dossiers and retrospective assessments has limited independent evaluation of whether refinement practices have improved over time. In short, refinement remains the most extensively discussed of the 3Rs, yet its implementation is often assumed rather than systematically demonstrated.

Replacement shows a similar asymmetry. While regulatory toxicology has adopted several validated non-animal methods,[69] some training programs have transitioned toward simulation-based and other animal-free approaches,[70] biomedical research—where most animals are used—still defaults to live models.[7] Contributing factors include weak enforcement of the hierarchical 3Rs (replacement before refinement), limited awareness, and scant incentives for deviating from legacy pipelines.[71]

The mitigation tools that should reduce suffering and bias in data are not consistently implemented; when they are, they cannot overcome fundamental species limits.

Where severe suffering persists—and how to target NAMs

When alternatives become binding regulatory standards, the welfare dividends dwarf incremental refinements. The monocyte-activation test and other in vitro pyrogen assays enabled the EU's phase-out of the rabbit pyrogen test—preventing more than 140,000 severe procedures annually.[40,72] Similarly, moving monoclonal-antibody production from mouse ascites to cell culture eliminated tens of thousands of severe cases each year in some countries.[8]

Yet severe classifications persist. Fish LD_{50} ecotoxicity, aggressive neuro-oncology models, and high-dose viral challenges remain major contributors in the EU and UK. Public datasets such as ALURES (ANIMALUSE REPORTING - EU SYSTEM) and the UK Home Office Portal now allow mapping of these pain points to prioritize NAM development and implement sunset clauses once validation is achieved.[42,73] The challenge is to pair intense refinement research with equally vigorous programs that convert validated alternatives into compulsory practice—realizing the ethical hierarchy envisioned by Russell and Burch in 1959.[26]

Outside Europe, severity frameworks are weaker, and outcome audits are rare. In the United States, the US Department of Agriculture's (USDA) pain categories C–E record only whether analgesia was provided and cover less than 10% of laboratory animals.[74] Canada's CCAC grades projected invasiveness but lacks compulsory outcome audits.[12] Australia's national A–E scale leaves retrospective review to state discretion, producing variable enforcement.[75] These gaps blunt incentives to implement refinement, obscure cumulative severity, and slow the shift to replacement. Measured reductions in headline "severe" numbers show that targeted replacement works.

However, persistent optimism in prospective assessments, under-analgesia, and the invisibility of lifetime burden mean the ethical ceiling is still far off—and the scientific one is already visible.

Funding bias, economic opportunity costs, and regulatory momentum

Public funding for biomedical research continues to favor animal models, despite growing recognition of their scientific and ethical shortcomings. In the European Union, only about 0.7% of the €7.3 billion Europe Health Pillar has been directed toward NAM-exclusive projects,[76] with initiatives such as the Animal-free Safety assessment of chemicals: Project cluster for the Implementation of novel Strategies (ASPIS) and the Partnership for the Assessment of Risks from Chemicals (PARC) partnership representing important but still marginal investments.[77,78] In the United States, the Government Accountability Office estimates that around US$ 5.5 billion annually supports animal research compared to just US$ 69 million for organoid and organ-on-chip work—a disproportion of roughly 80 to 1.[79] Australia shows a similar imbalance, with less than 0.2% of its combined public health research budget earmarked for NAMs.[13] Across these regions, systemic bias is clear: NAMs receive only a fraction of total funding, and in many cases, grant systems do not even track method use, further obscuring the true scale of investment.

The economic opportunity costs of this imbalance are significant. Animal-based pipelines underpin a 90% attrition rate in drug development, where most compounds that pass animal tests ultimately fail in humans.[3,50,80] The US Food and Drug Administration notes that non-human-primate safety studies can cost around US$ 50,000 per animal and delay programs by months.[21] By contrast, organ-on-chip systems, 3D-bioprinted tissues, and AI-driven in silico simulations can reduce timelines and direct costs while improving predictive accuracy.[81,82] Even modest reallocations—such as redirecting 5% of the ongoing animal-research budgets—could inject hundreds of millions annually into NAM science in the EU, the United States, and Australia. Modeling studies suggest that such reallocations could yield a five- to tenfold return on investment.[13]

Regulatory momentum is beginning to align with these arguments. In the EU, the One Substance, One Assessment package[83] anchors NAMs within a new cross-agency data platform to harmonize chemical-safety evaluations, while the European Medicines Agency has introduced annexes for organ-on-chip data in regulatory guidelines. In the United States, public and political pressure for change is mounting, with polls showing large majorities in favor of phasing out animal testing.[84] The FDA has pledged to reduce primate use in biologics development, while the NIH has created the Office of Research Innovation, Validation, and Application (ORIVA) to coordinate policy across agencies.[85] In June 2025, the NIH terminated funding for Harvard neuroscientist Margaret Livingstone's infant-macaque deprivation studies, citing its new strategic focus.[86] The same freeze also suspended US$ 2.2 billion in Harvard grants—including Don Ingber's flagship organ-on-chip program—prompting the university to sue the US Department of Health and Human Services.[87] Advocates hailed the termination of the primate studies as an ethical victory, yet the broader, politically charged cuts risk undermining the very NAM projects ORIVA is intended to support.

The EPA's 2019 plan to eliminate mammalian testing by 2035 was later rescinded before being reaffirmed in January 2026, underscoring how politically contingent animal-testing transition policies remain.[88] For the first time, the FDA,[89] NIH, and EPA now all have policy initiatives pointing toward reduced reliance on animal testing, offering a potential tipping point—if sufficient funding, implementation capacity, and clear milestones are sustained.

The transparency deficit

The much-publicized UK-based "Concordat on Openness on Animal Research" was drafted in 2014 and promoted by the lobbying group *Understanding Animal Research* (UAR)—an organization whose explicit mission is to defend and normalize animal experimentation.[90] Over 120 universities, charities, and industry bodies voluntarily signed the document, pledging clearer communication about their use of animals. Independent evaluations, however, show that most "openness" outputs celebrate refined husbandry practices and promised medical breakthroughs while omitting failed studies, unexpected deaths, and detailed adverse-event data.[18] Meanwhile, in the United Kingdom, Section 24 of the *Animals (Scientific Procedures) Act 1986 (as amended)*[22]—the Act's confidentiality clause—continues to criminalize the unauthorized release of project information, severely limiting external scrutiny despite the Concordat's rhetoric of transparency.

Across the EU, Directive 2010/63/EU requires the publication of nontechnical project summaries (NTS) and annual severity statistics. Yet, a review of 400 German and UK NTS found that only 44% adequately described predicted adverse effects, and 28% were rated "poor" or "very poor" overall.[38] Because severity data are aggregated only by species and overall purpose, it is impossible to hold individual projects accountable for the harms they inflict.

In North America, transparency gaps persist in different forms. Canada's Canadian Council on Animal Care issues annual national summaries, yet universities frequently redact license details when responding to provincial FOI requests, citing researcher safety and intellectual-property concerns.[12] In the United States, the Freedom of Information Act enables public access to USDA inspection reports, but more than 95% of laboratory animals—mice, rats, and birds—are excluded from the AWA and thus from routine public reporting. Furthermore, ClinicalTrials.gov registration rules do not cover preclinical studies, so negative animal results remain largely unpublished, perpetuating publication bias.[46]

Australia regulates animal research through state-based Animal Ethics Committees, whose minutes and license applications are rarely released. A review of public inquiries found that responses typically provided only high-level statistics with little contextual information or ethical justification.[91] The lack of national-level reporting impedes ethical oversight and prevents cross-jurisdictional benchmarking of scientific outcomes.

These patterns contribute to broader concerns about public trust. Sociologists describe this selective transparency as a "promissory discourse," in which institutions amplify hypothetical future benefits while downplaying present harms.[18] Public polling indicates that acceptance of animal research is conditional on trust that suffering is strictly necessary.[92] When institutional disclosures appear incomplete or self-congratulatory, that trust erodes, fueling citizen petitions, ballot initiatives, and calls for outright bans (e.g., the European Citizens' Initiatives "Stop Vivisection" and "Save Cruelty-free Cosmetics").

Regulatory track record

The EU's 2018–2022 implementation report shows that 20 Member States met the legal target of inspecting one-third of establishments annually, but 18 still missed the target in at least one year. Eight carried out no unannounced inspections across the entire five-year cycle.[93] Cyprus and Portugal have never conducted an unannounced visit since the Directive entered into force.

The UK Home Office recorded 169 breaches in 2023, involving approximately 155,000 animals; 57% related to basic husbandry failures—food deprivation, faulty ventilation, and overcrowding.[94] These numbers emerged despite mandatory training, designated veterinary care, and a public commitment to a "culture of care."

Approximately 15,000 project licenses are authorized each year across the EU and Norway; refusal rates hover at approximately 4.5%, and eight countries reported zero refusals in 2022.[93] Only Ireland requires that at least one evaluator possess demonstrable expertise in NAMs. Elsewhere, applicants often face panels lacking sufficient specialist knowledge to rigorously challenge the necessity of using animals.

Empirical analyses confirm that approval bias is entrenched: over 95% of submitted protocols are ultimately approved, raising concerns that review processes function more as technical checks than as genuine harm–benefit evaluations.[95] Oversight committees are frequently dominated by animal researchers, whose professional background and career incentives may predispose them toward accepting animal use rather than scrutinizing its necessity. In this "rubber-stamp" culture, discussions tend to focus on technical refinements or minor procedural adjustments, while claims of benefit are rarely interrogated with equal rigor.[95]

The lack of mandatory NAM expertise further compounds this bias, creating an uneven playing field in which proposals for human-relevant alternatives are disadvantaged from the outset. Requiring at least one accredited NAM specialist on every review and licensing panel would help counterbalance this bias and ensure that replacement options receive equitable consideration. Directive 2010/63/EU also allows each EU member state to determine how many licensing/inspection bodies—so-called competent authorities (CAs)—it will operate. The result is a patchwork: Denmark runs just one CA, Germany has 573, Poland 328, and Spain 112.[93] With hundreds of separate offices interpreting the same law, decisions on severity, housing standards, or project approval can vary substantially; multinational firms may therefore "forum-shop," submitting their protocols in whichever country is known for the quickest or most permissive review.

Together, approval bias and fragmented governance reinforce systemic inertia, ensuring that the Directive's stated principle of prioritizing replacement over refinement remains more aspirational than operational. This stands in direct contrast to Russell and Burch's original hierarchy of the 3Rs, which placed *replacement* as the first priority.[26] In practice, however, committees largely default to approving animal use with minor refinements, leaving the Directive's ethical foundation unfulfilled.

United States: Legal void post-Jarkesy ruling

In June 2024, the US Supreme Court's decision in *SEC v. Jarkesy* limited agencies' use of in-house proceedings to impose civil penalties, prompting reassessment of available enforcement pathways.[96] Although not about animals specifically, the ruling has affected AWA oversight: during this period of reassessment, USDA's Animal and Plant Health Inspection Service has not been issuing civil fines for AWA violations in research settings. Trends reported by the Animal Welfare Institute indicate long-standing concerns about weak deterrence and reduced penalties under the AWA, which predated but may now be compounded by this legal uncertainty.[96,97] Reporting on high-profile laboratories has likewise highlighted how weak enforcement cultures can manifest in practice, underscoring the need for transparent and legally durable sanction mechanisms.[98]

Ethical, scientific, and societal implications

The licensing calculus that permits animal experiments depends on two premises: (1) that suffering is minimized and (2) that any remaining harm is outweighed by reliable human benefit.[26] Evidence reviewed throughout this chapter undermines both claims. Severity is habitually underpredicted, pain undertreated, and cumulative distress underreported, while the anticpated benefits are undermined by irreproducible data, publication bias, and a ≥90% failure rate when animal-derived leads enter human trials.[2,3]

The scientific shortfall carries a measurable economic drag. The member countries of the Organisation for Economic Co-operation and Development spend an estimated US$10–12 billion each year on regulatory and exploratory animal studies, even though validated NAMs are already matching—or exceeding—predictive value at lower cost and with faster turnaround.[81] Capital and early-career talent continue to be channeled into an R&D architecture that generates high clinical attrition, while organoids, microphysiological systems, and AI-enabled modeling continue to receive only a small fraction of public funding.[6]

A further distortion is now visible: federal and industry roadmaps increasingly market NAMs as *complements* to legacy animal pipelines rather than replacements. Initiatives such as NIH's Complement-ARIE funding call[99] and FDA-endorsed "combinatorial NAMs"[100] actively promote murine or canine organ-on-chip systems and hybrid in silico platforms that are trained on animal data. Framed as synergy, the approach underwrites the animal-testing supply chain, siphons resources from fully human-based systems, and hard-wires the very translational failures NAMs were meant to solve.[101–105] Recent analysis has described this dynamic as a form of path dependency, in which institutional investment in animal-based infrastructures and regulatory paradigms perpetuates the integration of animal systems into emerging NAM frameworks rather than enabling a genuine transition toward fully human-relevant science.[106]

Public tolerance is already conditional. In the UK, fewer than half of respondents support animal use "for any research as long as it is well regulated"[92]; in the United States, 80% favor a federal plan to phase it out.[84] Secrecy provisions, such as the United Kingdom's retained Section 24 and the continued exclusion of purpose-bred mice, rats, and fish from US national statistics, erode that fragile social license to operate.[18]

Several policy measures are therefore critical for establishing a credible harm–benefit framework:

- Ring-fence ≥ 5% of public biomedical budgets for NAM development and validation.
- Legally define NAMs as human-derived or human-validated only; exclude animal-based "complementary" systems.
- Embed accredited NAM specialists on every review, funding, and regulatory panel.
- Publish outcome-based metrics—reproducibility indices, clinical translation rates, and surplus-breeding numbers—in open databases.
- Consolidate inspection into well-resourced agencies with ≥ 50% unannounced visits.
- Mandate colony-management tools (cryopreservation, genotype forecasting, regional sharing) to eliminate surplus kills.
- Require protocol-specific pain-recognition training and biomarker-supported humane endpoints.
- Attach binding sunset dates to each animal test once a NAM is validated (EU pyrogen-test precedent).

Almost seven decades after Russell and Burch coined the 3Rs, incremental refinement cannot compensate for structural scientific limits. Redirecting funds, embedding NAM expertise, enforcing outcome-based transparency, tightening enforcement, and inoculating human-based science against "complementary" co-option would steer biomedicine toward a paradigm that is humane by design and demonstrably more predictive for human health. The opportunity cost of inaction is no longer merely moral; it is scientific, economic, and societal.

References

[1] CSIRO. *Non-animal Models: A Strategy for Maturing Australia's Medical Product Development Capabilities*. Canberra, Australia: CSIRO; 2023. csiro:EP2023-3601. doi:10.25919/p5bt-3s39.

[2] Marshall LJ, Bailey J, Cassotta M, Herrmann K, Pistollato F. Poor translatability of biomedical research using animals - a narrative review. *Altern Lab Anim*. 2023;51(2):102–135. doi:10.1177/02611929231157756.

[3] Pound P, Ritskes-Hoitinga M. Is it possible to overcome issues of external validity in pre-clinical animal research? *J Transl Med*. 2018;16:304. doi:10.1186/s12967-018-1678-1.

[4] Hartung T, Tsatsakis AM. The state of the scientific revolution in toxicology. *ALTEX*. 2021;38(3):379–386. doi:10.14573/altex.2106101.

[5] Herrmann K, Jayne K, eds. *Animal Experimentation: Working Towards a Paradigm Change*. Leiden: Brill; 2019. doi:10.1163/9789004391192.

[6] Schmeisser S, Miccoli A, von Bergen M, et al. New approach methodologies in human regulatory toxicology – not if, but how and when! *Environ Int*. 2023;178:108082. doi:10.1016/j.envint.2023.108082.

[7] Taylor K. Trends in the use of animals and non-animal methods over the last 20 years. *ALTEX*. 2024;41(4):503–524. doi:10.14573/altex.2410111.

[8] European Commission. *Commission staff working document: Summary report on statistical information on the use of animals for scientific purpose in the Member States of the European Union and Norway in 2023*. Brussels: European Commission; 2024. Available at: https://circabc.europa.eu/ui/group/8ee3c69a-bccb-4f22-89ca-277e35de7c63/library/c51d5da3-b85b-4579-859f-8ae2969ef493/details. (Accessed 7 May 2026).

[9] U.S. Department of Agriculture, Animal and Plant Health Inspection Service (APHIS). *Standards for birds not bred for use in research under the Animal Welfare Act* (Final rule). *Fed Regist*. 88;2023(34):10654–10721. https://www.federalregister.gov/documents/2023/02/21/2023-03357/standards-for-birds-not-bred-for-use-in-research-under-the-animal-welfare-act.

[10] USDA – United States Department of Agriculture. FY 2024 research facility annual report summary. animal and plant health inspection service; 2025. https://www.aphis.usda.gov/sites/default/files/fy2024-research-animal-use-summary.pdf. Accessed 27 Jun 2025.

[11] Carbone L. Estimating mouse and rat use in American laboratories by extrapolation from Animal Welfare Act-regulated species. *Sci Rep*. 2021;11:493. doi:10.1038/s41598-020-79961-0.

[12] Canadian Council on Animal Care (CCAC). CCAC Animal Data Report 2024; 2025. Available at: https://ccac.ca/Documents/AUD/CCAC_Animal_Data_Report_2024.pdf. (Accessed 19 Jun 2026).

[13] AFSA. *Advancing Biomedical Research through Human-relevant Research: A Business Case for Funding Non-animal Methodologies*. Fitzroy, VIC: Animal-Free Science Advocacy; 2025. Available at: https://animalfreescienceadvocacy.org.au/wp-content/uploads/2025/02/AFSA-Business-Case-for-Funding-Non-Animal-Research-Methods.pdf. (Accessed 27 Jun 2025).

[14] CONCEA (Conselho Nacional de Controle de Experimentação Animal). *Relatório de Uso de Animais Em Ensino E Pesquisa — 2019-2023*. Brasília: Ministério da Ciência, Tecnologia e Inovações. n.d. Available at: https://www.gov.br/mcti/pt-br/composicao/conselhos/concea/paginas/Destaques/relatorio-de-uso-animal-concea-2019_2023-1-1.pdf. (Accessed 27 Jun 2025).

[15] Taylor K, Rego Alvarez L. An estimate of the number of animals used for scientific purposes worldwide in 2015. *Altern Lab Anim*. 2019;47(5-6):196–213. doi:10.1177/0261192919899853.

[16] European Commission. *Commission Staff Working Document — Summary Report on the Statistics on the Use of Animals for Scientific Purposes in the Member States of the European Union and Norway in 2022. SWD(2024) 185 Final*. Brussels: European Commission; 2024. Available at: https://data.consilium.europa.eu/doc/document/ST-12496-2024-INIT/en/pdf. Accessed 27 June 2025.

[17] European Commission. *ALURES Open Dataset: 2022 EU & Norway Animal-use Statistics (Directive 2010/63/EU)*. Brussels: DG ENV; 2024. Available at: https://webgate.ec.europa.eu/envdataportal/content/alures/section1_number-of-animals.html. Accessed 27 June 2025.

[18] Pound P, Blaug R. Transparency and public involvement in animal research. *ALTEX*. 2016;33(2):167–173. doi:10.1177/026119291604400210.

[19] Nelson CP, Brown P, Fitzpatrick S, et al. Advancing alternative methods to reduce animal testing. *Science*. 2024;386(6723):724–726. https://www.science.org/doi/10.1126/science.adg6228.

[20] Sewell F, Alexander-White C, Brescia S, et al. New approach methodologies (NAMs): identifying and overcoming hurdles to accelerated adoption. *Toxicol Res*. 2024;13(2):tfae044. doi:10.1093/toxres/tfae044.

[21] FDA. *Potential Approaches to Drive Future Integration of New Alternative Methods to Animal Testing*. Silver Spring, MD: FDA; 2024. Available at: https://www.fda.gov/media/182478/download (Accessed 19 Sept 2025).

[22] Animals (Scientific Procedures) Act 1986 (UK, as amended). Available at: https://www.legislation.gov.uk/ukpga/1986/14/contents. Accessed 27 Jun 2025.

[23] European Parliament & Council. Directive 2010/63/EU of 22 September 2010 on the protection of animals used for scientific purposes (Text with EEA relevance). *Off J Eur Union*. 276;2010:33–79. Available at: https://eur-lex.europa.eu/eli/dir/2010/63/oj.

[24] Home Office. *Animals in Science Regulation Unit (ASRU) Compliance Report 2022*. London: Home Office; 2023. Available at: https://assets.publishing.service.gov.uk/government/uploads/system/uploads/attachment_data/file/1171153/asru-annual-report-2022.pdf. Accessed 27 Jun 2025.

[25] Herrmann K, Flecknell PA. Severity classification of surgical procedures and application of health monitoring strategies in animal research proposals – a retrospective review. *ATLA*. 2018;46(5):273–289. doi:10.1177/026119291804600508.

[26] Russell WMS, Burch RL. *The Principles of Humane Experimental Technique*. Methuen, London. 1959.

[27] Herrmann K, Flecknell PA. Application of humane endpoints and humane killing methods in animal research proposals – a retrospective review. *ATLA*. 2018;46(6):1–17. doi:10.1177/026119291804600606.
[28] Herrmann K, Flecknell PA. Retrospective review of anaesthetic and analgesic regimens used in animal research proposals. *ALTEX*. 2019;36(1):65–80. doi:10.14573/altex.1804011.
[29] Pound P, Nicol CJ. Retrospective harm–benefit analysis of pre-clinical animal research for six treatment interventions. *PLOSONE*. 2018;13(3):e0193758. doi:10.1371/journal.pone.0193758.
[30] Hay M, Thomas DW, Craighead JL, Economides C, Rochon PA. Clinical development success rates for investigational drugs. *Nat Biotechnol*. 2014;32(1):40–51. doi:10.1038/nbt.2786.
[31] Mullard A. Parsing clinical success rates. *Nat Rev Drug Discov*. 2016;15:447. doi:10.1038/nrd.2016.136.
[32] Graham ML, Prescott MJ. The multifactorial role of the 3Rs in shifting the harm–benefit analysis in animal models of disease. *Eur J Pharmacol*. 2015;759:19–29. doi:10.1016/j.ejphar.2015.03.040.
[33] Caviola L, Everett JAC, Faber NS. The moral standing of animals: towards a psychology of speciesism. *J Pers Soc Psychol*. 2019;116(6):1011–1029. doi:10.1037/pspp0000182.
[34] Rollin BE. *Animal Rights & Human Morality*. 3rd. Amherst, NY: Prometheus Books; 2006.
[35] Rollin BE. Euthanasia and moral stress. *Loss Grief Care*. 1986;1(1–2):115–126.
[36] Lilley E, Hawkins P, Jennings M. A 'road map' toward ending severe suffering of animals used in research and testing. *Altern Lab Anim*. 2014;42(4):267–272. doi:10.1177/026119291404200408.
[37] Taylor K, Weber T, Rego Alvarez L. Have the non-technical summaries of animal experiments in Europe improved? An update. *ALTEX*. 2024;41(3):382–394. doi:10.14573/altex.2310181.
[38] Helmig S, Wenzel S. A new approach (CSA-Score) to assessing the cumulative overall stress load in the case of animal experiments and even beyond. *Discover Anim*. 2024;1:38. doi:10.1007/s44338-024-00035-w.
[39] European Commission. *Summary Report on the Statistics on the Use of Animals for Scientific Purposes in the Member States of the European Union and Norway in 2018*. Luxembourg: Publications Office of the European Union; 2020.
[40] European Directorate for the Quality of Medicines & HealthCare (EDQM), *European Pharmacopoeia to Put an End to the Rabbit Pyrogen Test. Press Release, 28 June 2021*. Strasbourg: Council of Europe; 2021. Available at: https://www.edqm.eu/en/-/european-pharmacopoeia-to-put-an-end-to-the-rabbit-pyrogen-test.
[41] RSPCA. *Latest statistics on severe suffering in the EU for 2021 and 2022*. Focus on Severe Suffering (FoSS); 2024, RSPCA. Available at: https://focusonseveresuffering.co.uk/latest/latest-statistics-on-severe-suffering-in-the-eu-for-2021-and-2022/. Accessed 27 Sept 2025.
[42] RSPCA. *Focus on Severe Suffering: UK Meeting Summary Report*. Horsham: RSPCA Science Group; 2023. Available at: https://focusonseveresuffering.co.uk/wp-content/uploads/2024/02/UK-FoSS-2023-Summary-Report.pdf. Accessed 27 Jun 2025.
[43] Home Office. *Annual Statistics of Scientific Procedures on Living Animals, Great Britain 2022*. London: Home Office; 2023. Available at: https://www.gov.uk/government/statistics/statistics-of-scientific-procedures-on-living-animals-great-britain-2022. Accessed 11 Jun 2025.
[44] Home Office. *Annual Statistics of Scientific Procedures on Living Animals, Great Britain 2023*. London: Home Office; 2024. Available at: https://www.gov.uk/government/statistics/statistics-of-scientific-procedures-on-living-animals-great-britain-2023. Accessed 11 Jun 2025.
[45] King H, et al. Anesthesia and analgesia for experimental craniotomy in mice and rats: a systematic scoping review comparing the years 2009 and 2019. *Front Neurosci*. 2023;17:1143109. doi:10.3389/fnins.2023.1143109. Accessed 12 Jun 2025.
[46] Bailey J. Does the stress of laboratory life and experimentation on animals adversely affect research data? A critical review. *Altern Lab Anim*. 2018;46(5):291–305. doi:10.1177/026119291804600501.
[47] Crabbe JC, Wahlsten D, Dudek BC. Genetics of mouse behaviour: interactions with laboratory environment. *Science*. 1999;284(5420):1670–1672. doi:10.1126/science.284.5420.1670.
[48] Wolfer DP, Litvin O, Morf S, et al. Cage enrichment and mouse behaviour. *Nature*. 2004;432(7019):821–822. doi:10.1038/432821a.
[49] Voelkl B, Altman NS, Forsman A, et al. Reproducibility of animal research in light of biological variation. *Nat Rev Neurosci*. 2020;21:384–393. doi:10.1038/s41583-020-0313-3.
[50] Seok J, Warren HS, Cuenca AG, et al. Genomic responses in mouse models poorly mimic human inflammatory diseases. *Proc Natl Acad Sci U S A*. 2013;110(9):3507–3512. doi:10.1073/pnas.1222878110.
[51] Wong CH, Siah KW, Lo AW. Estimation of clinical trial success rates and related parameters. *Biostatistics*. 2019;20(2):273–286. doi:10.1093/biostatistics/kxx069.

[52] Sun D, Gao W, Hu H, Zhou S. Why 90% of clinical drug development fails and how to improve it? *Acta Pharm Sin B*. 2022;12(7):3049–3062.

[53] Ford C. 'The harrowing lives of animal researchers: science depends on animal testing, but the work comes at a steep, hidden cost', *Vox – Future Perfect*; 2025. Available at: https://www.vox.com/future-perfect/400534/animal-researchers-lab-ethics-trauma-stress. Accessed 27 June 2025.

[54] Arluke A. Uneasiness among laboratory technicians. *Occup Med*. 0885-114X. 1999;14(2):305–316.

[55] Chang FT, Hart LA. Human–animal bonds in the laboratory: how animal behavior affects the perspectives of caregivers. *ILAR J*. 2002;43(1):10–18. doi:10.1093/ilar.43.1.10.

[56] Reeve CL, Rogelberg SG, Spitzmüller C, DiGiacomo N. The caring–killing paradox: euthanasia-related strain among animal-shelter workers. *J Appl Soc Psychol*. 2005;35(1):119–143. doi:10.1111/j.1559-1816.2005.tb02096.x.

[57] Johnson J, Smajdor A. Human wrongs in animal research: a focus on moral injury and reification. In: Herrmann K, Jayne K, (eds.), *Animal Experimentation: Working Towards a Paradigm Change*. Brill, Leiden;2019, pp. 305–317. doi:10.1163/9789004391192_014.

[58] Rohlf V, Bennett P. Perpetration-induced traumatic stress in persons who euthanize nonhuman animals in surgeries, animal shelters, and laboratories. *Soc Anim*. 2005;13(3):201–219. doi:10.1163/1568530054927763.

[59] Randall MS, Moody CM, Turner PV. Mental wellbeing in laboratory animal professionals: a cross-sectional study of compassion fatigue, contributing factors, and coping mechanisms. *J Am Assoc Lab Anim Sci*. 2021;60(1):54–63. doi:10.30802/AALAS-JAALAS-20-000039.

[60] Rumpel S, Kempen R, Merle R, Thoene-Reineke C. Psychological stress and strain in laboratory animal professionals: a systematic review. *Lab Anim*. 2023;57(4):396–411. doi:10.1177/00236772221129111.

[61] King M, Zohny H. Animal researchers shoulder a psychological burden that animal ethics committees ought to address. *J Med Ethics*. 2022;48(5):299–303. doi:10.1136/medethics-2020-106945.

[62] Hill EM, LaLonde CM, Reese LA. Compassion fatigue in animal care workers. *Traumatology*. 2020;26(1):96–108. doi:10.1037/trm0000218.

[63] Birke L, Arluke A, Michael M. *The Sacrifice: How Scientific Experiments Transform Animals and People*. West Lafayette, IN: Purdue University Press; 2007. ISBN: 9781557534572.

[64] Haslam SA, Reicher SD, Birney ME. Nothing by mere authority: evidence that in an experimental analogue of the Milgram paradigm participants are motivated not by orders but by appeals to science. *J Soc Issues*. 2014;70(3):473–488. doi:10.1111/josi.12072.

[65] Richardson CA, Flecknell PA. Anaesthesia and post-operative analgesia following experimental surgery in laboratory rodents: are we making progress? *Altern Lab Anim*. 2005;33(2):119–127. doi:10.1177/026119290503300207.

[66] Stokes EL, Flecknell PA, Richardson CA. Reported analgesic and anaesthetic administration to rodents undergoing experimental surgical procedures. *Lab Anim*. 2009;43(2):149–154. doi:10.1258/la.2008.008020.

[67] Carbone L, Austin J. Pain and laboratory animals: publication practices for better data reproducibility and better animal welfare. *PLoS ONE*. 2016;11(5):e0155001. doi:10.1371/journal.pone.0155001.

[68] Franco NH, Sandøe P, Olsson IAS. Researchers' attitudes to the 3Rs—An upturned hierarchy? *PLOS ONE*. 2018;13(8):e0200895. doi:10.1371/journal.pone.0200895.

[69] Basketter D, Clewell H, Kimber I, et al. A roadmap for the development of alternative (non-animal) methods for systemic toxicity testing. *ALTEX*. 2012;29(1):3–91. doi:10.14573/altex.2012.1.003.

[70] Pawlowski J, Feinstein D, Crandall ML, Gala S. Modernizing biomedical training: replacing live animal laboratories with human simulation. In: Herrmann K, Jayne K, eds. *Animal Experimentation: Working Towards a Paradigm Change, Brill Human Animal Studies*, Vol 22. Leiden: Brill; 2019, pp. 551–566. doi:10.1163/9789004391192_023.

[71] Herrmann K. Refinement on the way towards replacement: Are we doing what we can? In: Herrmann K., Jayne K. (Eds.), *Animal Experimentation: Working Towards a Paradigm Change*, Vol. 22. Leiden, The Netherlands: Brill; 2019, pp. 3–64. doi:10.1163/9789004391192_002.

[72] EPAA (European Partnership for Alternative Approaches to Animal Testing). *Annual Report 2023*. Brussels: European Commission; 2023. Directorate-General for Internal Market, Industry, Entrepreneurship and SMEs (DG GROW). See chapter 19. Available at: https://single-market-economy.ec.europa.eu/system/files/2023-11/Annual%20report-2023-WEB.pdf. Accessed 28 June 2025.

[73] European Commission. ALURES statistical EU database. n.d. Available from: https://webgate.ec.europa.eu/envdataportal/content/alures/section1_number-of-animals.html.

[74] USDA – United States Department of Agriculture. *Animal Welfare Act: Annual Report of Activities FY 2021*. Washington, DC: USDA. 2022.

[75] NHMRC. *Australian Code for the Care and Use of Animals for Scientific Purposes*, 9th ed. Canberra: National Health and Medical Research Council. 2021. https://www.nhmrc.gov.au/about-us/publications/australian-code-care-and-use-animals-scientific-purposes.

[76] European Commission. *Horizon Europe Work Programme 2023–2025, Cluster 1 "Health"*. Brussels: European Commission; 2024. Available at: https://ec.europa.eu/info/funding-tenders/opportunities/docs/2021-2027/horizon/wp-call/2023-2024/wp-4-health_horizon-2023-2024_en.pdf. Accessed 27 June 2025.

[77] ASPIS Cluster. *Animal-Free Safety Assessment of Chemicals: project factsheet*. European Commission; 2021. Available at: https://aspis-cluster.eu/wp-content/uploads/2022/08/ASPIS-fact-sheet_online-version.pdf. Accessed 27 Jun 2025.

[78] European Commission. *Summary Report on the Statistics on the Use of Animals for Scientific Purposes in the Member States of the European Union and Norway in 2020–2022. SWD(2023) 84 Final*. Brussels: European Commission; 2023. Available at: https://data.consilium.europa.eu/doc/document/ST-8083-2023-INIT/en/pdf. Accessed: 27 June 2025.

[79] U.S. Government Accountability Office (GAO). *National Institutes of Health: Assessing Efforts to Improve Animal Research Could Lead to Greater Human Health Benefits*. Washington, DC: GAO; 2024, pp. GAO–25–107140. Available at: https://www.gao.gov/assets/gao-25-107140.pdf. Accessed 24 June 2025.

[80] Kola I, Landis J. Can the pharmaceutical industry reduce attrition rates? *Nat Rev Drug Discov*. 2004;3(8):711–716. doi:10.1038/nrd1470.

[81] Meigs L, Smirnova L, Rovida C, Leist M, Hartung T. Animal testing and its alternatives – the most important omics is economics. *ALTEX*. 2018;35(3):275–305. doi:10.14573/altex.1807041.

[82] European Commission: Joint Research Centre. *Non-Animal Methods in Science and Regulation*. Luxembourg: Publications Office of the European Union; 2025. https://data.europa.eu/doi/10.2760/9648771. JRC141304.

[83] Council of the European Union. 'Council and Parliament agree on simplified and more efficient handling of chemical-assessment data'; 2025. Press release, 12 June. Available at: https://www.consilium.europa.eu/en/press/press-releases/2025/06/12/council-and-parliament-agree-on-simplified-and-more-efficient-handling-of-chemical-assessment-data/#:~:text=The%20Council%20and%20the%20European%20Parliament%20reached%20a,early%20detection%20and%20action%20on%20emerging%20chemical%20risks. Accessed 27 Jun 2025.

[84] Morning Consult. U.S. Gen Pop Beliefs on Animal Testing. Washington, DC; 2024. Available at: https://pcrm.widen.net/s/qzfxtfh7bw/animal-testing-survey. Accessed 24 Jun 2025.

[85] NIH (Extramural Nexus). 2025. "NIH funding announcements to align with NIH initiative to prioritize human-based research." July 10, 2025. Available at: https://grants.nih.gov/news-events/nih-extramural-nexus-news/2025/07/nih-funding-announcements-to-align-with-nih-initiative-to-prioritize-human-based-research (Accessed 24 Jun 2025.

[86] Bolotnikova M. The one thing the Trump administration got right. Too bad it's now sabotaging it., *Vox – Future Perfect*; 2025. Available at: https://www.vox.com/future-perfect/417127/trump-nih-harvard-defunding-monkey-research-livingstone. Accessed 27 Jun 2025.

[87] Harvard University. *President and Fellows of Harvard College v. U.S. Department of Health and Human Services: Complaint for Declaratory and Injunctive Relief* (U.S. District Court, D. Mass., Case No. 1:25-cv-11048, complaint filed 15 May 2025); 2025. Available at: https://www.harvard.edu/research-funding/wp-content/uploads/sites/16/2025/05/amended_complaint.pdf (Accessed 27 June 2025).

[88] U.S. Environmental Protection Agency. 22 January. Recommitment to reducing animal testing and continued support for new approach methods (NAMs). *Science*. 2026;(): https://www.epa.gov/system/files/documents/2026-01/recommitment-to-reducing-animal-testing-memo-1.22.26.pdf. doi: Accessed 21 January May 2026.

[89] FDA. *Roadmap to Reducing Animal Testing in Pre-clinical Safety Studies*. Silver Spring, MD: FDA; 2025. Available at: https://www.fda.gov/media/186092/download. Accessed 27 June 2025.

[90] Understanding Animal Research. *Concordat on Openness on Animal Research in the UK*. London: Understanding Animal Research; 2014. Available at: https://concordatopenness.org.uk/wp-content/uploads/2017/04/Concordat-Final-Digital.pdf. Accessed 27 June 2025.

[91] Merkes M, Buttrose R. Increasing the transparency of animal experimentation: an australian perspective. In: Herrmann, K., Jayne, K. (Eds.), *Animal Experimentation: Working Towards a Paradigm Change*. Brill. 2019;224–243. doi: 10.1163/9789004391192_010.

[92] Ipsos MORI. *Public Attitudes to Animal Research in the UK 2022*. London: Ipsos MORI. 2022.

[93] European Commission. *Union Overview on the Implementation of Directive 2010/63/EU on the Protection of Animals Used for Scientific Purposes in the Member States of the European Union in 2018–2022. Commission Staff Working Document SWD(2024) 183 Final*. Brussels: European Commission; 2024. Available at: https://circabc.europa.eu/ui/group/8ee3c69a-bccb-4f22-89ca-277e35de7c63/library/16a0a839-8f80-4aef-951b-d49c-fe21935e/details?download=true. Accessed 27 June 2025.

[94] Home Office. *Animals in Science Regulation Unit: Annual Report 2023*. London; 2024.

[95] Azilagbetor DM, Shaw D, Elger BS. Animal research regulation: improving decision-making and adopting a transparent system to address concerns around approval rate of experiments. *Animals*. 2024;14(6):846.

[96] *Securities and Exchange Commission v. Jarkesy*, 603 U.S. Slip Opinion No. 22–859; 2024. Available at: https://www.supremecourt.gov/opinions/23pdf/603us1r50_7kh7.pdf (Accessed 24 Jun 2025).

[97] Animal Welfare Institute. *Trends in Animal Welfare Act Enforcement*. Washington, DC: AWI; 2025. Available at: (PDF) https://awionline.org/sites/default/files/uploads/documents/awi-trends-animal-welfare-act-enforcement-report.pdf.

[98] Torrella K. Neuralink shows what happens when we bring "move fast and break things" to animal research. *Vox*; 2022. Available at: https://www.vox.com/future-perfect/2022/12/11/23500157/neuralink-animal-testing-elon-musk-usda-probe. (Accessed 28 Sept 2025).

[99] NIH. Complement-ARIE: advancing research through integrated efforts – notice of funding opportunity RFA-AR-25-001. Bethesda, MD. (No DOI available); 2024.

[100] NIH. *Report on combinatorial new approach methodologies*. Bethesda, MD. (No DOI available); 2023.

[101] Hartung T, Kleinstreuer N. Challenges and opportunities for validation of AI- based new approach methods. *ALTEX*. 2025;42(1):3–21. doi:10.14573/altex.2412291.

[102] Ingber DE. Human organs-on-chips for disease modelling, drug development and personalized medicine. *Nat Rev Genet*. 2022;23(8):467–491. doi:10.1038/s41576-022-00466-9.

[103] Pound P. Are animal models needed to discover, develop and test pharmaceutical drugs for humans in the 21st century? *Animals*. 2020;10(12). doi:10.3390/ani10122455.

[104] Nahle Z. A proof-of-concept study poised to remodel the drug development process: liver-Chip solutions for lead optimization and predictive toxicology. *Front Med Technol*. 2022;4:1053588. doi:10.3389/fmedt.2022.1053588.

[105] Nahle Z. Making NAMs a 'Complement' to animal experimentation is the Trojan Horse of the animal-industrial complex — a scheme to nullify the principles of the 3Rs and void the quest to 'replace' animals for good! *NAM J*. 3050-6204. 2025;1:100030. doi:10.1016/j.namjnl.2025.100030.

[106] Nahle Z. Path dependency and the rescuing of the biomedical research enterprise. Frontiers in Medical Technology. 2026;7:1683835. doi:10.3389/fmedt.2025.1683835.

CHAPTER

5

Case studies on the failures of animal research

Jarrod Bailey[a] *and Michael Balls*[b]

[a]Physicians Committee for Responsible Medicine, Washington, DC, United States

[b]University of Nottingham, Queen's Medical Center, Nottingham, United Kingdom

Abstract

Animals have borne the brunt of preclinical research, serving as "surrogate humans" for basic studies into how biological networks function in healthy and diseased individuals, the characterization of pathologies, the search for new drug-treatable targets, and the development of new human pharmaceuticals. This focus on animals harms not only the animals used but also humans, due to toxic effects not detected in animal tests, the abandonment of drugs incorrectly labeled as unsafe or ineffective following animal tests, and wasted research funds on poorly translatable animal experiments. This chapter describes case studies on the failures of animal research for Alzheimer's disease, Parkinson's disease, cancer, and drug development. Troubling disease burdens are discussed, along with translational failures that leave patients waiting for breakthroughs, and how human-specific methods can overcome the limitations of animal use. The chapter concludes with a discussion about the animal use approval process and how it contributes to both animal and human harms.

Keywords: Animal research; Ethics; Translation; Human harms; Clinical benefit; NAMs

Introduction

Animal research results in harm to humans and animals. It may be ironic that much attention is devoted to the harm caused to many tens of millions of animals in laboratories, while human harm may be passively accepted or unappreciated by many. This human harm may be direct, such as when new drugs cause serious side effects in people despite data from animal tests suggesting that they would likely *not* be toxic to humans. It can also be *indirect*, as in the case of new drugs that were halted in development due to signs of toxicity in animals but would not have been toxic to people, or due to the misdirection of valuable and limited research funds to poorly translatable methods. Both direct and indirect harm to humans augment the animal harm that is an obvious consequence of animal

Moving Toward Nonanimal Approaches in Medical Research and Testing
https://doi.org/10.1016/B978-0-443-26514-3.00011-4

research, increasing the gravity and urgency of the need to move away from animal research.

From basic research to drug development, animal failures are easy to find. One notable example in basic research involves some of the most severe suffering inflicted on animals—neuroscience research using monkeys—ostensibly intended to elucidate how the human brain and central nervous system function. The concordance of experimental data and biological functions between species is rare and superficial, and translation to clinical benefit is stark.[1,2] This is underpinned by significant and intractable interspecies differences in genetic complement, organization, and gene-expression, which influence the structure and function of the central nervous system and are associated with neurological diseases and conditions.[2,3] Structural, connective, and functional differences in diverse parts of the human and macaque brains affect the extrapolation of data from macaque neuroscientific investigations to human biology, including in prevalent research fields involving cortical functions.

In disease research, systematic reviews indicate that animal studies rarely result in clinical benefit.[4] To illustrate: it has been claimed that primate research was indispensable for the successful realization of clinical deep-brain stimulation for the control of involuntary movements and tremor in Parkinson's disease; that it enabled the development of functional magnetic resonance imaging; and that without it, we cannot hope to develop a vaccine for HIV/AIDS. In each case, in-depth reviews of the literature revealed that any advances made have been due to human-specific research.[1,5-7] There are numerous additional examples of research areas in which animal experimentation has translated poorly, including rheumatoid arthritis, respiratory diseases like asthma and chronic obstructive pulmonary disease,[8] multiple sclerosis, and vaccines for various diseases, including HIV/AIDS.[9]

This chapter presents case studies elaborating on the failures of animal research in Alzheimer's disease, Parkinson's disease, cancer, and drug development. Disease burdens and healthcare costs are discussed, along with translational failures that exacerbate these burdens and costs, and how human-specific approaches, including new approach methodologies (NAMs), can overcome the limitations of animal use. The chapter concludes with a discussion about the animal use approval process and how it contributes to both animal and human harms. A shift toward human-specific methods, including NAMs, is urgent. While there are positive signs of change, a better understanding of animal research failures and pressure points will accelerate this pace of change.

Alzheimer's disease

According to the World Health Organization (WHO), the number of people with dementia worldwide in 2023 was 57 million, with almost 10 million new cases each year, amounting to one every three seconds. Alzheimer's is the most common form, constituting 60%–70% of cases, and dementia is the seventh leading global cause of death and one of the major causes of disability and dependency.[10,11] The number of people with dementia is projected to double every twenty years globally, reaching an estimated 78 million by 2030 and 139 million by 2050.[12]

In the United States, an estimated 7.2 million people over 65 have Alzheimer's—one in nine people in that age group—and 200,000 Americans aged 30–64 have younger-onset dementia.[13] One in three older adults dies with dementia, and deaths from Alzheimer's have increased by 142% since 2000.[13] Statistics for the UK are similarly stark. Around one million UK citizens have dementia, and in a population of <70 million. This total will rise by almost half again by 2040, with one in three people born in the UK today developing dementia. More than 70,000 UK citizens have young-onset dementia.[11] Overall, Alzheimer's and other dementias are the *leading* cause of death in the UK and have been for a decade.[14]

The financial and economic impacts of these figures are considerable: in 2019, the estimated cost to the global economy was $1.3 trillion, with around half of those costs attributed to informal care from family members and friends.[10] This is forecast to grow to $2.8 trillion by 2030, including informal care, direct costs of social care, and medical care. If global dementia were a nation, it would be the 14th largest economy in the world.[12] In the United States, almost 12 million people provide unpaid care for people with dementia, amounting to over 19 billion unpaid hours. The costs to the US economy are calculated at $384 billion in 2025 and projected to rise to almost $1 trillion in 2050.[13] In the UK, costs amount to £42 billion per year, projected to more than double to £90 billion by 2040 (an equivalent of approximately $56.5 billion rising to $121 billion at the time of writing).[11]

Meanwhile, the very poor translation of animal research into Alzheimer's disease and other dementias has been described as "a graveyard for a large number of promising drugs," given its specific failure rate of 99.6%.[15,16] Alzheimer's research has focused for decades on the amyloid hypothesis, using genetically modified (GM) mice, GM primates, and aging primates. There are myriad problems with this hypothesis and research[17,18]; new amyloid-directed potential therapies have failed in clinical trials 99% of the time,[19] with just two approved in the United States—lecanemab and donanemab. Both these new therapies are considered controversial, with acknowledged issues surrounding adverse effects and efficacy.[20]

Concerns about the amyloid hypothesis and associated animal research point toward a future focused on human-specific investigations, such as advanced patient-derived stem-cell cultures, brain imaging, prospective biomarker studies, and computational biology.[18] This shift has already begun, with a growing research focus on human populations in the preclinical and early symptomatic stages of Alzheimer's, resulting in positive results for diagnostic biomarkers, a significant reduction in phase 3 clinical trials of anti-amyloid therapies, and phase 1 and phase 2 trials targeting more diverse therapeutic areas.[19] Furthermore, the use of live human brain slice cultures has detailed how endogenous levels of amyloid beta, tau, and other biomarkers are associated with patient characteristics, and demonstrated the effects of physiological and pathological amyloid on live human synapses.[21]

The use of NAMs to investigate Alzheimer's disease holds promising potential to overcome the limitations of animal approaches, leading to improved patient outcomes and more efficient research and development expenditures. Stem cell methods, such as organoids, have even been called the "preferred tools for Alzheimer's disease research" due to their ability to model multiple pathogenic factors, including immune disorders, impaired synapses, abnormal mitochondrial structure and function, and genetic variation, as well as their capacity to screen for and test new therapies.[22] Cerebral organoids can model the development and architecture of human neuronal tissues in a manner "unparalleled" by 2D cultures and animal approaches, offering "a bright outlook for more conclusive in-depth studies for [Alzheimer's disease]."[23] A review of neuropathogenesis-on-chips highlights their ability to measure functional changes via real-time human-specific electrophysiology, identify new drug candidates based on effects on neuropathophysiological features, and validate findings through multiorgan chip pharmacokinetics and pharmacodynamics.[24]

The development of a complex human stem-cell-based "vascularized neuroimmune" model facilitates the study of Alzheimer's in a human context. It has been validated via multiple pathologies and biomarkers, as well as the reduction of amyloid burden upon treatment with the approved drug lecanemab (the criticisms and arguments against the amyloid hypothesis above notwithstanding).[25] For genetic forms of Alzheimer's, advanced cerebral organoid models have been validated using multiple pathological phenotypes and are valuable platforms for real-time pathological analysis and drug screening.[26] Live human brain-slice cultures have also proven to be physiologically relevant models of Alzheimer's. Tiny samples of brain tissue from consenting patients undergoing surgery are cultured and kept alive for several weeks, enabling the study of human brain reactions to toxic Alzheimer's-associated proteins and facilitating the testing of new therapies in a human-specific manner.[21]

Parkinson's disease

The prevalence of Parkinson's Disease—the second most common neurodegenerative disease after Alzheimer's—has doubled in the past 25 years, with global estimates of more than 8.5 million people affected by the disease in 2019 and over 10 million people at present. This has resulted in almost 6 million disability-adjusted life years, representing the collective loss of the equivalent of 6 million years of full health—an increase of 81% since 2000.[27] In the United States, there will be an estimated 1.2 million people with Parkinson's by 2030, with a further 90,000 people diagnosed each year—an alarming increase from the previously estimated number of 60,000 annual diagnoses.[28] The total annual costs to US society were estimated at $61.5 billion in 2025.[29] Annual medication costs can amount to $2500 per person, with therapeutic surgery costs of up to $100,000 per patient.[29] The WHO reports that, like Alzheimer's, Parkinson's has a considerable impact on the population and on society, including caregivers, who suffer significant stress due to physical, emotional, and financial pressures. Patients are often subject to discrimination, including in their workplaces, and face a lack of opportunities to participate in their communities.[27]

Parkinson's research has suffered from a sustained focus on animal research, similar to Alzheimer's. There remains a poor understanding of the mechanisms of neurodegeneration, such as how alpha-synuclein (a protein central to Parkinson's pathology) becomes toxic, which hinders therapeutic development. Along with failures in understanding disease initiation and progression via modeling in animals, including monkeys, new preventive and curative therapies remain lacking. The main therapy, levodopa, entered clinical trials in the 1950s. The more than 23,000 animal studies investigating it between 1990 and 2018—including an estimated 5%–10% using primates—have failed to sufficiently replicate human pathology and the most important symptoms or to result in significant clinical benefit.[30,31]

Due to the complexity of the disease, the key to overcoming this will be human-focused research, including the use of ever-more physiologically relevant NAMs.[31] Human midbrain organoids and assembloids permit human-specific analysis of how alpha-synuclein pathology spreads from the hindbrain to the midbrain and dysregulates synapses, improving understanding of Parkinson's progression and aiding in the screening of new therapies.[32] They also permit the study of mitochondrial dysfunction.[33] These platforms have proven to be extremely reproducible and consistent.[34] They are regarded as "transformative" in their faithful modeling of the human midbrain, in the investigation of complex, multifaceted pathological mechanisms, and in the testing of new therapeutic approaches.[35] Increasingly, the importance of the interaction between the gut and the brain is being appreciated, including its potential influence on Parkinson's, and gut-brain organ-chips are helping to elucidate these in a human context and to identify new protective interventions and therapeutic targets.[36,37]

Cancer

The failure rate of new cancer drugs in clinical trials is one of the highest for any class of pharmaceuticals. The mean failure rate of all new drugs is 92%, but new drugs for cancer have a failure rate of 95%, with only cardiovascular and urology drugs faring worse out of 14 major disease areas.[38] Just 30% of approved cancer drugs extend patient survival beyond 2.5 months,[39] and $50–$60 billion every year is devoted to failed oncology drug trials. This is exemplified by the failure of 16 insulin-like growth factor-1 receptor (IGF-1R) compounds, all of which failed in 183 clinical trials involving more than 12,000 patients at a cost of more than $1.6 billion.[40] The causes for these and other failures have been laid squarely on the predictive validity of preclinical models and a lack of rigorous analyses of those failures.[40,41]

Despite these failures, it is claimed by some that cancer research and drug development are generally successful and are saving lives. Yet cancer remains one of the leading causes of death worldwide, with almost 10 million reported deaths (one in every six deaths) annually and 20 million new cancer cases every year. These figures are projected to rise to almost 30 million new cases and 15.3 million deaths annually by 2040.[42] In the United States, while there has been a welcome decrease in the rates of cancer incidence and deaths from 2015–2019, the mean decrease in death rate for men, women, and children was just 1.9%, and for the period 2018–2022, this was 1.5%.[43] Every year, an estimated 2 million US citizens will be diagnosed with cancer, and more than 600,000 will die from cancer. Care costs for cancer patients in the United States alone were an estimated $209 billion per year in 2020, with projections of almost one-quarter of a trillion dollars annually by 2030.[44]

An estimated 40%–50% of cancers worldwide and in the United States are preventable, being linked to tobacco use, alcohol consumption, obesity, low intake of fruits and vegetables, and insufficient physical activity.[45–47] It also seems clear that public health (focusing on populations, epidemiology, or behavioral sciences) is underfunded, with estimates suggesting it receives less than 10% of funds dedicated to cancer research.[48] A contemporary analysis of the cancer prevention research landscape in Europe was similarly damning: just 14% of European cancer research papers focus on prevention research, and only 16% of all European cancer research funders show an interest in prevention research, with less than 4% of them showing an interest in primary prevention.[49]

While a greater focus on prevention research and policies is needed, shifting from animals to NAMs and other human-specific methods will also drive progress. Advanced 3D human cell cultures are already moving the needle for many different cancers. Derived from resections and metastatic biopsies, they retain physiologically relevant tumor microenvironments, replicate tumor-specific variants and histopathology, and demonstrate long-term viability.[50] The incorporation of complex vascular networks and angiogenesis to build organ-chip models has been crucial to their increased fidelity, facilitating real-time analysis of proliferation, migration, and intravasation, all in a patient and tumor-specific manner.[50] Organoids can accurately represent the tumor microenvironment in 3D, which will be pivotal in the development of personalized cancer immunotherapies. This is expected to improve even further with enhanced culture stability and the integration of artificial intelligence, microfluidics, immune and stromal components, 3D bioprinting, and genetic modification.[51] Metastasis-on-a-chip technology enables investigations of the regulation of cancer cell invasion and evaluations of anti-metastatic drugs, such as for bone metastasis and melanoma.[52] Reviews have specifically praised new NAMs-focused initiatives at the US National Institutes of Health (NIH) and Food and Drug Administration (FDA; see the section "Conclusions and discussion"), noting that they will mark "a major stride toward more accurate, ethical, and efficient research methods" in many areas, including cancer specifically.[52]

Drug development

International guidelines have historically required that, when preclinical data are submitted to regulatory agencies to support progression to clinical trials, those data must include results from at least one rodent species (typically mice or rats) and one non-rodent species (typically dogs or monkeys). For many years, there was a paucity of information to demonstrate that these animal tests were predictive of human responses and risks. Today, empirical evidence suggests otherwise. The failure rate of new drugs in clinical trials has consistently been 90% or more, with the latest statistic being 92%.[38] This indicates that preclinical tests are not predictive for humans.

To elucidate this, the authors of this chapter conducted their own study of 2336 drugs with publicly available data, assessing the evidential weight provided by animal tests for and against human safety and toxicity.[2,53–57] They found that data from animals suggesting a lack of toxicity provides little or no evidential weight for a similar lack of toxicity in humans. This can be quantified: a new drug with a prior probability of a lack of toxic effects in humans of 70%, based on in silico and in vitro tests, will have that probability raised to just 70.4% by a lack of toxic effects in monkey tests or 72% by dog tests. Positive toxicity prediction fared little better, with a median value lower than the acknowledged diagnostic value of a "useful" test. Subsequent industry studies claimed that animal tests *were* predictive for humans, but this assertion was *not* supported by their data—which aligned with our studies—nor by their admission that their studies compelled an agreement with our salient finding on the prediction of a lack of toxicity.[53] This lack of evidential weight from animal tests explains the extraordinary failure rate of new drugs in clinical trials, most of which are due to toxicity or a lack of efficacy not indicated by animal testing.

Incorrect positive toxicity predictions, which lead to the termination of drugs due to unacceptable toxicities that would have been free from significant toxicities in humans, are an often-overlooked cost of animal-focused testing. As would be expected, the identities of many such compounds are unknown due to a lack of human data. However, there are several examples of compounds that revealed animal toxicities post-approval or for which a combination of animal toxicities with a lack of toxicity in human in vitro testing led to continued development. These have been summarized in reviews and editorials and include penicillin, paracetamol, aspirin, tamoxifen, and Gleevec/Glivec (imatinib).[58,59]

Ultimately, people bear the financial burden of this failure through the cost of their pharmaceuticals. Estimates of the average cost of developing a typical new drug vary, ranging from $314 million to $4.5 billion per compound, depending on the therapeutic area and modeling assumptions.[60] Another estimate, which includes the cost of failures, proposed a median capitalized R&D investment for each new drug at just under one billion dollars, with a mean total investment of $1.34 billion.[61] In the United States, the annual average R&D expenditure of leading biopharmaceutical research companies amounted to $100 billion worldwide for 2021–2023.[62] These losses are recouped in the form of exorbitant costs for new drugs that successfully make it through the development process and clinical trials.[41] Globally, only a small proportion of patients who could benefit from these drugs are able to afford them, leading to calls for these costs to be borne by pharmaceutical companies rather than patients.

These costs and losses will be mitigated by the increased replacement of animal-based preclinical tests with NAMs. This has been elegantly illustrated for drug-induced liver injury (DILI), which is directly responsible for many failures of drugs in clinical trials and post-approval.[63] Human liver-chip-based testing of 27 small-molecule drugs, selected to challenge complex liver models in benchmarking tests by a pharmaceutical consortium, showed a sensitivity of 87% and specificity of 100%. In other words, the chips predicted 7 out of 8 toxic drugs correctly and showed a lack of liver toxicity in all drugs known to be non-hepatotoxic.[64] This provides confidence in two ways: eleven of these 27 drugs had caused at least 242 deaths despite prior animal tests that did not flag DILI as a risk, and they showed zero false positives, suggesting that valuable drugs might not be lost in development due to false indications of toxicity, an acknowledged issue with animal data.[64] These and other similar platforms are now being incorporated into drug development pipelines to reduce attrition and have been estimated to be worth around $3 billion per annum in increased R&D productivity to the pharmaceutical industry, or $24 billion if extended to other human organ chips.[64,65] In 2024, Emulate's Liver-Chip S1 was accepted into the FDA's Innovative Science and Technology Approaches for New Drugs (ISTAND) program—a new pilot initiative to qualify NAMs for regulatory use.[66]

The approval processes of animal research contribute to human harm

Some argue that the human harms of poorly translatable animal research ought to be factored into much more rigorous harm-benefit analyses (HBAs). These analyses weigh the objectives of animal research studies against their potential benefits and are required by legislation to be conducted as part of the approval process for animal research in the EU and the UK. Although not required by law, they are conducted *de facto* by Institutional Animal

Care and Use Committees (IACUCs) in the United States.[67] Notably, these analyses have been questioned regarding their rigor and critical nature. For example, an unprecedented *retrospective* HBA of more than 200 animal studies worldwide (in contrast to *prospective* HBAs conducted for regulatory reasons) showed that fewer than 7% of them could have met guidelines for prospective approval. If "alternative" nonanimal methods had been considered as well, this proportion would have been even lower.[4]

The evidence, therefore, suggests that the approval process, including at the very least the HBA, is in dire need of being made considerably more rigorous, and that one way to help improve these processes would be to incorporate human harms such as those discussed above. Emerging publications have examined the function of IACUCs and similar ethics committees outside the United States.[68–70] Among numerous salient findings concerning the decision-making processes of animal ethics committees and other stakeholders, they noted that, while US-based researchers are required to *consider* replacements for animal methods, they are not required to *use* them, and further, that ethics committee members are not expected to review those considerations of alternative methods.

While the process may be more involved in Europe, there is considerable room for improvement everywhere. In many European countries (including the UK), there is a legal stipulation that "scientifically satisfactory" alternatives to animals should be used wherever possible; however, there are widespread, established concerns that these regulations are almost worthless because they are not enforced. A report by the UK National Center for the 3Rs concluded that replacement is not well served by any existing process, including ethical review bodies and the UK's Animals in Science Regulation Unit.[68] The report found that replacements are rarely advised, that inspectors and committee members do not and cannot have sufficient expert knowledge of each area and application of research they cover, and that there is an assumption that both researchers and funders have fully considered replacement before deciding that animal use is unavoidable. In addition, the report concluded that funders do not focus on replacement but instead may simply ask if the researchers' chosen approaches will allow them to answer their research questions.

This echoes another report, which highlights that the legal articles "lack teeth" and provide easy opt-outs for researchers, and that protocols are often complex, making it hard to judge replacement options.[71] It also notes significant problems with the HBA, including a lack of clarity in guidelines, its subjective nature and lack of transparency, and conflicts of interest.[72–74] Collectively, these papers noted significant issues with expertise, the conduct of HBAs, alternative searches, the rigor of the review process (e.g., in the United States, this may be conducted by few committee members or even one), lack of training, education, and guidance for the process, conflicts of interest, committee composition and professional diversity, and more. Overall, ethics committee approval of animal research needs considerable improvement and cannot, and should not, be used as a yardstick for adherence to ethical guidelines beyond the minimum standards legally required.

Conclusions and discussion

The human harms from biomedical research with poor human relevance are substantial, widespread, comprehensive, and unmistakable. Though this can include the use of poor-quality human in vitro models, it is clear that the predominant cause is reliance on animals as models. Human harms are primarily a consequence of the poor translation of the results from animal studies to human biology and disease due to intractable species differences, compounded by welfare issues that are endemic to animal research, which affect many crucial biological systems and, therefore, the quality and value of the experimental data. These compounding factors cannot be avoided, and the poor translation from animal to human cannot be significantly improved. It is reasonable to conclude that "those who defend animal experimentation at almost any cost must bear part of the responsibility for the human harms which result."[75]

International initiatives are gathering momentum, building on the arguably well intentioned but fairly vague and generic roadmaps of the past. In 2025, the new director of the NIH announced an initiative to expand innovative, human-based science while reducing animal use in research and detailed actions to deliver this, including the establishment of the Office of Research Innovation, Validation, and Application. The Complement Animal Research in Experimentation (Complement-ARIE) program seeks to advance a comprehensive and diverse array of NAMs and increase their acceptance, use, and appreciation of their value.[76] This includes funding opportunities for researchers and NAM developers.[77,78] The FDA is investing millions of dollars toward expediting NAMs into regulatory use in place of animals, including through its ISTAND program,[79] collaborating with industrial and academic partners internationally, and developing NAMs guidance to help those applying for new drug approvals. The new FDA commissioner announced in 2025 that the agency would phase out animal use in favor of human-relevant methods.[80] This announcement came on the heels of the reintroduction of the FDA Modernization Act 3.0[81] aimed at actively encouraging and facilitating the qualification and widespread use of human-specific NAMs in the testing of new drugs.[82] Outside the United States, the EU's Horizon Program is funding and driving innovation to reduce and replace animal use in research and testing.[83] The Netherlands' Transition Program for Innovation is creating educational, training, collaborative, and funding ventures to accelerate the shift toward animal-free research,[84] and the UK's NC3Rs oversees a NAMs Network and Hub of NAMs researchers, developers, and industry and regulatory end-users, aiming to foster collaborations and facilitate greater use of NAMs.[85]

This burgeoning sea of change is undoubtedly driven by many factors, not least economics and the unprecedented power of NAMs, but also by a growing appreciation and frustration in the face of myriad human harms resulting from animal research, such as those exemplified in this chapter. It is the end-product that matters, not only for the tens of millions of animals used in laboratories each year but also for the billions of people affected by illness, disease, and shortened lifespans. A shift away from animal research is happening, and the greater focus on human-based methods, driven by the advent and ongoing development of NAMs, will deliver a monumental reduction in the suffering of animals and humans alike.

References

[1] Bailey J, Taylor K. Non-human primates in neuroscience research: the case against its scientific necessity. *Altern Lab Anim*. 2016;44:43–69.

[2] Bailey J. Arguments against Using Nonhuman Primates in Research Nonhuman Primate Welfare: From History, Science, and Ethics to Practice. Robinson LM, Weiss A eds. Nonhuman Primate Welfare: From History, Science, and Ethics to Practice. Cham: Springer International Publishing; 2023, pp. 559–588.

[3] Bailey J. Monkey-based research on human disease: the implications of genetic differences. *Altern Lab Anim*. 2014;42:287–317.

[4] Pound P, Nicol CJ. Retrospective harm benefit analysis of pre-clinical animal research for six treatment interventions. *PLoS One*. 2018;13:e0193758.
[5] Bailey J. Letter to the Editor[letter]. *Altern Lab Anim*. 2017;45(3):165–166.
[6] Bailey J. Letter to the Editor. ATLA. 2015;43:205–207.
[7] Bailey J. Letter to the Editor. *ATLA*. 2015;43:428–431.
[8] Marshall LJ, Bailey J, Cassotta M, Herrmann K, Pistollato F. Poor Translatability of Biomedical Research Using Animals — a Narrative Review. *Altern Lab Anim*. 2023;51:102–135. Available at: https://doi.org/10.1177/02611929231157756.
[9] Bailey J, Balls M. Clinical impact of high-profile animal-based research reported in the UK national press. *BMJ Open Sci*. 2020;4(1):4e100039.
[10] WHO. Dementia fact sheet, 2025. https://www.who.int/news-room/fact-sheets/detail/dementia 2025, Accessed May 30 2025).
[11] Alzheimer's Society (UK). Facts about dementia, https://www.alzheimersresearchuk.org/dementia-information/need-to-know-more/facts-stats/10-facts-about-dementia/ (2025, Accessed May 30 2025).
[12] Alzheimer's Disease International. Dementia statistics, https://www.alzint.org/about/dementia-facts-figures/dementia-statistics/ (2020, Accessed May 30 2025).
[13] Alzheimer's Association. Alzheimer's disease facts and figures, https://www.alz.org/getmedia/ef8f48f9-ad36-48ea-87f9-b74034635c1e/alzheimers-facts-and-figures.pdf (2025, Accessed Jun 6 2025).
[14] Office for National Statistics (ONS). ONS website, statistical bulletin, Deaths registered in England and Wales: 2023, https://www.ons.gov.uk/peoplepopulationandcommunity/birthsdeathsandmarriages/deaths/bulletins/deathsregistrationsummarytables/2023#leading-causes-of-death (2024, Accessed Jun 9 2025).
[15] Burke M. Why Alzheimer's drugs keep failing. Sci. Am. 2014. https://www.scientificamerican.com/article/why-alzheimer-s-drugs-keep-failing/ (Accessed 30 May 2025).
[16] Balls M, Combes R. Animal experimentation and alternatives: revealed preferences [editorial]. *Altern Lab Anim*. 2017;45(1):1.
[17] Morris GP, Clark IA, Vissel B. Questions concerning the role of amyloid-β in the definition, aetiology and diagnosis of Alzheimer's disease. *Acta Neuropathol*. 2018;136:663–689.
[18] Pistollato F, Ohayon EL, Lam A, et al. Alzheimer disease research in the 21st century: past and current failures, new perspectives and funding priorities. *Oncotarget*. 2016;7:38999–39016.
[19] Huang L-K, Kuan Y-C, Lin H-W, Hu C-J. Clinical trials of new drugs for Alzheimer disease: a 2020-2023 update. *J Biomed Sci*. 2023;30:83.
[20] Espay AJ, Kepp KP, Herrup K. Lecanemab and donanemab as therapies for Alzheimer's disease: an illustrated perspective on the data. *eNeuro*. 2024;11:ENEURO.0319-23.2024.
[21] McGeachan RI, Meftah S, Taylor LW, et al. Divergent actions of physiological and pathological amyloid-β on synapses in live human brain slice cultures. *Nat Commun*. 2025;16:3753.
[22] Cao Z, Kong F, Ding J, et al. Promoting Alzheimer's disease research and therapy with stem cell technology. *Stem Cell Res Ther*. 2024;15:136.
[23] Sreenivasamurthy S, Laul M, Zhao N, Kim T, Zhu D. Current progress of cerebral organoids for modeling Alzheimer's disease origins and mechanisms. *Bioeng Transl Med*. 2023;8:e10378.
[24] Amartumur S, Nguyen H, Huynh T, et al. Neuropathogenesis-on-chips for neurodegenerative diseases. *Nat Commun*. 2024;15:2219.
[25] Ji Y, Chen X, Wang Z, et al. Alzheimer's disease patient brain extracts induce multiple pathologies in novel vascularized neuroimmune organoids for disease modeling and drug discovery. *Mol Psychiatry*. 2025. doi:10.1038/s41380-025-03041-w. Epub ahead of print. PMID: 40316675.
[26] Choe MS, Yeo HC, Kim JS, et al. Simple modeling of familial Alzheimer's disease using human pluripotent stem cell-derived cerebral organoid technology. *Stem Cell Res Ther*. 2024;15:118.
[27] World Health Organization (WHO). Parkinson disease, https://www.who.int/news-room/fact-sheets/detail/parkinson-disease (2023, Accessed Jun 6 2025).
[28] Willis AW, Roberts E, Beck JC, et al. Incidence of Parkinson disease in North America. *NPJ Parkinsons Dis*. 2022;8:170.
[29] Parkinson's Foundation. Understanding Parkinson's, https://www.parkinson.org/understanding-parkinsons/statistics (2025, Accessed Jun 6 2025).
[30] Marshall LJ, Willett C. Parkinson's disease research: adopting a more human perspective to accelerate advances. *Drug Discov Today*. 2018;23:1950–1961.

[31] Cassotta M, Geerts H, Harbom L, et al. The future of Parkinson's disease research: a new paradigm of human-specific investigation is necessary … and possible. *ALTEX*. 2022;39:694–709.
[32] Gomez-Giro G, Frangenberg D, Vega D, et al. α-Synuclein pathology spreads in a midbrain-hindbrain assembloid model. *Adv Sci*. 2025;12(20):12e2409040.
[33] Sabaté Soler S, Rosety I, Giro G, et al. Released mitochondrial DNA and neurofilament light chain as Parkinson's disease phenotypes in patient-specific midbrain assembloids. 2025. *bioRxiv* 2025.03.28.645921. doi:10.1101/2025.03.28.645921.
[34] Zuccoli E, Al Sawaf H, Tuzza M, et al. *Reproducibility of PD Patient-specific Midbrain Organoid Data for in Vitro Disease Modelling*. 2025. bioRxiv 2025.04.17.649331. doi:10.1101/2025.04.17.649331.
[35] Qin S, Gao J, Ding M, Vicuna LH, Wang X. Parkinson's disease in a dish: the emerging role of organoids in research and therapy. *Organoid Res*. 2025;1:025040006.
[36] Reiner O, Sapir T, Parichha A. Using multi-organ culture systems to study Parkinson's disease. *Mol Psychiatry*. 2020;26:725–735.
[37] Jones EJ, Skinner BM, Parker A, et al. An in vitro multi-organ microphysiological system (MPS) to investigate the gut-to-brain translocation of neurotoxins. *Biomicrofluidics*. 2024;18:054105.
[38] Thomas DW, Chancellor D, Micklus A, et al. Clinical development success rates and contributing factors 2011-2020. Available at: https://www.bio.org/clinical-development-success-rates-and-contributing-factors-2011-2020 2021 pp 34.
[39] Sun D, Macedonia C, Chen Z, et al. Can machine learning overcome the 95% failure rate and reality that only 30% of approved cancer drugs meaningfully extend patient survival? *J Med Chem*. 2024;67:16035–16055.
[40] Jentzsch V, Osipenko L, Scannell JW, Hickman JA. Costs and causes of oncology drug attrition with the example of insulin-like growth factor-1 receptor inhibitors. *JAMA Netw Open*. 2023;6:e2324977.
[41] Fojo T. Journeys to failure that litter the path to developing new cancer therapeutics. *JAMA Netw Open*. 2023;6:e2324949.
[42] International Agency for Research on Cancer. Global cancer observatory, cancer tomorrow, https://gco.iarc.fr/e (2025, Accessed Jun 9 2025).
[43] Sherman RL, Firth AU, Henley SJ, et al. Annual Report to the Nation on the Status of Cancer, featuring state-level statistics after the onset of the COVID-19 pandemic. *Cancer*. 2025;131:e35833.
[44] Mariotto AB, Enewold L, Zhao J, Zeruto CA, Yabroff KR. Medical care costs associated with cancer survivorship in the United States. *Cancer Epidemiol Biomarkers Prev*. 2020;29:1304–1312.
[45] Islami F, Marlow EC, Thomson B, et al. Proportion and number of cancer cases and deaths attributable to potentially modifiable risk factors in the United States, 2019. *CA Cancer J Clin*. 2024;74:405–432.
[46] Siegel RL, Miller KD, Fuchs HE, Jemal A. Cancer statistics, 2022. *CA Cancer J Clin*. 2022;72:7–33.
[47] World Health Organization (WHO). Preventing cancer, https://www.who.int/activities/preventing-cancer (2025, Accessed Jun 9 2025).
[48] McIntosh SA, Alam F, Adams L, et al. Global funding for cancer research between 2016 and 2020: a content analysis of public and philanthropic investments. *Lancet Oncol*. 2023;24:636–645.
[49] Schmutz A, Matta M, Cairat M, et al. Mapping the European cancer prevention research landscape: a case for more prevention research funding. *Eur J Cancer*. 2023;195.
[50] Liu K, Chen X, Fan Z, et al. From organoids to organoids-on-a-chip: current applications and challenges in biomedical research. *Chin Med J*. 2025;138:792–807.
[51] Wang Q, Yuan F, Zuo X, Li M. Breakthroughs and challenges of organoid models for assessing cancer immunotherapy: a cutting-edge tool for advancing personalised treatments. *Cell Death Discov*. 2025;11:222.
[52] Li B, Tang Y, Ma L, Song J, Xue L. Synergistic innovation in organ-on-a-chip and organoid technologies: reshaping the future of disease modeling, drug development and precision medicine. *Protein Cell*. 2025:pwaf058. doi:10.1093/procel/pwaf058Epub ahead of print. PMID: 40653861.
[53] Bailey J, Balls M. Recent efforts to elucidate the scientific validity of animal-based drug tests by the pharmaceutical industry, pro-testing lobby groups, and animal welfare organisations. *BMC Med Ethics*. 2019;20:16.
[54] Bailey J, Thew M, Balls M. Predicting human drug toxicity and safety via animal tests: can any one species predict drug toxicity in any other, and do monkeys help? *Altern Lab Anim*. 2015;43:393–403.
[55] Bailey J, Thew M, Balls M. An analysis of the use of animal models in predicting human toxicology and drug safety. *Altern Lab Anim*. 2014;42:181–199.
[56] Bailey J, Thew M, Balls M. An analysis of the use of dogs in predicting human toxicology and drug safety. *Altern Lab Anim*. 2013;41:335–350.

[57] Bailey J. A response to the ABPI's letter to the use of dogs in predicting drug toxicity in humans[letter]. *Altern Lab Anim.* 2014;42(2):149–153.
[58] Editorial. Follow the yellow brick road. *Nat Rev Drug Discov.* 2003;2:167.
[59] Van Norman GA. Limitations of animal studies for predicting toxicity in clinical trials: is it time to rethink our current approach? *JACC Basic Transl Sci.* 2019;4:845–854.
[60] Sertkaya A, Beleche T, Jessup A, Sommers BD. Costs of drug development and research and development intensity in the US, 2000-2018. *JAMA Netw Open.* 2024;7:e2415445.
[61] Wouters OJ, McKee M, Luyten J. Estimated research and development investment needed to bring a new medicine to market, 2009-2018. *JAMA.* 2020;323:844–853.
[62] Pharmaceutical research and manufacturers of America (PhRMA). 2024 PhRMA Annual Membership Survey, https://cdn.aglty.io/phrma/global/resources/import/pdfs/PhRMA_2024 Annual Membership Survey.pdf (2025, Accessed Jun 9 2025).
[63] Lee S, Yoo S. InterDILI: interpretable prediction of drug-induced liver injury through permutation feature importance and attention mechanism. *J Cheminform.* 2024;16:1.
[64] Ewart L, Apostolou A, Briggs SA, et al. Performance assessment and economic analysis of a human Liver-Chip for predictive toxicology. *Commun Med.* 2022;2:154.
[65] Taylor K, Ram R, Ewart L, et al. Perspective: how complex in vitro models are addressing the challenges of predicting drug-induced liver injury. *Front Drug Discov.* 2025;5:1536756.
[66] Emulate. Emulate Liver-Chip S1 Accepted into FDA's ISTAND Pilot Program. What's Next?, https://emulatebio.com/liver-chip-accepted-into-fda-istand-pilot-program-lorna-ewart-interview/ (2024, Accessed Sep 2 2025).
[67] Brønstad A, Newcomer CE, Decelle T, et al. Current concepts of harm-benefit analysis of animal experiments - report from the AALAS-FELASA working group on harm-benefit analysis - Part 1. *Lab Anim.* 2016;50:1–20.
[68] Rawle F. The role of review and regulatory approvals processes for animal research in supporting implementation of the 3Rs - A report by Dr Frances Rawle, commissioned by the NC3Rs, https://nc3rs.org.uk/role-review-and-regulatory-approvals-processes-animal-research-supporting-implementation-3rs-2023 (2023, Accessed May 25 2025).
[69] Milford A, De Clercq E, Louis-Maerten E, Geneviève LD, Elger BS. How animal ethics committees make decisions - a scoping review of empirical studies. *PLoS One.* 2025;20:e0318570.
[70] Pippin JJ, Bailey J, Kennedy M, et al. Institutional animal care and use committees and the challenges of evaluating animal research proposals. *Res Integr Peer Rev.* 2025;10(11). doi:https://doi.org/10.1186/s41073-025-00169-9.
[71] Cruelty Free Europe. Reducing and replacing animal experiments: europe needs an action plan, https://crueltyfreeeurope.org/cruelty-free-europe-reducing-and-replacing-animal-testing-europe-needs-action-plan (2022, Accessed May 23 2025.
[72] Taylor KD. Harms versus benefits: a practical critique of utilitarian calculations. In: Linzey A, Linzey C, eds. *The Ethical Case against Animal Experiments. Oxford Centre for Animal Ethics.* Champaign, IL, USA: University of Illinois Press; 2018, pp. 148–159.
[73] Grimm H. Turning apples into oranges? The harm-benefit analysis and how to take ethical considerations into account. *Altern Lab Anim.* 2015;43:P22–4.
[74] Grimm H, Eggel M, Deplazes-Zemp A, Biller-Andorno N. The road to hell is paved with good intentions: why harm-benefit analysis and its emphasis on practical benefit jeopardizes the credibility of research. *Animals.* 2017;7:70.
[75] Balls M. It's time to include harm to humans in harm-benefit analysis - but how to do it, that is the question. *Altern Lab Anim.* 2021;49:182–196.
[76] NIH. Complement animal research in experimentation (complement-ARIE) program, https://commonfund.nih.gov/complementarie (2025, Accessed 20 Feb 2025).
[77] NIH. Complement-ARIE funding opportunities, https://commonfund.nih.gov/complementarie/fundingopportunities (2025, Accessed 20 Feb 2025).
[78] NIH. Complement-ARIE new approach methodologies (NAMs) technology development centers (UM1 Clinical Trial Optional), https://grants.nih.gov/grants/guide/rfa-files/RFA-RM-24-010.html (2025, Accessed 20 Feb 2025).
[79] US Food and Drug Administration. Innovative science and technology approaches for new drugs (ISTAND) program, https://ntp.niehs.nih.gov/sites/default/files/2025-10/04_Sekar_SACATM2025_508.pdf (2025, Accessed Sep 2 2025).

[80] Makary MA. FDA announces plan to phase out animal testing requirement for monoclonal antibodies and other drugs, https://www.fda.gov/news-events/press-announcements/fda-announces-plan-phase-out-animal-testing-requirement-monoclonal-antibodies-and-other-drugs (2025, Accessed Jun 13 2025).
[81] U.S. Congress FDA Modernization Act 3.0, https://www.booker.senate.gov/imo/media/doc/fdamodernization302025.pdf (2025, Accessed 20 Feb 2025).
[82] Carratt SA, Zuch de Zafra CL, Oziolor E, et al. An industry perspective on the FDA Modernization Act 2.0/3.0: potential next steps for sponsors to reduce animal use in drug development. *Toxicol Sci*. 2025;203:28–34.
[83] European Commission. Accelerating the implementation of New Approach Methodologies and other innovative non-animal approaches for the development, testing and production of health technologies. HORIZON-JU-IHI-2023-05-01, https://ec.europa.eu/info/funding-tenders/opportunities/portal/screen/opportunities/topic-details/horizon-ju-ihi-2023-05-01 (2023).
[84] TPI. Transition programme for innovation, https://www.animalfreeinnovationtpi.nl (2025, Accessed Sep 2 2025).
[85] NC3Rs. NAMs Network, https://nc3rs.org.uk/3rs-resource-library/nams-network (2025, Accessed 20 Feb 2025).

PART II

Advances in nonanimal methods for research and testing

CHAPTER

6

Microphysiological systems

Konrad Schmidt[a], *Alessandro Bentivogli*[a], *and Sarah Hedtrich*[b,c]

[a]Berlin Institute of Health at Charité—Universitätsmedizin Berlin, Berlin, Germany
[b]School of Biomedical Engineering, The University of British Columbia, Vancouver, BC, Canada
[c]Centre for Blood Research & Life Sciences Institute, University of British Columbia, Life Sciences Centre, Vancouver, BC, Canada

Abstract

Microphysiological systems (MPS) are complex human-based in vitro models that aim to closely replicate human (patho)physiology. This chapter explores the current state-of-the-art and limitations of organoids and organ-on-chip (OoC) systems as preeminent MPS technologies. Organoids, self-assembled 3D cellular structures, closely mimic tissue-specific architecture and functions, advancing biomedical research in disease modeling, drug discovery, and personalized medicine. Complementing organoids, OoC systems integrate microfluidics and mechanical cues to recreate dynamic biological environments, addressing organ-specific and inter-organ interactions. Key aspects of OoC systems, including design principles, cell sources, and in situ monitoring, are highlighted for their potential to model complex diseases and pharmacokinetics. Despite their promise, MPS technologies face challenges such as limited complexity, scalability, and standardization, which hinder broader application. Nevertheless, by overcoming these constraints, MPS have the potential to improve biomedical research and reduce the usage of animal models in the future.

Keywords: Microphysiological systems; Organoids; Organ-on-chip; Disease modeling; Drug development

Introduction to microphysiological systems

Microphysiological systems (MPS) are human-based in vitro models that aim to recapitulate physiological and pathophysiological aspects of the human body ex vivo. While this is the overarching purpose of MPS, they differ widely in their complexity. Examples of MPS include organoids, bioengineered or bioprinted tissue models, as well as organ-on-chips (OoC). In this chapter, we provide an overview of organoids and OoC setups as commonly applied and increasingly used MPS systems.

Moving Toward Nonanimal Approaches in Medical Research and Testing
https://doi.org/10.1016/B978-0-443-26514-3.00002-3

Organoids

Definition and background

Historically, simplistic 2D monolayers formed by primary or immortalized cells have represented the main alternative to animal models in biomedical research.[1,2] These cells are fairly easy to maintain, low-cost, and offer the possibility of easy genetic manipulation. However, they do not exhibit the tissue-relevant spatial organization and differentiation found in humans, and cell lines mostly represent a cancerous, pathological cell state and do not account for the genetic variation of individuals.[3] Finally, simplistic 2D monolayers do not emulate organ-specific 3D microenvironments, which is critical when striving to recapitulate key features of human tissue.

A major breakthrough was achieved in 2009 when Sato et al. reported the formation of 3D structures in vitro.[4] Using murine intestinal stem cells seeded in an extracellular matrix (ECM) substitute, they observed the formation of self-assembled 3D structures composed of differentiated intestinal cell types and stem cells. This landmark paper paved the way for organoid research.

By definition, organoids are 3D cell assemblies that have the unique ability of self-renewal and self-organization.[3] These structures are obtained by embedding stem cells of various origins into an ECM and treating them with a cocktail of growth factors.[5] A key element to induce organoid formation is the ability to mimic the stem cell niche and its microenvironment.[6] The latter guides stem cell proliferation and differentiation and, thus, tissue homeostasis through biochemical, biophysical, and mechanical cues.[7]

Building on this initial success, there have been continuous attempts to engineer solutions to obtain organoid cultures from other tissues and achieve greater control over organogenesis. Researchers continue to optimize the cell sources, the ECM, as well as biochemical and physical cues to guide organoid formation and differentiation.

Cell source

Organoids can be obtained either from adult stem cells (ASCs) or induced pluripotent stem cells (iPSCs) (Fig. 1). The origin of the cells is very important as it determines batch-to-batch variability, cellular heterogeneity, as well as the functionality of the organoids.[8] ASCs are tissue-resident stem or progenitor cells obtained from a biopsy and dissociated through enzymatic digestion.[8,9] Subsequently, the stem cells are selected by fluorescence-activated or magnetic-activated cell sorting and resuspended in an ECM substrate of choice, followed by the addition of tissue-specific growth and differentiation factors.[8–10]

The choice of ECM represents a critical step as it needs to provide a specific niche for stem cell growth, proliferation, and differentiation.[3] Among currently available ECM substitutes, Matrigel is most commonly used.[3] Matrigel is a laminin-111-rich basement membrane extracted from the Engelbreth-Holm-Swarm mouse sarcoma, which provides valuable biochemical and mechanical cues for organoid culture in vitro.[11] While Matrigel has significantly

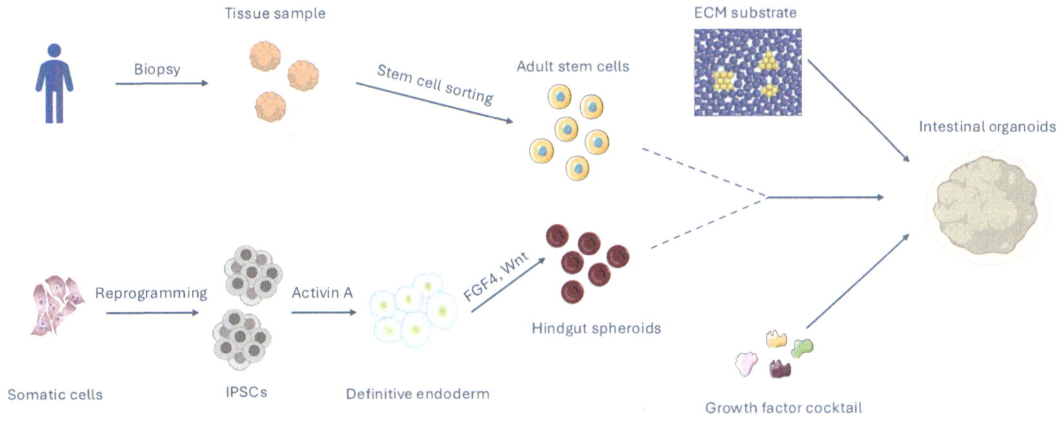

FIG. 1 Generation of intestinal organoids from adult stem cells (ASCs; top) and induced pluripotent stem cells (iPSCs; bottom). To generate ASCs, tissue samples are obtained through biopsies, the cells are then sorted, and only stem cells are selected. For iPSCs, after the somatic cells have been reprogrammed into pluripotent stem cells, they are pushed to definitive endoderm and later to form hindgut spheroids. Subsequently, either the ASCs or the hindgut spheroids are integrated into a niche-mimicking ECM substitute and supplied with a growth factor cocktail to induce the formation of intestinal organoids.

aided the advancement of organoid research, its exact composition is poorly defined, and high batch-to-batch variabilities hamper clinical translation.[3]

Consequently, there have been numerous attempts to move past Matrigel using alternative, customizable, and animal-free matrices.[12] These alternatives are either natural or synthetic products composed of biomaterials such as collagen, fibrin, or hyaluronic acid. Hydrogels with synthetic backbones offer the advantage of being easily functionalized to improve cell adherence and to tune properties such as stiffness and degradation for tissue-specific applications.

While ASC-derived organoids were established before iPSC-derived organoids and can be differentiated into many cell types found in vivo, they lack mesenchymal precursor cells.[5] Furthermore, their use is contingent on the availability of biological samples. Here, iPSC-based organoids offer distinct advantages, such as operating with a better-defined and infinitely expandable cell source, thus reducing biological variability. iPSCs are stem cells obtained by reprogramming mature adult cells (mostly fibroblasts) into a pluripotent state. As such, they can address critical gaps for tissues where clinical samples cannot be easily obtained, such as the liver, where it is challenging to generate and maintain ASC-derived liver organoids, or the brain.[13–17]

Finally, iPSCs, as well as ASCs, allow the generation of patient-specific organoids, which can be leveraged for testing and predicting treatment strategies in vitro. However, dealing with pluripotent stem cells also means that it can be difficult to obtain pure tissue-specific cultures. Often, differentiation protocols result in a heterogeneous mixture of various cell types.[17] In addition, iPSC-derived models often exhibit immature functional characteristics, emulating embryonic or fetal traits.[18]

Organoids in biomedical research

The first organoids were derived from murine intestinal crypts.[4] A few years later, the first human ASC-derived intestinal organoids were generated, followed by organoids from other parts of the GI tract.[10,19,20] As of today, human ASC-derived organoids of the lungs, liver, pancreas, endometrium, mammary gland, prostate, and many other tissues have been reported.[21–26] Similarly, iPSCs are widely used, including to generate organoids that cannot be obtained via ASCs, such as brain organoids.[27–32]

In both cases, the ability to expand the organoids indefinitely has led to the creation of organoid biobanks, which play an increasingly important role in biomedical research.[10,33–35] In fact, organoids can help us better understand human biology and investigate developmental events, including stem cell differentiation and tissue development.[36,37] Diseased models can be obtained by altering the expression and activity of certain genes or by generating patient-specific organoids to replicate a disease in vitro. For example, Lancaster et al. first showed that brain development can be modeled through iPSC organoids and were able to produce a disease model of microcephaly by mutating the *CDK5RAP2* gene.[38]

Organoids can also aid in the understanding of infectious diseases by overcoming the inadequacy of many animal models regarding host specificity.[1] A prime example was the use of brain organoids to identify the Zika virus as causative for the rise in newborns with microcephaly. Several other studies verified the effect of the Zika virus on human brain development using brain organoids, proving that these models serve as relevant platforms to study pathogenic infections.[39,40] Another example is the first successful in vitro cultivation of the norovirus in human intestinal organoids.[41] This study showed that certain viral strains require the presence of specific bile components and the activity of the host enzyme galactosidase 2-alpha-L-fucosyltransferase 2 for infections.[41] Intestinal organoids also played a key role in unraveling the pathomechanisms of human rotavirus (HRV), where models infected with the virus enterotoxin displayed luminal swelling, a characteristic of rotavirus-induced diarrhea.[42] Organoids can also serve as drug screening platforms. In the case of HRV, organoids facilitated the identification of mycophenolic acid, IFNγ, and anti-rotavirus VP7 antibodies as potent rotavirus replication inhibitors.[43]

In addition to infectious diseases, organoids are proving increasingly valuable in unraveling other disease mechanisms, as demonstrated in Alzheimer's and Parkinson's diseases.[44,45] Patient-derived organoids can also be harnessed to predict treatment responses. One prime example is the use of intestinal organoids derived from cystic fibrosis patients to test the restoration efficacy of the cystic fibrosis transmembrane conductance regulator (CFTR) activity following gene therapies or other drugs, such as CFTR modulators.[46,47]

Finally, patient-specific organoids are having an increasingly significant impact in the cancer field. The ability to grow human cancer organoids enables the generation of a diverse set of tumor models, including those of the colon, brain, prostate, pancreas, liver, breast, bladder, stomach, esophagus, endometrium, and lungs.[22,48–54] Such patient-specific cancer organoids represent an important leap forward in the context of personalized medicine, facilitating drug screening and testing ad personam. For this purpose, Vlachogiannis et al. compared drug responses between patients and their cancer organoid models, revealing a 90% concordance.[55]

In addition to biomedical research, organoids have proven useful in drug discovery and toxicology studies.[56] For example, Winkler et al. used human ASC-derived lung organoids to evaluate the potential toxic effects of microplastic fibers on our airways, showing a tendency of the organoids to incorporate the fibers over time during co-culture.[57] Moreover, Zhu et al. employed iPSC-derived brain organoids to explore the mechanisms underlying neural dysfunctions in prenatal alcohol exposure.[58] Interestingly, the organoids showed decreased neurite outgrowth and skewed neural maturation following ethanol exposure, highlighting novel pathways involved in this condition. The relatively low cost and ease of culture have also made these models ideal for high-throughput drug testing using more or less complicated systems.[59,60] For example, Shinozawa et al. leveraged iPSC-derived liver organoids to develop an organoid-based assay measuring viability, cholestatic and/or mitochondrial toxicity, which allowed them to investigate the potential risk of causing drug-induced liver injury for 238 different drugs at 4 different concentrations with high predictive values.[61]

Organ-on-chips

OoCs represent an innovative approach with the potential to revolutionize biomedical research by enabling the close emulation of human (patho)physiology in vitro. OoCs address the limitations of conventional in vitro cell culture models, which fail to capture the complex microenvironment of human tissues.[62] The development of the first lung-on-chip system in 2010 launched the rapid evolution of OoCs[62] over the past decade. Many OoC setups have been described, including OoCs of the lungs,[62–64] gastrointestinal (GI) tract,[65,66] liver,[67,68] heart,[69,70] and kidneys.[71] OoCs are primarily used in disease modeling,[72–74] to understand (patho)physiological processes and elucidate possible treatments. OoCs are promising tools for preclinical and clinical drug development and testing,[67,68,75,76] as they provide a highly controlled and scalable microenvironment based on human physiology within miniaturized devices. Applications such as personalized medicine or clinical diagnostics are desirable but remain in the early stages due to limitations[77] discussed later in this chapter.

Design principles of OoCs

OoCs are microfabricated devices, usually the size of a microscopy slide, designed to mimic (patho)physiological organ function at the microscale level by integrating bioengineered tissue models. Through innovative design strategies, these devices aim to recapitulate tissue-specific functionalities and (patho)physiological mechanisms. The OoC design enables the mimicking of a dynamic biological environment, which is a major improvement over conventional static in vitro models and a missing link toward alternatives to animal models. In general, an OoC is composed of microfluidic, structural, mechanical, and biological components, which are discussed in more detail below (Fig. 2).

Microfluidics: The dynamic features of OoCs originate from microfabricated channels within the chip. These channels interconnect different compartments on the chip and can be perfused with fluids, thus emulating systemic blood flow or lymphatic drainage. The primary

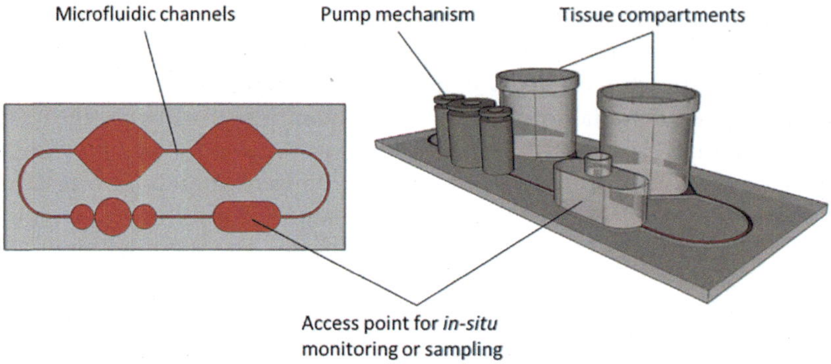

FIG. 2 Design principles of OoC. A schematic illustration of a basic OoC system includes essential features such as the pump mechanism, microfluidic channels, tissue compartments, and access points for in situ monitoring and sampling.

function of this perfusion is the supply of nutrients and oxygen, as well as the removal of metabolic waste products. Furthermore, the resulting fluid dynamics can directly impact the tissue through shear forces. Such forces play a pivotal role in vivo, for example, in vascular biology, where flow and shear stress guide functions such as the polarization and homeostasis of endothelial cells.[78,79] The flow rate is measured as volume per unit of time and can range from a few picoliters to several milliliters per minute.

Different types of flow can be induced, such as constant or pulsatile flow. Pulsatile flow is more suitable for mimicking blood flow in large blood vessels, such as arteries, whereas steady flow is more comparable to the flow observed in small capillaries within a tissue.[80] Hence, these parameters must be tailored to the microfluidic device, the scientific question, and the physiology of the organ of interest.

The flow can be either induced actively through peristaltic pumps,[74,81] syringes,[72,75,82] or microvalve-driven actuators, or passively by diffusion or gravity.[73,83,84] While the latter often merely allows one-way perfusion, it reduces the complexity of the OoC setup. Furthermore, even though one-way perfusion secures constant replenishment of nutrients, it is poorly suited for studies focused on inter-organ crosstalk. A circulatory setup enables inter-organ crosstalk, where tissue models are continuously exposed to the same media conditioned by mediators released from other tissue models. For this reason, a circulatory setup is the more popular approach.[85]

Materials: The material selection is critical for the OoC fabrication of the microfluidic components and tissue cultivation. The most commonly used material is polydimethylsiloxane (PDMS), which is widely applied in bioengineering. Due to its biocompatibility, flexibility, and elastic properties, it is well suited as a structural and cell-interacting component in OoCs.[86] Further, PDMS is compatible with many microfabrication methods, resulting in short OoC production times. Nevertheless, PDMS has distinct drawbacks, including the absorption of small hydrophobic molecules, which can be a limiting factor in drug screening assays.[63] Hence, ongoing research explores alternative materials to address its limitations.[86]

Hydrogels that exhibit native ECM-like properties might be a suitable alternative. This includes synthetic and biological hydrogels like gelatin, alginate, or fibrin, which offer high flexibility and applicability to a wide range of fabrication methods, including 3D printing, allowing rapid prototyping.[87] Hydrogels can also complement PDMS in microfluidic devices, for example, as coatings of the microfluidic channels to prevent molecule absorption. However, their comparatively low mechanical stability can lead to degradation over time.

Thermoplastics such as polycarbonate, polystyrene, or poly(methyl methacrylate)are also cost-effective alternatives that are widely applied in standard lab cell culture. They possess high mechanical and thermal stability, rendering them useful for high-throughput fabrication methods.[88] Nevertheless, their limited gas permeability and flexibility prove disadvantageous.

Taken together, in-house and commercially available organ chips typically combine different materials.[64,71,73] The materials for the different organ chip parts are chosen based on their properties. For example, glass is commonly used as an optically clear base for microscopy, thermoplastics for the housing of the microfluidic channels, and PDMS or hydrogels for the cultivation of biological tissues.

Mechanical cues

OoC systems yield high biomimicry, which includes the biomechanical forces present in the human body and the organ(s) of interest. For example, mechanical forces are critical for bone formation and remodeling, as the latter relies on compaction and bone load.[89] Another example is the breathing mechanics of the lungs. Through inhalation and exhalation, especially in the terminal respiratory units, the lung tissue is continuously stretched. These cyclic forces influence the proliferation, differentiation, and migration of lung epithelial cells,[90] as well as surfactant production in the alveoli.[91] The lungs also serve as an interesting example of the effects of unbalanced mechanical tissue stimulation, as occurs during mechanical ventilation in patients with respiratory failure. The application of unphysiological pressure results in overstretching of the alveoli, causing ventilator-associated lung injury.[92] Hence, when striving to mimic biomechanical forces in an OoC, knowledge about (patho)physiologically relevant parameters is key. To emulate the breathing mechanics of human lungs, for instance, different approaches are being used,[62,63,93] such as the application of a vacuum inside channels of the organ chip, which results in the stretching of an integrated membrane where the cells are cultivated.[64] Other examples of biomechanical cues integrated into OoC systems include the peristaltic movement of the GI tract,[74,82] electrical stimulation of heart muscle tissue,[94] and fluid shear forces, which are pivotal for the vasculature and the kidneys.[95]

In situ monitoring

An exciting opportunity when working with OoC technology is the possibility of in situ monitoring of the tissue cultures. In fact, OoCs have greatly benefited from achievements in microengineering, which allow the production of very small sensors and electrodes that can be implemented into OoC setups, thus providing real-time analysis of the system. For example, integrated electrodes can record the transepithelial electrical resistance as a measure of epithelial barrier integrity[96] or the action potentials of cells[97] in a dynamic environment. Furthermore, sensors can be implemented into the microfluidic network, allowing real-time analysis of media components such as pH, oxygen, or glucose,[68,97,98] which provides detailed information about the metabolic state and viability of the biological samples. OoCs can also be coupled with high-end imaging tools, including live cell imaging. While in vivo monitoring of standard parameters like glucose levels is also possible in animals, the development of more advanced analyte sensors is still in its infancy.[99] Yet, in OoCs, the combination of real-time monitoring with endpoint assays to investigate histology, transcriptomics, or proteomics, and offline measurement of soluble factors by media sampling, is essential to allow in-depth analysis and requires further research.

Tissue models

A variety of biological samples can be integrated into an OoC setup, such as simple cell monolayers,[62,63] spheroids,[71] organoids,[72–74] and complex 3D bioengineered tissues.[67,75,76] When designing an OoC setup, a critical point is the choice of the cell source for the tissue model, which must be selected in a context-based manner. Commonly used cell types include immortalized cell lines, primary human cells, iPSCs, and ASCs.[100] Cell lines are well suited for high-throughput screening approaches as they are cost-effective, easy to handle, and readily available; however, they have limited predictive value. Primary cells exhibit the highest similarity to in vivo conditions, but their limited availability and often restricted proliferative capacity remain challenges. Additionally, when expanded outside of the human body, primary cells may lose their tissue-specific phenotype and alter their functionality.[101,102] iPSCs and ASCs are increasingly utilized cell sources, harboring the potential to generate all (iPSCs) or many (ASCs) somatic cell types. Currently, variability in differentiation efficiencies, which leads to immaturity, as well as limited functionality and standardization, are obstacles that restrict their usability.

Multi-organ-on-chips

A major advantage of OoCs is that multiple tissue models can be combined into a multi-organ-on-chip (multiOoC), which allows for studying inter-tissue crosstalk and mimicking more complex interactions in the human body. This offers a significant advantage over other MPS, such as classic organoid cultures, which do not offer that option per se. The majority of available multi-OoC systems are composed of two to three tissue models, although highly complex systems composed of up to ten organs have also been described.[71,81,103] Multi-OoCs offer great potential for basic and biomedical research, enabling the assessment of drug pharmacokinetics, pharmacodynamics, and toxicity across multiple organ models. For example, it is possible to study drug absorption through relevant biological barriers found in the airways, intestine, or skin and monitor its distribution, metabolism, and excretion through other tissue models. This may help in identifying potential side effects on other organs in a human-based system.[76] Furthermore, multi-OoC setups may facilitate a deeper understanding of (patho)physiological processes that involve several organs, as is the case for infectious diseases (e.g., COVID-19),[104] inflammatory diseases (e.g., atopic diseases),[105] or inter-organ crosstalk, as evident in the gut-liver[106] or brain-gut axis.[107]

Challenges

Organoids

Although major strides have been achieved in the context of organoids, several limitations remain. A major drawback of the setups is the high experimental variability of organoid cultures, which ultimately hampers reproducibility and data robustness.[1,8] A variety of different methods are used to generate and maintain organoid cultures, rendering the inter-lab variability high. At the same time, once the organoids have formed, there is limited control over how they develop due to the stochastic nature of self-organization and cell fate outcomes.[3] This lack of reproducibility poses another limitation to the full scope at which organoid models can be used.

In addition, organoids fail to reproduce the entirety of tissue functions. While many studies have been able to obtain relatively complex models, most are still merely comprised of the epithelial layer, not accounting for the mesenchyme, immunological components, vasculature, and the microbiome. Organoids do not account for the crosstalk between different organs in the human body, only allowing researchers to study tissues in an isolated manner. A major limitation to multitissue organoids is the differing proliferation rates and growth factor requirements of the cells, which may yield functional impairment and de-differentiation in certain cases due to the co-culture conditions.[8] At the same time, organoids are intrinsically static, thus failing to recapitulate the dynamic environment of the human body, such as mechanical stress, perfusion, and inter-organ communication. For example, being embedded in a matrix prevents the cells of an intestinal organoid from experiencing the mechanical stress they would face in the human body due to, for example, peristalsis. Vascularization is also largely absent in these model systems.[1] Probably the biggest drawback at the moment is the lack of inter-tissue communication, which only allows researchers to have a reductionist view of tissue behavior.[108]

iPSC-based organoids often do not reach full cellular maturity and fail to recapitulate adult tissue structure and function.[17] Another challenge is the closed geometry of organoids, which limits nutrient access and leads to the accumulation of dead cells that cannot be removed. Finally, access to the inner lumen of the organoids is limited, though it may be required in some investigations, such as when studying certain secreted factors.

Organ-on-chips

The overall goal for the development of human-based OoC setups is to routinely generate humans-on-chip that include all major organs of the body.[109] While the field has experienced substantial growth over the past decade, this ambitious goal remains a future vision for the following reasons.

Cell source and tissue complexity: The fidelity of OoC models relies on human cells. While primary human cells emulate in vivo-like functions best, the existing culture conditions often don't allow them to remain in a tissue-specific and differentiated state long-term. iPSCs may help fill this gap, especially when striving for autologous multi-OoC systems. However, to unlock the full potential of iPSCs, the development of standardized and robust differentiation protocols is warranted. Furthermore, emulating the full complexity of the human body in vitro is extremely challenging, if not unrealistic. For example, the great variety of immune components is missing in the available OoCs. Even though certain immune cell types can be successfully incorporated into OoCs,[62,65] this is far from mimicking the complete adaptive and innate immune systems. This also holds true for organ complexity. The respiratory system includes approximately 61 cell types,[110] but lung MPS typically entail up to ten.[62–64] Functional and long-term tissue vascularization is another bottleneck that has not been satisfactorily addressed yet. Finally, it should be noted that increasing OoC complexity will also make experimentation and handling more challenging. Hence, a commonly applied approach is to define the minimal OoC complexity in the context of the scientific question.[85]

Scalability: Most of the available OoC setups (commercially available and in-house products) are not compatible with high-throughput approaches due to limitations regarding the generation of complex organ models and the production of organ chips. The available high-throughput devices include simple models that are easy and fast to generate to reach a certain number of replicates.[67,73] In addition, PDMS, which is commonly used for OoC production, is not compatible with mass fabrication methods, and its molecule absorption properties require the search for alternative materials that can circumvent limitations.

Accessibility and standardization: Some OoC systems are commercially available, which provides easier access to end users. Nevertheless, these setups are still very expensive, limiting their wide-spread application. Companies generally provide their own microphysiological (pump) setups and predefined organ chips that may not necessarily match the requirements of every research question. Besides, most organ chips are costly and designed for single-use. This ultimately reduces accessibility for many academic research groups that cannot afford commercial OoCs in adequate amounts or have special demands on their structure. Therefore, groups oftentimes develop in-house-produced chips that lack standardization and are typically used for proof-of-concept studies. These custom-made devices require very specific handling, rendering it challenging to establish them in other laboratories. Overall, the definition of standards for OoC design and operation would potentially facilitate the validation and wide-spread acceptance of OoCs. See Chapter 17 for more on standardization.

Media composition: Another bottleneck is the distinct media requirements of different organ models. In a multiorgan setup, inadequate media composition can impair the validity of the outcomes, particularly when working with highly differentiated cells and iPSC-derived models, as inadequate media composition can initiate cell de-differentiation. Furthermore, while certain media supplements, like corticosteroids, are required for some tissue models, they may exert toxic effects on other cell types.[84,111] This highlights how important it is to identify a media composition that is suitable for all included cells and tissue types. Media optimization, however, is a time-consuming process. So far, mostly 1:1 media mixtures are being used in multi-OoCs. Here, design-of-experiment approaches can help to study the necessity and impact of different media components to pinpoint those critical for the scientific question.[112]

Future directions

The advancement of organoid research has been instrumental in expanding our understanding of the human body and driving biomedical innovation. However, as discussed, there is a pressing need to develop more complex and biomimetic human models. Engineering strategies aimed at optimizing biochemical and mechanical cues, as well as cell-ECM interactions, have sought to simplify and automate organoid generation—enhancing control over morphology and maturity. Efforts to recreate inter-tissue crosstalk have led to the development of multitissue organoids.[113] However, these models face challenges, as each tissue requires distinct culture conditions, making it difficult to establish tailored microenvironments that support multiple tissue types equally.

To address these limitations, OoC models have gained traction. These systems introduce technical complexity while enabling human-based models to be cultured under dynamic conditions, incorporating mechanical forces (e.g., flow, shear stress, pressure) and environmental stimuli such as hypoxia.

While MPS hold significant promise and will play an increasing role in biomedical research, fully replacing animal models remains unrealistic in the near future. Certain aspects, such as memory, nociception, and systemic drug biodistribution, are not yet replicable in human-based models. However, integrating these models can mitigate interspecies discrepancies and enhance the translatability of preclinical data. To maximize their impact, it is crucial to validate MPS against human physiological benchmarks—an often-overlooked step that requires greater emphasis as the field progresses.

References

[1] Kim J, Koo BK, Knoblich JA. Human organoids: model systems for human biology and medicine. *Nat Rev Mol Cell Biol*. 2020;21(10):571–584.

[2] Lopez-Escalera S, Wellejus A. Evaluation of Caco-2 and human intestinal epithelial cells as in vitro models of colonic and small intestinal integrity. *Biochem Biophys Rep*. 2022;31:101314–101314.

[3] Hofer M, Lutolf MP. Engineering organoids. *Nat Rev Mater*. 2021;6(5):402–420.

[4] Sato T, Vries RG, Snippert HJ, et al. Single Lgr5 stem cells build crypt-villus structures in vitro without a mesenchymal niche. *Nature*. 2009;459(7244):262–265.

[5] Loewa A, Feng JJ, Hedtrich S. Human disease models in drug development. *Nat Rev Bioeng*. 2023;1(8):545–559.

[6] Voog J, Jones DL. Stem cells and the niche: a dynamic duo. *Cell Stem Cell*. 2010;6(2):103–115.

[7] Jones DL, Wagers AJ. No place like home: anatomy and function of the stem cell niche. *Nat Rev Mol Cell Biol*. 2008;9(1):11–21.

[8] Zhao Z, Chen X, Dowbaj AM, et al. Organoids. *Nat Rev Methods Primers*. 2022;2(1):1–21.

[9] Hu P, Zhange W, Xin H, Deng G. Single cell isolation and analysis. *Front. Cell Dev Biol*. 2016;4(OCT):116–116.

[10] Sato T, Stange DE, Ferrante M, et al. Long-term expansion of epithelial organoids from human colon, adenoma, adenocarcinoma, and Barrett's epithelium. *Gastroenterology*. 2011;141(5):1762–1772.

[11] Kleinman HK, Martin GR. Matrigel: basement membrane matrix with biological activity. *Semin Cancer Biol*. 2005;15(5):378–386.

[12] Kratochvil MJ, Seymour AJ, Li TL, et al. Engineered materials for organoid systems. *Nat Rev Mater*. 2019;4(9):606–622.

[13] Takahashi K, Tanabe K, Ohnuki M, Yamanaka S. Induction of pluripotent stem cells from adult human fibroblasts by defined factors. *Cell*. 2007;131(5):861–872.

[14] Lancaster MA, Knoblich JA. Organogenesis in a dish: modeling development and disease using organoid technologies. *Science*. 2014;345(6194):1247125–1247125.

[15] McCauley HA, Wells JM. Pluripotent stem cell-derived organoids: using principles of developmental biology to grow human tissues in a dish. *Development*. 2017;144(6):958–962.

[16] Eichmüller OL, Knoblich JA. Human cerebral organoids—a new tool for clinical neurology research. *Nat Rev Neurol*. 2022;18(11):661–680.

[17] Xu Z, Yang J, Xin X, et al. Merits and challenges of iPSC-derived organoids for clinical applications. *Front Cell Dev Biol*. 2023;11:1188905.

[18] Youssef AA, Ross EG, Bolli R, et al. The promise and challenge of induced pluripotent stem cells for cardiovascular applications. *JACC Basic Transl Sci*. 2016;1(6):510–523.

[19] Bartfeld S, Bayram T, van de Wetering M, et al. In vitro expansion of human gastric epithelial stem cells and their responses to bacterial infection. *Gastroenterology*. 2015;148(1):126–136.e6.

[20] Yui S, Nakamura T, Sato T, et al. Functional engraftment of colon epithelium expanded in vitro from a single adult Lgr5+ stem cell. *Nat Med*. 2012;18(4):618–623.

[21] Huch M, Gehart H, van Boxtel R, et al. Long-term culture of genome-stable bipotent stem cells from adult human liver. *Cell*. 2015;160(1-2):299–312.

[22] Kim M, Mun M, Sung CO, et al. Patient-derived lung cancer organoids as in vitro cancer models for therapeutic screening. *Nat Commun*. 2019;10(1):3991.

[23] Kim M, Bonfanti P, Boj SF, et al. Unlimited in vitro expansion of adult bi-potent pancreas progenitors through the Lgr5/R–spondin axis. *EMBO J*. 2013;32(20):2708–2721.

[24] Chua CW, Shibata M, Lei M, et al. Single luminal epithelial progenitors can generate prostate organoids in culture. *Nat Cell Biol*. 2014;16(10):951–961.

[25] Linnemann JR, Miura H, Meixner LK, et al. Quantification of regenerative potential in primary human mammary epithelial cells. *Development*. 2015;142(18):3239–3251.

[26] Turco MY, Gardner L, Hughes J, et al. Long-term, hormone-responsive organoid cultures of human endometrium in a chemically defined medium. *Nat Cell Biol*. 2017;19(5):568–577.

[27] Spence JR, Mayhew CN, Rankin SA, et al. Directed differentiation of human pluripotent stem cells into intestinal tissue in vitro. *Nature*. 2011;470(7332):105–109.

[28] McCracken KW, Catá EM, Crawford CM, et al. Modelling human development and disease in pluripotent stem-cell-derived gastric organoids. *Nature*. 2014;516(7531):400–404.

[29] Takebe T, Sekine K, Enomura M, et al. Vascularized and functional human liver from an iPSC-derived organ bud transplant. *Nature*. 2013;499(7459):481–484.

[30] Wong AP, Bear CE, Chin S, et al. Directed differentiation of human pluripotent stem cells into mature airway epithelia expressing functional CFTR protein. *Nat Biotechnol*. 2012;30(9):876–882.

[31] Lancaster MA, Renner M, Martin CA, et al. Cerebral organoids model human brain development and microcephaly. *Nature*. 2013;501(7467):373–379.

[32] Huang L, Hotzinger A, Jagan I, et al. Ductal pancreatic cancer modeling and drug screening using human pluripotent stem cell- and patient-derived tumor organoids. *Nat Med*. 2015;21(11):1364–1371.

[33] Bruna A, Rueda OM, Greenwood W, et al. A biobank of breast cancer explants with preserved intra-tumor heterogeneity to screen anticancer compounds. *Cell*. 2016;167(1) 260–274.

[34] Yan HHN, Siu HC, Law S, et al. A comprehensive human gastric cancer organoid biobank captures tumor subtype heterogeneity and enables therapeutic screening. *Cell Stem Cell*. 2018;23(6):882–897.

[35] Sachs N, de Ligt, Kopper O, et al. A living biobank of breast cancer organoids captures disease heterogeneity—pubmed. *Cell*. 2018;172(1–2):373–386.

[36] Fatehullah A, Tan SH, Barker N. Organoids as an in vitro model of human development and disease. *Nat Cell Biol*. 2016;18(3):246–254.

[37] Clevers H. Modeling development and disease with organoids. *Cell*. 2016;165(7):1586–1597.

[38] Lancaster MA, Huch M. *Disease modelling in human organoids*. *Dis Model Mech*. 2019;12(7):dmm039347–dmm039347.

[39] Dang J, Tiwari SK, Lichinchi G, et al. Zika virus depletes neural progenitors in human cerebral organoids through activation of the innate immune receptor TLR3. *Cell Stem Cell*. 2016;19(2):258–265.

[40] Garcez P, Loiola EC, da Costa RM, et al. Zika virus impairs growth in human neurospheres and brain organoids. *Science*. 2016;352(6287):816–818.

[41] Ettayebi K, Crawford SE, Murakami K, et al. Replication of human noroviruses in stem cell-derived human enteroids. *Science*. 2016;353(6306):1387–1393.

[42] Saxena K, Blutt SE, Ettayebi K, et al. Human intestinal enteroids: a new model to study human Rotavirus infection, host restriction, and pathophysiology. *J Virol*. 2015;90(1):43–56.

[43] Chen S, Li P, Wang Y, et al. Rotavirus infection and cytopathogenesis in human biliary organoids potentially recapitulate biliary atresia development. *mBio*. 2020;11(4):1–6.

[44] Choi SH, Kim YH, Hebisch M, et al. A three-dimensional human neural cell culture model of Alzheimer's disease. *Nature*. 2014;515(7526):274–278.

[45] Kim H, Park HJ, Choi H, et al. Modeling G2019S-LRRK2 sporadic Parkinson's disease in 3D midbrain organoids. *Stem Cell Rep*. 2019;12(3):518–518.

[46] Dekkers JF, Wiegerinck CL, de Jonge HR, et al. A functional CFTR assay using primary cystic fibrosis intestinal organoids. *Nat Med*. 2013;19(7):939–945.

[47] Schwank G, Koo BK, Sasselli V, et al. Functional repair of CFTR by CRISPR/Cas9 in intestinal stem cell organoids of cystic fibrosis patients. *Cell Stem Cell*. 2013;13(6):653–658.

[48] van de Wetering M, Francies HE, Francis JM, et al. Prospective derivation of a living organoid biobank of colorectal cancer patients. *Cell*. 2015;161(4):933–945.

[49] Gao D, Vela I, Sboner A, et al. Organoid cultures derived from patients with advanced prostate cancer. *Cell*. 2014;159(1):176–187.

[50] Lee SH, Hu W, Matulay JT, et al. Tumor evolution and drug response in patient-derived organoid models of bladder cancer. *Cell*. 2018;173(2):515–528.

[51] Boretto M, Maenhoudt N, Luo X, et al. Patient-derived organoids from endometrial disease capture clinical heterogeneity and are amenable to drug screening. *Nat Cell Biol*. 2019;21(8):1041–1051.

[52] Li X, Francies HE, Secrier M, et al. Organoid cultures recapitulate esophageal adenocarcinoma heterogeneity providing a model for clonality studies and precision therapeutics. *Nat Commun*. 2018;9(1):2983.

[53] Seidlitz T, Merker SR, Rothe A, et al. Human gastric cancer modelling using organoids. *Gut*. 2019;68(2):207–217.

[54] Jacob F, Salinas RD, Zhang DY, et al. A patient-derived glioblastoma organoid model and biobank recapitulates inter- and intra-tumoral heterogeneity. *Cell*. 2020;180(1):188–204.e22.

[55] Vlachogiannis G, Hedavat S, Vatsiou A, et al. Patient-derived organoids model treatment response of metastatic gastrointestinal cancers. *Science*. 2018;359(6378):920–926.

[56] Hu C, Yang S, Zhang T, et al. Organoids and organoids-on-a-chip as the new testing strategies for environmental toxicology-applications & advantages. *Environ Int*. 2024;184:108415.

[57] Winkler AS, Cherubini A, Rusconi F, et al. Human airway organoids and microplastic fibers: a new exposure model for emerging contaminants. *Environ Int*. 2022;163:107200.

[58] Zhu Y, Wang L, Yu F, et al. Probing impaired neurogenesis in human brain organoids exposed to alcohol. *Integr Biol*. 2017;9(12):968–978.

[59] Kozuka K, He Y, Koo-McCoy S, et al. Development and characterization of a human and mouse intestinal epithelial cell monolayer platform. *Stem Cell Rep*. 2017;9(6):1976–1990.

[60] Cil O, Phuan P, Gillespie AM, et al. Benzopyrimido-pyrrolo-oxazine-dione CFTR inhibitor (R)-BPO-27 for antisecretory therapy of diarrheas caused by bacterial enterotoxins. *FASEB J*. 2017;31(2):751–760.

[61] Schinozawa T, Kimura M, Cai Y, et al. High-fidelity drug-induced liver injury screen using human pluripotent stem cell-derived organoids. *Gastroenterology*. 2021;160(3):831–846.e10.

[62] Huh D, Matthes BD, Mammoto A, et al. Reconstituting organ-level lung functions on a chip. *Science*. 2010;328(5986):1662–1668.

[63] Zamprogno P, Wüthrich S, Achenbach S, et al. Second-generation lung-on-a-chip with an array of stretchable alveoli made with a biological membrane. *Commun Biol*. 2021;4(1):168.

[64] Nawroth JC, Roth D, van Schadewijk A, et al. Breathing on chip: dynamic flow and stretch accelerate mucociliary maturation of airway epithelium in vitro. *Mater Today Bio*. 2023;21:100713.

[65] Nikolaev M, Mitrofanova O, Broguiere N, et al. Homeostatic mini-intestines through scaffold-guided organoid morphogenesis. *Nature*. 2020;585(7826):574–578.

[66] Kasendra M, Tovaglieri A, Sontheimer-Phelps A, et al. Development of a primary human small intestine-on-a-chip using biopsy-derived organoids. *Sci Rep*. 2018;8(1):2871.

[67] Bircsak KM, DeBiasio R, Miedel M, et al. A 3D microfluidic liver model for high throughput compound toxicity screening in the OrganoPlate. *Toxicology*. 2021;450:152667.

[68] Bavli D, Prill S, Ezra E, et al. Real-time monitoring of metabolic function in liver-on-chip microdevices tracks the dynamics of mitochondrial dysfunction. *Proc Natl Acad Sci U S A*. 2016;113(16):E2231–E2240.

[69] Sakamiya M, Fang Y, Mo X, Shen J, Zhang T. A heart-on-a-chip platform for online monitoring of contractile behavior via digital image processing and piezoelectric sensing technique. *Med Eng Phys*. 2020;75:36–44.

[70] Kujala VJ, Pasqualini FS, Goss JA, Nawroth JC, Parker KK. Laminar ventricular myocardium on a microelectrode array-based chip. *J Mater Chem B*. 2016;4(20):3534–3543.

[71] Maschmeyer I, Lorenz AK, Schimek K, et al. A four-organ-chip for interconnected long-term co-culture of human intestine, liver, skin and kidney equivalents. *Lab Chip*. 2015;15(12):2688–2699.

[72] Workman MJ, Gleeson JP, Troisi EJ, et al. Enhanced utilization of induced pluripotent stem cell–derived human intestinal organoids using microengineered chips. *Cell Mol Gastroenterol Hepatol*. 2017;5(4):669–677.e2.

[73] Rajasekar S, Lin DSY, Abdul L, et al. IFlowPlate—a customized 384-well plate for the culture of perfusable vascularized colon organoids. *Adv Mater*. 2020;32(46):2002974.

[74] Lee KK, McCauley HA, Broda TR, et al. Human stomach-on-a-chip with luminal flow and peristaltic-like motility. *Lab Chip*. 2018;18(20):3079–3085.

[75] Paek J, Park SE, Lu Q, et al. Microphysiological engineering of self-assembled and perfusable microvascular beds for the production of vascularized three-dimensional human microtissues. *ACS Nano*. 2019;13(7):7627–7643.
[76] Hübner J, Raschke M, Rütschle I, et al. Simultaneous evaluation of anti-EGFR-induced tumour and adverse skin effects in a microfluidic human 3D co-culture model. *Sci Rep*. 2018;8(1):15010.
[77] Zhou C, Li Z, Lu K, et al. Advances in human organs-on-chips and applications for drug screening and personalized medicine. *Fundam Res*. 2024;5(3):1258–1272.
[78] Sonmez UM, Cheng YW, Watkins SC, Roman BL, Davidson LA. Endothelial cell polarization and orientation to flow in a novel microfluidic multimodal shear stress generator. *Lab Chip*. 2020;20(23):4373–4390.
[79] Vion AC, Perovic T, Petit C, et al. Endothelial cell orientation and polarity are controlled by shear stress and VEGF through distinct signaling pathways. *Front Physiol*. 2021;11:623769.
[80] Westerhof BE, Noordegraaf AV, Fukumitsu M. *Pulsatile Hemodynamics and Ventricular–Arterial Interactions in the Pulmonary Circulation: Physiologic Concepts*. Elsevier; 2022:883–897.
[81] Ronaldson-Bouchard K, Teles D, Yeager K, et al. A multi-organ chip with matured tissue niches linked by vascular flow. *Nat Biomed Eng*. 2022;6(4):351–371.
[82] Kim HJ, Huh D, Hamilton G, Ingber DE. Human gut-on-a-chip inhabited by microbial flora that experiences intestinal peristalsis-like motions and flow. *Lab Chip*. 2012;12(12):2165–2174.
[83] Ong LJY, Chong LH, Jin L, et al. A pump-free microfluidic 3D perfusion platform for the efficient differentiation of human hepatocyte-like cells. *Biotechnol Bioeng*. 2017;114(10):2360–2370.
[84] Chen HJ, Miller P, Shuler ML. A pumpless body-on-a-chip model using a primary culture of human intestinal cells and a 3D culture of liver cells. *Lab Chip*. 2018;18(14):2036–2046.
[85] Leung CM, de Haan P, Ronaldson-Bouchard K, et al. A guide to the organ-on-a-chip. *Nat Rev Methods Primers*. 2022;2(1):33.
[86] Campbell SB, Wu Q, Yazbeck J, et al. Beyond polydimethylsiloxane: alternative materials for fabrication of organ-on-a-chip devices and microphysiological systems. *ACS Biomater Sci Eng*. 2021;7(7):2880–2899.
[87] Liu H, Wang Y, Cui K, et al. Advances in hydrogels in organoids and organs-on-a-chip. *Adv Mater*. 2019;31(50):1902042.
[88] Rodríguez CF, Andrade-Pérez V, Vargas MC, et al. Breaking the clean room barrier: exploring low-cost alternatives for microfluidic devices. *Front Bioeng Biotechnol*. 2023;11:1176557.
[89] Jiang JX, Siller-Jackson AJ, Burra S. Roles of gap junctions and hemichannels in bone cell functions and in signal transmission of mechanical stress. *Front Biosci*. 2007;12(1):1450.
[90] Doryab A, TAS S Taskin MB, et al. Evolution of bioengineered lung models: recent advances and challenges in tissue mimicry for studying the role of mechanical forces in cell biology. *Adv Funct Mater*. 2019;29(39):1903114.
[91] Wirtz HRW, Dobbs LG. Calcium mobilization and exocytosis after one mechanical stretch of lung epithelial cells. *Science*. 1990;250(4985):1266–1269.
[92] Pinhu L, Whietehead T, Griffiths M. Ventilator-associated lung injury. *Lancet*. 2003;361(9354):332–340.
[93] Stucki AO, Stucki JD, Hall SRR, et al. A lung-on-a-chip array with an integrated bio-inspired respiration mechanism. *Lab Chip*. 2015;15(5):1302–1310.
[94] Park SM, Eom S, Hong H, et al. Reconstruction of in vivo-like in vitro model: enabling technologies of microfluidic systems for dynamic biochemical/mechanical stimuli. *Microelectron Eng*. 2019;203–204:6–24.
[95] Wang D, Gust M, Ferrell N. Kidney-on-a-chip: mechanical stimulation and sensor integration. *Sensors*. 2022;22(18):6889.
[96] Henry OYF, Villenave R, Cronce MJ, et al. Organs-on-chips with integrated electrodes for trans-epithelial electrical resistance (TEER) measurements of human epithelial barrier function. *Lab Chip*. 2017;17(13):2264–2271.
[97] Aydogmus H, Hu M, Ivancevic L, et al. An organ-on-chip device with integrated charge sensors and recording microelectrodes. *Sci Rep*. 2023;13(1):8062.
[98] Bussooa A, Tubbs E, Revol-Cavalier F, et al. Real-time monitoring of oxygen levels within thermoplastic organ-on-chip devices. *Biosens Bioelectron X*. 2022;11:100198.
[99] Flynn CD, Chang D, Mahmud A, et al. Biomolecular sensors for advanced physiological monitoring. *Nat Rev Bioeng*. 2023;1(8):560–575.
[100] Singh D, Mathur A, Arora S, Roy S, Mahindroo N. Journey of organ on a chip technology and its role in future healthcare scenario. *Appl Surf Sci Adv*. 2022;9:100246.
[101] Herbst CJ, Lopez-Rodriguez E, Gluhovic V, et al. Characterization of commercially available human primary alveolar epithelial cells. *Am J Respir Cell Mol Biol*. 2024;70(5):339–350.
[102] Verma A, Verma M, Singh A. Animal Tissue Culture Principles and Applications. Elsevier; 2020:269–293.

[103] Edington CD, Chen WLK, Geishecker E, et al. Interconnected microphysiological systems for quantitative biology and pharmacology studies. *Sci Rep*. 2018;8(1):4530.
[104] Shahabipour F, Satta S, Mahmoodi M, et al. Engineering organ-on-a-chip systems to model viral infections. *Biofabrication*. 2023;15(2):022001.
[105] Graff P, Woerz D, Wilzopolski J, et al. Extracellular matrix remodeling in atopic dermatitis harnesses the onset of an asthmatic phenotype and is a potential contributor to the atopic march. *J Invest Dermatol*. 2024;144(5):1010–1021.e23.
[106] Yang J, Hirai Y, Iida K, et al. Integrated-gut-liver-on-a-chip platform as an in vitro human model of non-alcoholic fatty liver disease. *Commun Biol*. 2023;6(1):310.
[107] Kim M-H, Kim D, Sung JH. A gut-brain axis-on-a-chip for studying transport across epithelial and endothelial barriers. *J Ind Eng Chem*. 2021;101:126–134.
[108] Park G, Rim YA, Sohn Y, Nam Y, Ju JH. Replacing animal testing with stem cell-organoids: advantages and limitations. *Stem Cell Rev Rep*. 2024;20(6):1375–1386.
[109] Sikkema L, Ramírez-Suástegui C, Strobl DC, et al. An integrated cell atlas of the lung in health and disease. *Nat Med*. 2023;29:1563–1577.
[110] Marx U, Walles H, Hoffmann S, et al. 'Human-on-a-chip' developments: a translational cutting-edge alternative to systemic safety assessment and efficiency evaluation of substances in laboratory animals and man? *Altern Lab Anim*. 2012;40(5):235–257.
[111] Trapecar M, Communal C, Velazquez J, et al. Gut-liver physiomimetics reveal paradoxical modulation of IBD-related inflammation by short-chain fatty acids. *Cell Syst*. 2020;10(3):223–239.e9.
[112] Galbraith SC, Bhatia H, Liu H, Yoon S. Media formulation optimization: current and future opportunities. *Curr Opin Chem Eng*. 2018;22:42–47.
[113] Koike H, Iwasawa K, Ouchi R, et al. Modelling human hepato-biliary-pancreatic organogenesis from the forgut-midgut boundary. *Nature*. 2019;574(7776):112–116.

CHAPTER

7

The application of generative adversarial networks (GANs) as new approach methods (NAMs) toward nonanimal testing

Xi Chen, Ting Li, Weida Tong, and Minjun Chen

National Center for Toxicological Research, U.S. Food and Drug Administration, Jefferson, AR, United States

Abstract

Animal models have historically played a critical role in biomedical research and regulatory submissions. However, there is a growing movement toward nonanimal approaches driven by ethical concerns, cost-effectiveness, and limitations in predicting human outcomes. This chapter explores the transformative potential of artificial intelligence (AI) applications, particularly generative adversarial networks (GANs), in advancing nonanimal testing methodologies. It discusses how GANs can leverage existing data from traditional animal models, offering an alternative to animal studies as future avenues for integrating AI into the risk and safety assessment of drugs and chemicals. The chapter elucidates the principles of GANs with examples in toxicology through three case studies: ToxGAN, AnimalGAN, and TransOrGAN. In addition, it discusses the existing challenges and potential trajectories for leveraging this technology to enhance preclinical safety studies.

Keywords: Machine learning; Data science; Drug; Generative adversarial networks; Artificial intelligence; ToxGAN; Animalgan; Transorgan

Introduction

In the landscape of biomedical research and regulatory submissions, animal models have long been the cornerstone for understanding complex biological processes and assessing potential hazards of pharmaceutical compounds.[1–3] However, a paradigm shift is underway, driven by mounting ethical concerns, economic considerations, efficiency demands, and the persistent challenge of accurately predicting human responses.[1,4,5] In recent years, as

Moving Toward Nonanimal Approaches in Medical Research and Testing
https://doi.org/10.1016/B978-0-443-26514-3.00019-9

the limitations of traditional animal testing methodologies have become more evident, the pursuit of alternatives to traditional animal testing methodologies has become increasingly imperative within the biomedical research community.[6–12]

Among these alternatives, in silico methods have emerged as powerful tools for predicting the safety and efficacy of drugs, chemicals, and other substances.[13–16] In silico methods leverage computational models and algorithms to simulate biological processes and predict their outcomes, offering a cost-effective, time-efficient, and ethically superior approach compared to animal experimentation. This shift toward in silico methodologies is driven by the consensus among public, private, and non-profit sectors that hazard identification and risk assessment should capitalize on the advancements in computational screening and modeling to move away from in vivo animal testing toward more human-relevant methods.[17] The development and application of these new tools, including the use of big data and artificial intelligence (AI), are often termed new approach methods (NAMs).[18] These methodologies encompass a broad spectrum of techniques, including in vitro tests or assays, in chemico assays, and in silico algorithms, which collectively aim to reduce, refine, or replace animals in testing.[19] The adoption of in silico methods is not only a response to ethical concerns but also expands our capability beyond animal models in predicting human responses, thereby enhancing the robustness of safety assessments.[20] Furthermore, the integration of in silico methods into regulatory frameworks is evidenced by the increasing acceptance of computational models and simulations by authoritative bodies such as the Food and Drug Administration (FDA) and the European Medicines Agency (EMA), underscoring the critical role of these methodologies in decision-making in science, engineering, and medicine.[21]

One advancement in in silico methods is the emergence of generative adversarial networks (GANs), a class of AI algorithms that have garnered widespread attention for their ability to generate synthetic data that closely mimics real-world phenomena. GANs, introduced by Ian Goodfellow and his colleagues in 2014,[22] consist of two neural networks—a generator and a discriminator—engaged in a competitive game where the generator aims to produce realistic samples and the discriminator strives to distinguish between real and synthetic data. Through adversarial training, GANs can learn to capture the underlying distribution of a dataset and generate in silico samples that exhibit similar characteristics. This innovative approach to data generation has revolutionized various domains beyond its original application in computer vision.[23–27] The transformative potential of GANs in the field of nonanimal testing methodologies lies in their capacity to address several critical challenges. First, GANs offer a means to augment limited datasets,[28–30] particularly in domains where collecting large amounts of labeled data is impractical or expensive. By generating synthetic data that captures the complexity of biological systems, GANs can alleviate the data scarcity problem and enable the development of robust predictive models. Moreover, GANs facilitate the generation of diverse and representative samples, allowing researchers to explore a broader range of chemical space and environmental conditions. This directly supports the replacement of animal testing by enabling the creation of virtual testing environments that can simulate the biological responses of animals to various substances without requiring live subjects, thereby adhering to the 3Rs principles (replacement, reduction, and refinement of animal testing) while maintaining or even enhancing predictive accuracy.

However, despite their transformative potential, GANs pose several challenges and limitations that must be addressed for their widespread adoption in nonanimal testing methodologies. This chapter aims to explore the transformative role of GANs in nonanimal testing methodologies, with a specific focus on their applications in toxicology and safety assessment. By examining the principles, applications, advantages, challenges, and future directions of GANs in this context, we aim to provide insights into how this innovative technology can revolutionize the field of preclinical safety prediction and pave the way for more ethical, efficient, and reliable testing practices.

Principles of GANs

Since its inception in 2014,[22] the GANs model has garnered significant attention and recognition for its remarkable ability to generate synthetic data that closely resembles real-world observations. At the core of GANs lies a novel framework comprised of two neural networks—the generator and the discriminator—which engage in an adversarial training process to produce high-quality synthetic data. The concept of generative modeling predates GANs, with approaches such as variational autoencoders (VAEs)[31] and Boltzmann machines[32] being prominent in the field. However, GANs introduced a novel paradigm by framing the generation task as a two-player game, leading to significant advancements in generative modeling.

A GAN consists of two interconnected neural networks: the generator and the discriminator (as shown in Fig. 1). The generator network is tasked with generating synthetic data samples,

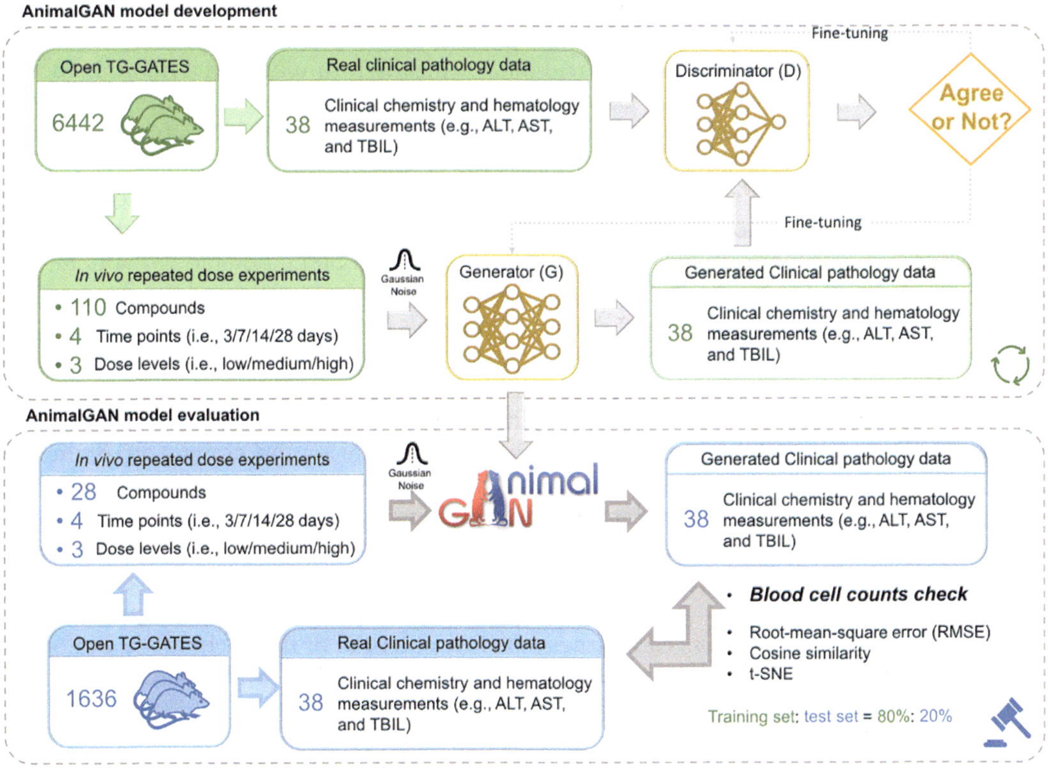

FIG. 1 Basic GAN architecture and learning process. The GAN consists of two main components: the generator and the discriminator. The generator creates fake samples from a latent vector (a mathematical representation in multidimensional space that encodes random noise or specific features, which serve as input to the generator), while the discriminator distinguishes between real samples and fake samples generated by the generator. The model is trained through iterative updates, where the discriminator's feedback is used to improve the generator, aiming to minimize the loss function.

such as images, text, or numerical sequences, that resemble authentic data from a given distribution. Conversely, the discriminator network aims to distinguish between real and synthetic data samples by assigning a probability score indicating the likelihood of a sample being real. Through iterative training, the generator learns to produce increasingly realistic samples, while the discriminator improves its ability to differentiate between real and synthetic data.

The essence of GANs lies in their adversarial training regime, wherein the generator and discriminator networks engage in a minimax game. During training, the generator seeks to minimize the distinguishability of its synthetic samples from real data, while the discriminator aims to maximize its ability to discriminate between the two. This adversarial interplay drives the iterative refinement of both networks, leading to the generation of increasingly realistic synthetic data.

The generator network typically takes random noise or latent vectors as input and transforms them into synthetic data samples through a series of learnable transformations. In contrast, the discriminator network receives input sample data and outputs a probability score indicating the likelihood of the input being real. Both networks are trained concurrently, with the generator learning to produce realistic samples that deceive the discriminator and the discriminator learning to distinguish between real and synthetic data.

To guide the training process, GANs utilize specific loss functions (mathematical metrics that quantify how well the model is performing its task) tailored to the objectives of the generator and discriminator networks. The generator's loss function incentivizes it to produce samples that are indistinguishable from real data, typically formulated as a minimization of the cross-entropy loss between the generated samples and the discriminator's predictions. Conversely, the discriminator's loss function encourages it to correctly classify real and synthetic samples, often defined as a binary classification loss.

GANs have undergone significant innovations and extensions to improve their stability, convergence properties (how reliably and efficiently the networks reach an optimal state during training), and performance across different domains. Some notable variants include Conditional GANs (cGANs),[33] which condition the generation process on additional information, such as class labels or input data; Wasserstein GANs (WGANs),[34] which use Wasserstein distance to improve training stability; Progressive GANs (ProGANs),[35] which gradually increase the resolution of generated images during training; and CycleGAN,[36] which introduces a cycle-consistency loss term to facilitate unsupervised translation between two domains, making it particularly useful for tasks like style transfer and domain adaptation. In addition, extensions like StyleGAN[37] and StyleGAN2[38] offer techniques for controlling the style and appearance of generated data, yielding high-quality, realistic outcomes.

In summary, GANs represent a groundbreaking approach to generative modeling, leveraging adversarial training to produce realistic data samples. By understanding the fundamental principles underlying GANs, including their architecture, training process, and variants/extensions, researchers can harness the full potential of this technology in nonanimal testing methodologies, thereby fostering more ethical, efficient, and reliable approaches to medical research and testing.

Applications of GANs in toxicology

GANs are a powerful class of deep generative models that can learn the distribution of real data and generate new synthetic data points that are statistically indistinguishable from the training data. This unique capability enables GANs to produce in silico data as an alternative to in vivo animal studies. This section showcases a series of pioneering studies that exemplify the transformative role of GANs in advancing toxicological research and testing while reducing reliance on animal models.

ToxGAN: Generating in vivo transcriptomic profiles

Toxicogenomics (TGx) integrates genomics data into toxicology studies to understand underlying biological mechanisms and develop biomarkers for predictive toxicology.[39–41] However, profiling thousands of compounds using traditional animal TGx studies is impractical due to time, cost, and ethical constraints. ToxGAN,[26] developed by Chen et al., aims to address this challenge by using GANs to generate high-quality in vivo transcriptomic profiles for rats treated with given compounds under different doses and treatment durations, without requiring new animal experiments. Transcriptomics refers to the study and analysis of RNA transcripts produced in cells or tissues. Transcriptomic profiles capture gene expression patterns, indicating which genes are active or suppressed under specific conditions, such as exposure to potentially toxic compounds. These data can be generated using technologies like microarrays (which ToxGAN models and generates synthetic data for) or RNA sequencing (which TransOrGAN models and generates synthetic data for), offering insights into molecular responses to various treatments or exposures. It does this by leveraging a deep GAN architecture to learn from existing animal TGx datasets to model the relationship between treatment conditions (the combinations of a compound's chemical structure, dose, and duration) and the resulting transcriptomic response.

As shown in Fig. 2, ToxGAN was trained on a large preexisting rat liver TGx dataset, in which the rats were treated by repeatedly dosing compounds under different treatment durations from the Open Toxicogenomics Project-Genomics-Assisted Toxicity Evaluation System (Open TG-GATES)[42] database. By employing a combination of cGANs and WGANs with Gradient Penalty (WGAN-GP),[43] ToxGAN learns to capture the intricate

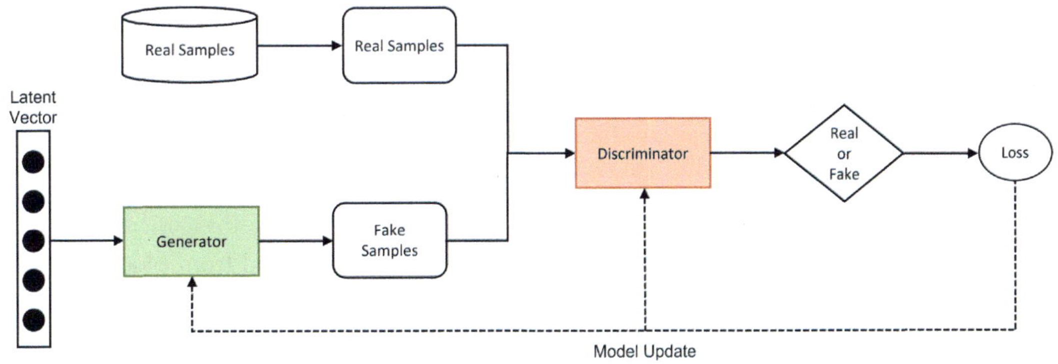

FIG. 2 ToxGAN overview. ToxGAN was trained using 80% of the data (A), evaluated with the remaining 20% of the data (B), and the model's application was then assessed in mechanistic evaluation, biomarker development, and read-across analysis (C). *Modified from Chen X, Roberts R, Tong W, Liu Z. Tox-GAN: an artificial intelligence approach alternative to animal studies—a case study with toxicogenomics. Toxicol Sci. 2022;186(2):242–259. doi:10.1093/toxsci/kfab157.*

relationships between chemical exposures and gene expression patterns in rat livers. The generator takes a compound's representation for molecular structure, dose level, treatment duration, and random noise as inputs to produce a simulated transcriptomic profile that closely resembles real profiles obtained from legacy animal studies. The discriminator attempts to distinguish the generated profiles from real ones, providing feedback to improve the generator iteratively. Once trained, ToxGAN can infer transcriptomic responses for new compounds based solely on their chemical structure and intended treatment regimen.

To validate the effectiveness of ToxGAN, researchers assessed the concordance between the generated transcriptomic profiles and real profiles at both the gene and pathway levels. Remarkably, ToxGAN demonstrated high similarity between the generated and real transcriptomic profiles. Specifically, the average Pearson correlation was 0.997 for gene expression intensities and 0.740 for fold changes in gene expression between the compound-treated group and the matched vehicle-treated controls. This indicates that the synthetic profiles captured genomic responses across different pathological findings and treatment conditions.

Beyond just replicating TGx data, ToxGAN showcased several valuable applications. (1) It can facilitate toxicity mechanism exploration by accurately capturing genomic responses across different pathological findings and treatment regimens, enabling Gene Ontology and pathway analysis. Over 87% agreement in enriched Gene Ontology terms between ToxGAN-generated results and their corresponding real gene expression data highlights its potential for elucidating toxicity pathways and facilitating mechanistic insights. (2) It can enhance gene expression-based biomarker development for predictive toxicology by identifying important gene signatures from the generated profiles comparable to real data. ToxGAN-generated profiles performed comparably to real gene expression data in predicting liver necrosis and identifying relevant biomarker genes, demonstrating the predictive power of ToxGAN-generated transcriptomic profiles for liver-related toxicity prediction. (3) It can aid chemical read-across approaches by incorporating high-fidelity biological data, e.g., TGx, alongside chemical structure information. Importantly, ToxGAN could generalize to an external dataset (DrugMatrix) and different therapeutic categories beyond its training space, demonstrating robustness.

AnimalGAN: Simulating clinical pathology measurements

The AnimalGAN[25] model takes a complementary approach by targeting the simulation of multi-dimensional clinical pathology data commonly collected in traditional animal toxicology studies. This rich, contextual data on hematology, clinical chemistry, and other systemic measures provide critical insights into toxicity assessment that go beyond single-endpoint predictions.

AnimalGAN was developed using clinical pathology data acquired from blood samples collected from the abdominal aorta of rats involved in in vivo tests within the Open TG-GATEs database. This repository hosts an extensive collection of in vivo rat toxicology studies, comprising 38 different clinical pathology measurements, including 21 hematology and 17 biochemistry parameters, across 8078 rats treated with 138 compounds under various dose and duration conditions (Fig. 3).

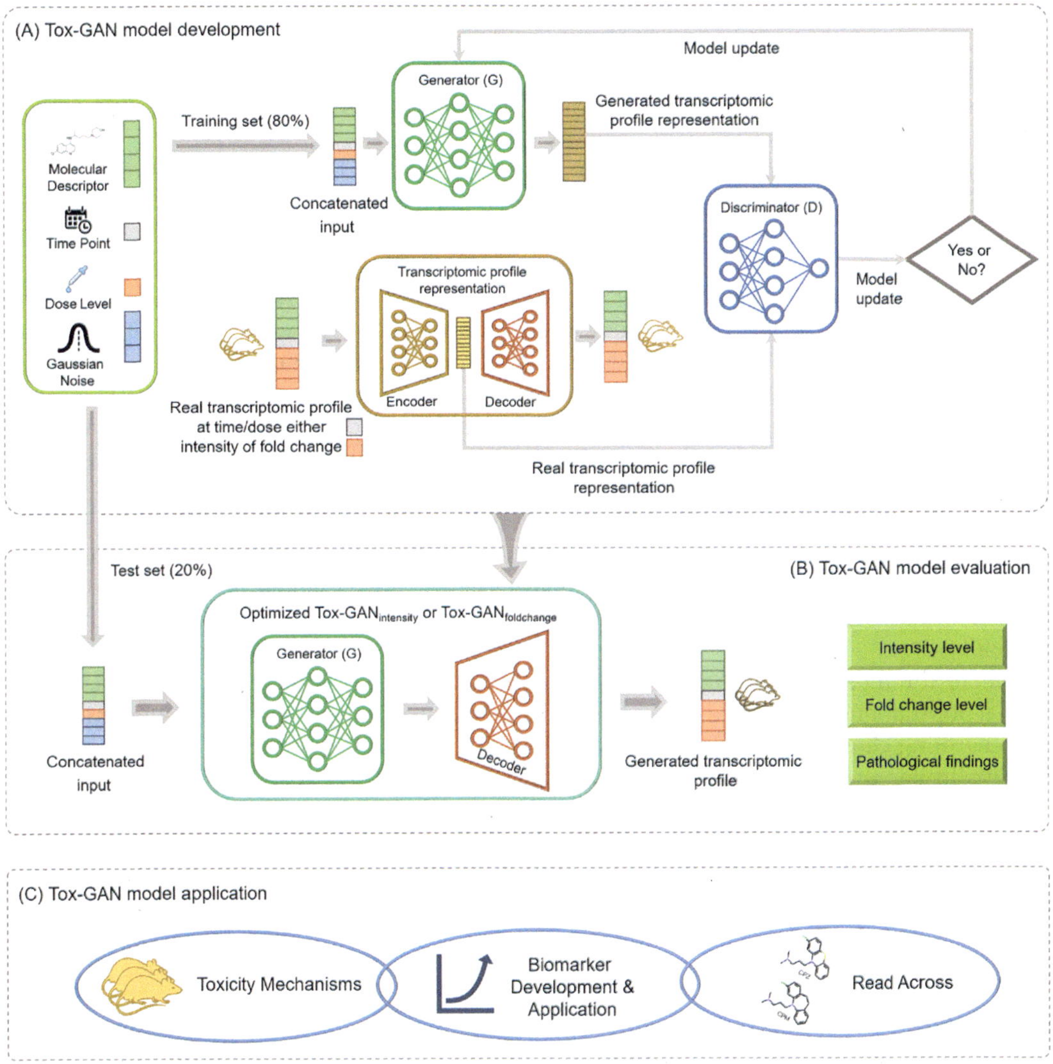

FIG. 3 AnimalGAN overview. AnimalGAN was trained on 80% of the clinical pathology data in the open TG-GATEs and evaluated on the remaining 20% of the data. The generator produced synthetic clinical pathology measurements from chemical descriptors, time points, dose levels, and Gaussian noise, which were evaluated against real data by the discriminator. The model was considered established once the discriminator could no longer distinguish between synthetic and real data. *Modified from Chen X, Roberts R, Liu Z, et al. A generative adversarial network model alternative to animal studies for clinical pathology assessment. Nat Commun. 2023;14(1):7041. doi:10.1038/s41467-023-42933-9.*

AnimalGAN employs a cGAN architecture, where the generator takes the molecular descriptors of a compound, dose level, treatment duration, and random noise as inputs. It then produces a 38-dimensional vector representing the simulated clinical pathology profile for those specific exposure conditions. The discriminator tries to distinguish the generated profiles from the real data, driving the generator to improve iteratively.

To enhance the stability and performance of cGAN training, the AnimalGAN framework incorporated techniques from WGAN-GP regularization. These enhancements helped the generator learn to faithfully reproduce the underlying data distribution, even with the small sample sizes and high-dimensional treatment conditions typical of toxicology datasets.

A crucial aspect of the AnimalGAN approach was the addition of post-optimization criteria. Beyond just training the GAN to generate statistically similar data, the researchers implemented checks to ensure the synthetic clinical pathology measures adhered to biological constraints (e.g., valid blood cell count ratios).

The results demonstrated the robustness and versatility of the AnimalGAN model. Even when challenged with test compounds having significantly different chemical structures, drug classes, or FDA approval timelines compared to the training data, the model maintained high concordance with the real animal data. Remarkably, AnimalGAN outperformed 12 conventional quantitative structure-activity relationship models in predicting all 38 clinical pathology endpoints, highlighting the advantages of the generative approach over traditional predictive modeling techniques.

The researchers further showcased AnimalGAN's utility in toxicity assessment by comparing its results to real animal studies. AnimalGAN demonstrated 96%–100% consistency with animal data in determining statistically significant changes in clinical pathology measures between treatment and control groups, which is a critical step in safety evaluation.

One of the most compelling applications of AnimalGAN was its ability to simulate large virtual populations of animal data, enabling the detection of rare, idiosyncratic adverse events. By generating clinical pathology profiles for 100,000 hypothetical rats treated with three structurally similar diabetes drugs—troglitazone, pioglitazone, and rosiglitazone—AnimalGAN could recapitulate the different frequencies of drug-induced liver injury observed in human populations. This unprecedented scale of synthetic data generation enhances statistical power, allowing for the detection of subtle toxicity signals that may be missed in real-world experiments due to limited sample sizes.

TransOrGAN: Mapping transcriptomic profiles between organs, ages, and sexes

In contrast to ToxGAN, which generates transcriptomic profiles specifically for the liver, TransOrGAN[27] has the capability to synthesize transcriptomic profiles in rats across nine organs. This broad scope is crucial for conducting comprehensive studies to elucidate the complex biological processes occurring across different organs. As a result, this approach enables the synthesis of transcriptomic profiles for organs with extensive data, such as the liver and kidney, and addresses the challenge of limited data availability for organs like the spleen, heart, and lung.

TransOrGAN was developed using the comprehensive rat RNA-seq transcriptomic BodyMap dataset.[44] This dataset includes nine non-sex-specific organs (i.e., adrenal gland, brain, heart, kidney, liver, lung, muscle, spleen, and thymus) and different experimental conditions: two sexes (i.e., male and female) and four developmental stages (juvenile [2 weeks old], adolescence [6 weeks old], adult [21 weeks old], and aged [104 weeks old]). In total, there are 64,800 pairwise transcriptomic profiles between organs under different sex and developmental stage combinations.

TransOrGAN combines two GAN models, cGAN[45] and Cycle-consistent GAN (CycleGAN),[36] to facilitate the synthesis of transcriptomic data across rat organs. While CycleGAN was originally designed for image translation between two different domains or types, TransOrGAN adapts it for translating transcriptomic data across nine organs. In the original approach, a separate model was needed for each organ pair (e.g., liver and kidney, liver and heart, liver and spleen). This was challenging due to the limited sample size available for training. However, by integrating CycleGAN with cGAN, which allows for the incorporation of conditional experimental information (organ, sex, and developmental stage) into the training process, TransOrGAN can translate transcriptomic profiles between different experimental conditions. This approach overcomes the limitation of sample size and improves efficiency, enabling the translation of transcriptomic profiles across multiple organs within a single model.

Specifically, TransOrGAN consists of a generator, a discriminator, and a profile enhancer. The generator is trained to generate transcriptomic profiles specific to an organ, sex, and developmental stage, with cycle-consistency regulation through the profile enhancer. The discriminator evaluates the generated profiles against the measured profiles under the same experimental conditions. This iterative process continues until the discriminator can no longer differentiate between synthetic and measured transcriptomic profiles. The well-trained generator in TransOrGAN can then be used to translate transcriptomic profiles across organs under different experimental conditions.

The generator in TransOrGAN takes three inputs: the transcriptomic profile from the source organ, the experimental condition of the source organ (including organ, sex, developmental stage) for the source, and the experimental condition of the target organ. A Gaussian noise vector of the same size, sampled from a normal distribution, is added to the inputs to improve model robustness. The generator is then trained to produce a transcriptomic profile for the target organ under a specific sex and developmental stage. On the other hand, the discriminator in TransOrGAN takes either a generated or measured transcriptomic profile and distinguishes whether the input profile is real or generated.

To assess the performance of TransOrGAN, Li et al.[27] employed three evaluation methods: cosine similarity, RMSE, and the dimensionality reduction tool Uniform Manifold Approximation and Projection (UMAP).[46] All three indicated a high level of agreement between the synthetic and real profiles. In addition, this study demonstrated high similarities in toxicity-related pathways between the generated and measured transcriptomic profiles.

TransOrGAN is capable of translating female profiles from male profiles within the same organ or across organs, using both the full gene set and a set of genes related to sex hormones. This could help tackle the issue of females being underrepresented in preclinical and clinical studies, despite known differences in physical characteristics, behavior, biological regulation, susceptibility to adverse drug reactions, and lifestyle. In addition, TransOrGAN could translate transcriptomic profiles between age groups—predicting profiles of young and old rats from adolescent data—within and across organs. This presents the opportunity to comprehensively assess the potential toxicity in younger and older animals without the need for additional animal testing.

Challenges and limitations in implementing GANs for nonanimal testing

GANs offer immense potential in revolutionizing nonanimal testing methodologies in biomedical research. However, their implementation presents several challenges that must be addressed to fully leverage their capabilities in this domain.

A critical limitation of GAN-based methods for nonanimal testing is their reliance on animal-derived training data. While GANs can generate synthetic data closely correlated with real animal data, extrapolating these findings to human outcomes remains a significant challenge due to species-specific physiological differences. Even if animal studies are perfectly simulated, there are inherent translational gaps across species that can affect the capability of predicting human toxicities or drug responses accurately using animal data. In addition, if biases or limitations exist in the animal data used for training, these may be perpetuated or potentially amplified in the synthetic outputs, potentially affecting downstream applications. Addressing this challenge will require integrating more diverse sources of data, including human-relevant in vitro and in silico data, clinical data, and mechanistic toxicology insights. While direct GAN training on human data remains limited due to factors such as data availability, privacy concerns, and regulatory barriers, advances in data curation, transfer learning, and hybrid modeling could progressively lessen reliance on animal data, paving the way for more human-relevant, noanimal testing methodologies.

A primary challenge in implementing GANs for nonanimal testing is the generation of high-quality and diverse synthetic data. GANs are designed to learn and generate data that mimic real samples from a given dataset. However, ensuring the generated data accurately represent the complex real-world biological data of interest is a formidable task. GANs are susceptible to "mode collapse," where the generator fails to capture the full diversity of the data distribution, resulting in unrealistic and biased samples.[47,48] Addressing this challenge requires the development of advanced augmentation techniques and optimization strategies to enhance the diversity and quality of the generated data.

In addition, GANs require large amounts of high-quality labeled data to learn meaningful representations of the underlying data distribution. However, acquiring labeled data for training GANs in biomedical research applications can be challenging due to privacy concerns, data scarcity, and class imbalance issues (e.g., rare toxicity events). Developing techniques for effective data augmentation, semi-supervised learning, and transfer learning is essential for overcoming these challenges and enabling the robust training of GANs for nonanimal testing tasks.[47]

Furthermore, GANs are sensitive to hyperparameters and training configurations, making them prone to training instabilities and convergence problems.[48] Tuning GAN parameters to achieve optimal performance can be time-consuming and computationally intensive, requiring expertise and computational resources. Addressing these challenges requires the development of advanced optimization algorithms and architectural modifications to improve the stability and convergence properties of GANs in nonanimal testing applications.

While GANs offer tremendous potential in transforming nonanimal testing methodologies in biomedical research and replacing animal tests, their implementation poses several challenges that must be overcome. By addressing these challenges through interdisciplinary research, collaboration, and innovation, researchers can unlock the full potential of GANs for efficient and reliable medical research and testing.

Integration of GANs with other AI techniques

In recent years, there has been growing interest in integrating GANs with other AI techniques, which may improve their effectiveness in nonanimal testing methodologies. By combining the strengths of different AI approaches, researchers aim to overcome the limitations of individual methods and achieve superior performance in biomedical research applications.[49]

One promising avenue of integration is the combination of GANs with machine learning techniques such as supervised learning and semi-supervised learning.[50] By leveraging labeled data in conjunction with GAN-generated samples, researchers can train more robust models for tasks such as drug toxicity prediction and compound screening. In addition, semi-supervised GANs enable the utilization of both labeled and unlabeled data, thereby maximizing the efficiency of data utilization in nonanimal testing methodologies.

Deep learning techniques, including convolutional neural networks (CNNs) and recurrent neural networks (RNNs), can also be integrated with GANs to enhance their capabilities in analyzing complex biomedical data.[28] CNNs excel at extracting hierarchical features from images and other structured data, making them well-suited for tasks such as image-based toxicity assessment. Similarly, RNNs are proficient in modeling sequential data, offering opportunities for GANs to generate realistic time-series data for chemical toxicity prediction and drug response modeling.

Reinforcement learning (RL) presents another avenue for integrating GANs with other AI techniques in nonanimal testing methodologies. By formulating nonanimal testing tasks as sequential decision-making problems, researchers can employ RL algorithms to optimize the generation process of GANs. This integration enables adaptive and interactive data generation, allowing GANs to adapt to changing environmental conditions and user preferences in real-time.

Overall, the integration of GANs with other AI techniques holds significant promise in advancing nonanimal testing methodologies in biomedical research. By leveraging the complementary strengths of different approaches, researchers can overcome challenges and unlock new opportunities for innovation in drug discovery, toxicity assessment, and beyond. For more details regarding AI applications, see Chapter 9.

Acceptance and regulatory implications of GAN-based nonanimal testing

The utilization of GANs in nonanimal testing methodologies presents not only scientific challenges but also implications for acceptance and regulation that warrant careful consideration.

One of the primary considerations surrounding GAN-based nonanimal testing is the reliability and validity of the generated data. Unlike traditional experimental approaches that rely on empirical evidence obtained from animal studies, GANs generate synthetic data based on learned patterns from existing datasets. This raises questions about the reliability and validity of the generated data in accurately representing complex biological phenomena. Ensuring that GAN-generated data faithfully reflects real-world scenarios is essential for fostering acceptance of nonanimal testing methodologies.

Transparency and interpretability are also critical considerations in the adoption of GAN-based nonanimal testing methodologies. GANs operate as black-box models, making it challenging to understand the underlying mechanisms driving their predictions. Concerns arise regarding the transparency and interpretability of GAN-generated results, particularly in critical decision-making processes such as drug development and toxicity assessment. Efforts to enhance the interpretability of GAN architectures and establish standards for transparent reporting of model performance are crucial for gaining acceptance of nonanimal testing practices.

From a regulatory perspective, the use of GANs in nonanimal testing introduces novel challenges for validation and approval processes. Regulatory agencies, such as the US FDA and the EMA, typically require robust evidence of safety and efficacy before approving new drugs and medical devices. However, the evaluation of GAN-based nonanimal testing methodologies poses unique challenges due to the lack of established standards and benchmarks for assessing model performance and predictive accuracy.

To navigate these challenges, stakeholders need to engage in collaborative efforts to establish ethical guidelines and regulatory frameworks for the responsible use of GANs in nonanimal testing. This includes defining criteria for evaluating the reliability, reproducibility, and generalizability of GAN-generated data, as well as developing protocols for transparent reporting and independent validation of GAN-based models. By fostering a culture of ethical conduct and regulatory compliance, researchers can harness the transformative potential of GANs in advancing nonanimal testing methodologies while upholding the highest standards of scientific integrity and patient safety.

Future directions and opportunities

The substantial progress made in recent years has laid a solid foundation for the application of AI/machine learning in nonanimal testing, with studies such as ToxGAN, AnimalGAN, and TransOrGAN representing typical examples of using GANs as NAMs to generate synthetic data as alternatives to animal testing. These frameworks demonstrate GANs' transformative potential to produce multi-dimensional, context-rich data mimicking complex toxicological processes traditionally evaluated through animal studies. Notably, opportunities exist to expand and refine GAN-based approaches further. A major direction is integrating data from diverse sources, including in vitro, in silico, and human studies, to enhance the performance, applicability, and translatability of GAN-generated synthetic data. Moving beyond animal-derived datasets to incorporate human-relevant data will be critical for improving their predictive value in human health risk assessment. Achieving this requires the inclusion of multimodal datasets such as in vitro experimental results (e.g., high-throughput screening assays, organ-on-a-chip models, and induced pluripotent stem cell-based assays), human biomarker data (e.g., clinical biobank records, transcriptomic and metabolomic profiles), and computational modeling outputs (e.g., quantitative structure-activity relationship models and physiologically based pharmacokinetic simulations). Integrating these diverse data sources into GAN frameworks can mitigate species differences and improve their generalizability, ultimately bridging the gap between preclinical findings and human outcomes.

Beyond data integration, continued exploration and refinement of GAN architectures hold promise for enhancing model performance and robustness. Innovations in architecture design, such as attention mechanisms, self-attention, and graph neural networks, could enable GANs to capture more complex patterns and relationships in biological data, further advancing their utility in nonanimal testing methodologies.

A critical step in advancing GAN-based NAMs is the development of well-curated, standardized databases that integrate comprehensive animal study data, in vitro experimental results, human clinical information, and mechanistic toxicology insights. Such databases will be pivotal for training robust GAN models applicable to real-world scenarios. In addition, collaborative efforts to establish benchmarks for evaluating GAN performance can facilitate comparative studies, improve transparency, and support regulatory acceptance of nonanimal testing methodologies.

As computational resources and data availability continue to improve, GAN-based methodologies could enable virtual, animal-free evaluations of thousands of new compounds routinely, providing critical toxicity information to prioritize testing and streamline product development pipelines.

In summary, the utilization of GANs in nonanimal testing methodologies holds promise for producing high-quality in silico data, reducing reliance on animal testing, and ultimately improving human health outcomes. By embracing these opportunities and overcoming the associated challenges, researchers can pave the way for a more ethical, efficient, and innovative approach to toxicological practices, driving progress in biomedical research and advancing human health.

Disclaimer

This manuscript reflects the views of the authors and does not necessarily reflect those of the Food and Drug Administration. Any mention of commercial products is for clarification only and is not intended as an approval, endorsement, or recommendation.

References

[1] Thomas S. Animal research and the search for understanding. *Nat Genet*. 2006;38(8):859.

[2] Mukherjee P, Roy S, Ghosh D, et al. Role of animal models in biomedical research: a review. *Lab Anim Resh*. 2002;38(1):18. doi:10.1186/s42826-022-00128-1.

[3] Barré-Sinoussi F, Montagutelli X. Animal models are essential to biological research: issues and perspectives. *Future Sci OA*. 2015;1(4):FSO63. doi:10.4155/fso.15.63.

[4] Rice J. Animal models: not close enough. *Nature*. 2012;484:S9. doi:10.1038/nature11102.

[5] Peers IS, South MC, Ceuppens PR, Bright JD, Pilling E. Can you trust your animal study data? *Nat Rev Drug Discov*. 2014;13(7):560. doi:10.1038/nrd4090-c1.

[6] U.S. Food and Drug Administration. FDA's predictive toxicology RoadMap; 2017. Available at: https://www.fda.gov/media/109634/download. Accessed 9 August 2026.

[7] Congress.Gov. H.R. 2565- FDA modernization act; 2021. Available at: https://www.congress.gov/index.php/bill/117th-congress/house-bill/2565. Accessed 9 August 2026.

[8] Doke SK, Dhawale SC. Alternatives to animal testing: a review. *Saudi Pharm J*. 2015;23(3):223–229. doi:10.1016/j.jsps.2013.11.002.

[9] Han JJ. FDA modernization act 2.0 allows for alternatives to animal testing. *Artif Organs*. 2023;47(3):449–450. doi:10.1111/aor.14503.

[10] Löwa A, Jevtić M, Gorreja F, Hedtrich S. Alternatives to animal testing in basic and preclinical research of atopic dermatitis. *Exp Dermatol*. 2018;27(5):476–483. doi:10.1111/exd.13498.
[11] Meigs SL, Rovida C, et al. Animal testing and its alternatives—the most important omics is economics. *Altex*. 2018;35(3):275–305. doi:10.14573/altex.1807041.
[12] Reddy N, Lynch B, Gujral J, Karnik K. Alternatives to animal testing in toxicity testing: current status and future perspectives in food safety assessments. *Food Chem Toxicol*. 2023;179:113944. doi:10.1016/j.fct.2023.113944.
[13] Raies AB, Bajic VB. In silico toxicology: computational methods for the prediction of chemical toxicity. *Wiley Interdiscip Rev Comput Mol Sci*. 2016;8(3):147–172. doi:10.1002/wcms.1352.
[14] Cronin MTD, Madden JC, Yang C, Worth AP. Unlocking the potential of in silico chemical safety assessment—a report on a cross-sector symposium on current opportunities and future challenges. *Comput Toxicol*. 2019;10:38–43. doi:10.1016/j.comtox.2018.12.006.
[15] Madden JC, Enoch S, Paini A, Cronin MTD. A review of in silico tools as alternatives to animal testing: principles, resources and applications. *Altern Lab Anim*. 2020;48(4):146–172. doi:10.1177/0261192920965977.
[16] Yang H, Sun L, Li W, Liu G, Tang Y. In silico prediction of chemical toxicity for drug design using machine learning methods and structural alerts. *Front Chem*. 2018;6. doi:10.3389/fchem.2018.00030.
[17] Andersen ME, McMullen PD, Phillips MB, et al. Developing context appropriate toxicity testing approaches using new alternative methods (NAMs). *Altex*. 2019;36(4):523–534. doi:10.14573/altex.1906261.
[18] Ram RN, Gadaleta D, Allen TEH. The role of 'big data' and 'in silico' new approach methodologies (NAMs) in ending animal use—a commentary on progress. *Comput Toxicol*. 2022;23:100232. doi:10.1016/j.comtox.2022.100232.
[19] Carvalho C, Varela SAM, Bastos LF, et al. The relevance of in silico, in vitro and non-human primate based approaches to clinical research on major depressive disorder. *Altern Lab Anim*. 2019;47(3–4):128–139. doi:10.1177/0261192919885578.
[20] Kahrass H, Pietschmann I, Mertz M. Why do I choose an animal model or an alternative method in basic and preclinical biomedical research? A spectrum of ethically relevant reasons and their evaluation. *Animals*. 2024;14(4):651. doi:10.3390/ani14040651.
[21] Tatka LT, Smith LP, Hellersstein J, Sauro HM. Adapting modeling and simulation credibility standards to computational systems biology. *J Transl Med*. 2023;21(1):501. doi:10.1186/s12967-023-04290-5.
[22] Goodfellow IJ, Pouget-Abadie J, Mirza M, et al. Generative adversarial nets. In: *Advances in Neural Information Processing Systems 27 (NIPS 2014)*, Vol. 27. 2014:2672–2680.
[23] Dash A, Ye J, Wang G. A review of generative adversarial networks (GANs) and its applications in a wide variety of disciplines: from medical to remote sensing. *IEEE Access*. 2024;12:18330–18357. Available at: https://ieeexplore.ieee.org/stamp/stamp.jsp?tp=&arnumber=10372211. doi:10.1109/access.2023.3346273.
[24] Rizvi SKJ, Azad MA, Fraz MM. Spectrum of advancements and developments in multidisciplinary domains for generative adversarial networks (GANs). *Arch Comput Methods Eng*. 2021;28(7):4503–4521. doi:10.1007/s11831-021-09543-4.
[25] Chen X, Roberts R, Liu Z, Tong W, et al. A generative adversarial network model alternative to animal studies for clinical pathology assessment. *Nat Commun*. 2023;14(1):7141. doi:10.1038/s41467-023-42933-9.
[26] Chen X, Roberts R, Tong W, Liu Z. Tox-GAN: an artificial intelligence approach alternative to animal studies—a case study with toxicogenomics. *Toxicol Sci*. 2022;186(2):242–259. doi:10.1093/toxsci/kfab157.
[27] Li T, Roberts R, Liu Z, Tong W. TransOrGAN: an artificial intelligence mapping of rat transcriptomic profiles between organs, ages, and sexes. *Chem Res Toxicol*. 2023;36(6):916–925. doi:10.1021/acs.chemrestox.3c00037.
[28] Frid-Adar M, Diamant I, Klang E, et al. GAN-based synthetic medical image augmentation for increased CNN performance in liver lesion classification. *Neurocomputing*. 2018;321:321–331. doi:10.1016/j.neucom.2018.09.013.
[29] Shin HC, Tenenholtz NA, Rogers JK, et al. Medical image synthesis for data augmentation and anonymization using generative adversarial networks. In: *International Workshop on Simulation and Synthesis in Medical Imaging*. Cham: Springer International Publishing; 2018: 1–11. doi:10.1007/978-3-030-00536-8_1.
[30] Salehinejad H, Valaee S, Dowdell T, et al. Generalization of deep neural networks for chest pathology classification in x-rays using generative adversarial networks. arXiv. Canada: arXiv. 2017. doi:10.48550/arxiv.1712.01636.
[31] Kingma DP, Welling M. Auto-encoding variational bayes. arXiv e-prints. 2013. doi:10.48550/arXiv.1312.6114.
[32] Ackley DH, Hinton GE, Sejnowski TJ. A learning algorithm for boltzmann machines. *Cogn Sci*. 1985;9(1):147–169. doi:10.1016/S0364-0213(85)80012-4.
[33] Mirza M, Osindero S. Conditional generative adversarial nets. arXiv preprint. 2014. doi:10.48550/arXiv.1411.1784.

[34] Arjovsky M, Chintala S, Bottou L. Wasserstein GAN. arXiv e-prints. 2017. doi:10.48550/arXiv.1701.07875.

[35] Karras T, Aila T, Laine S, Lehtinen J. Progressive growing of gans for improved quality, stability, and variation. arXiv e-prints. 2017. doi:10.48550/arxiv.1710.10196.

[36] Zhu JY, Park T, Isola P, Efros AA Unpaired image-to-image translation using cycle-consistent adversarial networks. In: *Proceedings of the IEEE International Conference on Computer Vision*. 2017;2223–2232.

[37] Karras T, Laine S, Aila T. A style-based generator architecture for generative adversarial networks. arXiv e-prints. 2018. doi:10.48550/arxiv.1812.04948.

[38] Karras T, Laine S, Aittala M, et al. Analyzing and improving the image quality of StyleGAN. arXiv preprint. 2019. doi:10.48550/arxiv.1912.04958.

[39] Liu Z, Huang R, Roberts R, Tong W. Toxicogenomics: a 2020 vision. *Trends Pharmacol Sci*. 2019;40(2):92–103. doi:10.1016/j.tips.2018.12.001.

[40] Podtelezhnikov AA, Monroe J, Aslamkhan AG, et al. Quantitative transcriptional biomarkers of xenobiotic receptor activation in rat liver for the early assessment of drug safety liabilities. *Toxicol Sci*. 2020;175(1):98–112. doi:10.1093/toxsci/kfaa026.

[41] Monroe JJ, Tanis KQ, Podtelezhnikov A, et al. Application of a rat liver drug bioactivation transcriptional response assay early in drug development that informs chemically reactive metabolite formation and potential for drug-induced liver injury. *Toxicol Sci*. 2020;177(1):281–299. doi:10.1093/toxsci/kfaa088.

[42] Igarashi Y, Nakatsu N, Yamashita T, et al. Open TG-GATEs: a large-scale toxicogenomics database. *Nucleic Acids Res*. 2015;43(Database issue):D921–7. doi:10.1093/nar/gku955.

[43] Gulrajani I, Ahmed F, Arjovsky M, et al. Improved training of wasserstein GANs. In: *NIPs'17: Proceedings of the 31st International Conference on Neural Information Processing Systems. Advances in Neural Information Processing Systems*. 2017;5769–5779.

[44] Yu Y, Fuscoe JC, Zhao C, et al. A rat RNA-seq transcriptomic BodyMap across 11 organs and 4 developmental stages. *Nat Commun*. 2014;5:230. doi:10.1038/ncomms4230.

[45] Isola P, Zhu JY, Zhou T, et al. Image-to-image translation with conditional adversarial networks. In: *Proceedings of the IEEE Conference on Computer Vision and Pattern Recognition*. 2017;1125–1134.

[46] McInnes L, Healy J, Melville J. Umap: uniform manifold approximation and projection for dimension reduction. arXiv preprint arXiv:1802.03426. 2018. Accessed 12 March 2025. doi:10.48550/arXiv.1802.03426.

[47] Saxena D, Cao J. Generative adversarial networks (GANs): challenges, solutions, and future directions. arXiv. Hong Kong: arXiv. 2020. doi:10.48550/arXiv.2005.00065.

[48] Kodali N, Abernethy J, Hays J, Kira Z. On convergence and stability of GANs. arXiv preprint. 2017. doi:10.48550/arXiv.1705.07215.

[49] Iglesias G, Talavera E, Díaz-Álvarez A. A survey on GANs for computer vision: recent research, analysis and taxonomy. arXiv e-prints. 2022. doi:10.48550/arXiv.2203.11242.

[50] Odena A. Semi-supervised learning with generative adversarial networks. arXiv e-prints. 2016. doi:10.48550/arXiv.1606.01583.

CHAPTER

8

3D bioprinting as a promising alternative to animal models: Technologies and applications

Ahmed S.M. Ali, Dongwei Wu, and Jens Kurreck

Applied Biochemistry, Berlin University of Technology, Berlin, Germany

Abstract

Three-dimensional (3D) bioprinting has emerged as an advanced technology at the intersection of biology, engineering, and materials science, offering unprecedented opportunities for improving healthcare, drug development, and regenerative medicine. This chapter explores core technologies—inkjet, extrusion, light-based systems, and volumetric printing—detailing their mechanisms, bioink requirements, and applications in cancer modeling, drug screening, and regenerative medicine. Natural, synthetic, and decellularized extracellular matrix-based bioinks are analyzed for their biocompatibility, mechanical properties, and ethical considerations, emphasizing the shift toward xeno-free models. Case studies show how bioprinted tissues better mimic human physiology than traditional models, including vascularized constructs and patient-specific disease platforms. Challenges such as scalability, vascular integration, and regulatory hurdles are addressed, alongside emerging innovations like artificial intelligence-driven optimization.

Keywords: 3D Bioprinting; bioinks; tissue engineering; xeno-free models; nonanimal methods

Introduction

The journey of additive manufacturing, commonly known as three-dimensional (3D) printing, began in the early 1980s and was primarily employed for rapid prototyping. Over time, the technology has evolved to play a pivotal role in diverse industries. Today, 3D printing is not only integral to the production of complex, end-use parts in aerospace, automotive, and consumer goods but also finds applications in fields as varied as construction, healthcare, and even fashion.[1] One particularly promising branch of this technology is bioprinting, which leverages the technology for tissue engineering and regenerative medicine.[2] In essence, 3D bioprinting uses the same principles as traditional 3D printing but substitutes standard materials with bioink, a formulated mixture of living cells, biomaterials, and

https://doi.org/10.1016/B978-0-443-26514-3.00018-7

bioactive molecules, to build functional, 3D tissue constructs that mimic the architecture and function of natural tissues.

The history of bioprinting can be traced back to the early 2000s, when significant milestones were achieved in the development of bioinks and cell deposition techniques. One of the earliest breakthroughs occurred when Thomas Boland at Clemson University modified a conventional inkjet printer to dispense living cells instead of ink, marking the birth of inkjet bioprinting.[3] This technological leap paved the way for further innovations in the field, as scientists began to experiment with various bioinks and printing methods to create increasingly complex tissue-like structures. Over the years, bioprinting evolved from producing simple 2D cellular patterns to fabricating functional 3D tissue constructs. Today, bioprinting stands at the intersection of several disciplines, including cell biology, materials science, mechanical engineering, and computer science, making it a highly interdisciplinary field.[4]

3D bioprinting is a promising advancement with the potential to improve how we understand, diagnose, and treat diseases. This technology aims to address critical challenges in modern healthcare, from organ shortages to ineffective drug testing models, ultimately advancing human-relevant medical research, improving patient outcomes, and reducing animal testing. At its core, 3D bioprinting involves the precise deposition of bioinks to create 3D structures. It is characterized by the presence of living cells and other biological components in the printed material during the printing process. However, printing a scaffold that is subsequently populated with cells is also sometimes considered bioprinting. In the present chapter, we focus on bioprinting processes involving living cells.

Types of 3D bioprinting technologies

The development of various 3D bioprinting technologies has been a cornerstone of the field's progress. Each technology has its unique mechanism for depositing bioinks, along with distinct advantages and limitations. The selection of technology is often dictated by the specific application, the type of tissue being printed, and the desired resolution or structural complexity. Prominent methods include inkjet-based bioprinting, extrusion-based bioprinting, and light-based 3D printing technologies, such as stereolithography (SLA), digital light processing (DLP), and volumetric bioprinting (Table 1). None of these methods can be considered the best or optimal technology; each has its own strengths and limitations, and the most suitable approach must be chosen based on the specific application.

Inkjet-based bioprinting

Inkjet-based bioprinting, one of the earliest bioprinting technologies, originated from conventional inkjet printing used in the paper and graphics industry.[5] In inkjet bioprinting, droplets of bioink containing living cells are ejected from a nozzle onto a substrate. The ejection of these droplets can be controlled by several mechanisms. The thermal and piezoelectric techniques are the most prominent approaches. In thermal inkjet bioprinting, heat is used to create a small vapor bubble inside the print head, propelling a droplet of bioink out of the nozzle.[6] In piezoelectric inkjet bioprinting, an electric current is applied to a piezoelectric crystal, generating pressure that forces the ejection of a droplet of bioink.[7]

TABLE 1 Comparative overview of major 3D bioprinting technologies. Primary bioprinting methods are categorized into four groups—inkjet-based (thermal, piezoelectric, electromagnetic), extrusion-based (pneumatic, piston-driven, screw-driven), light-based bioprinting technologies (liquid crystal display (LCD), DLP, and SLA), and volumetric bioprinting (computed axial lithography (CAL) and xolography). For each subtype, working principles, key advantages, and representative applications are summarized.

Technology	Classification	Working principle	Advantages	Application examples
Inkjet-based	Thermal	A heating element briefly raises the temperature to generate vapor bubbles, which eject droplets of bioink	• Fast • Low cost • High resolution droplet deposition	• Hydrogel microparticles[8] • Direct human cartilage repair[9]
	Piezoelectric	A piezoelectric actuator produces pressure pulses that eject droplets	• Precise control over droplet size and placement • Better cell viability	• Bioprinting microvessels[10] • Precise tissue construction[11]
	Electro-magnetic	Electromagnetic forces actuate the ejection of droplets	• Precise droplet control • Rapid actuation	• Vascularized tumor-environment model[12]
Extrusion-based	Pneumatic	Uses pressurized air to force bioink through a nozzle	• Simple setup • Low cost • High material versatility	• 3D lung model for infection studies[13] • Human neural tissues[14]
	Piston-driven	A mechanical piston pushes bioink through a nozzle	• Good control over deposition volume and speed • Suitable for moderately viscous inks	• Perfusable vascularized organ models[15] • Tubular structures[16]
	Screw-driven	A rotating screw (auger) delivers bioinks through a nozzle	• Handles highly viscous materials • Consistent • Continuous extrusion for large-scale constructs	• Real-time mixing and 3D bioprinting[17]
Light-based	LCD-based	An LCD panel selectively masks light (typically UV or visible) to cure a photoresponsive bioink layer-by-layer from a vat	• Relatively low cost • High resolution • Fast curing across an entire layer simultaneously	• Bone tissue engineering[18]
	DLP-based	Uses a digital micromirror device to project light patterns onto a vat of photocurable bioink, solidifying the exposed layer	• High speed • High resolution • Uniform layer curing	• Multivascular networks[19] and • Vascularized tissues[20]
	SLA	A focused laser beam scans over a vat of photoactive bioink to cure precise patterns layer by layer	• Very high resolution • Excellent surface finish	• Human corneal stroma equivalent[21]

Continued

TABLE 1 Comparative overview of major 3D bioprinting technologies. Primary bioprinting methods are categorized into four groups—inkjet-based (thermal, piezoelectric, electromagnetic), extrusion-based (pneumatic, piston-driven, screw-driven), light-based bioprinting technologies (liquid crystal display (LCD), DLP, and SLA), and volumetric bioprinting (computed axial lithography (CAL) and xolography). For each subtype, working principles, key advantages, and representative applications are summarized—cont'd

Technology	Classification	Working principle	Advantages	Application examples
Volumetric bioprinting	CAL	Projects patterned light into a rotating vat of bioink, curing an entire 3D volume simultaneously from multiple angles	• Ultrarapid printing (seconds rather than hours). • Smooth surfaces • Minimal layer-by-layer defects; high cell viability due to uniform curing	• Bioprinting of dECM[22] • Modeling exocrine pancreatic units[23]
	Xolography	Employs a dual-wavelength light system: one wavelength activates a photoinitiator in the resin, while the other defines the 3D shape, enabling precise volumetric solidification of the resin	• Fast • High resolution • Can create intricate 3D architectures	• Aggregate bioprinting[24]

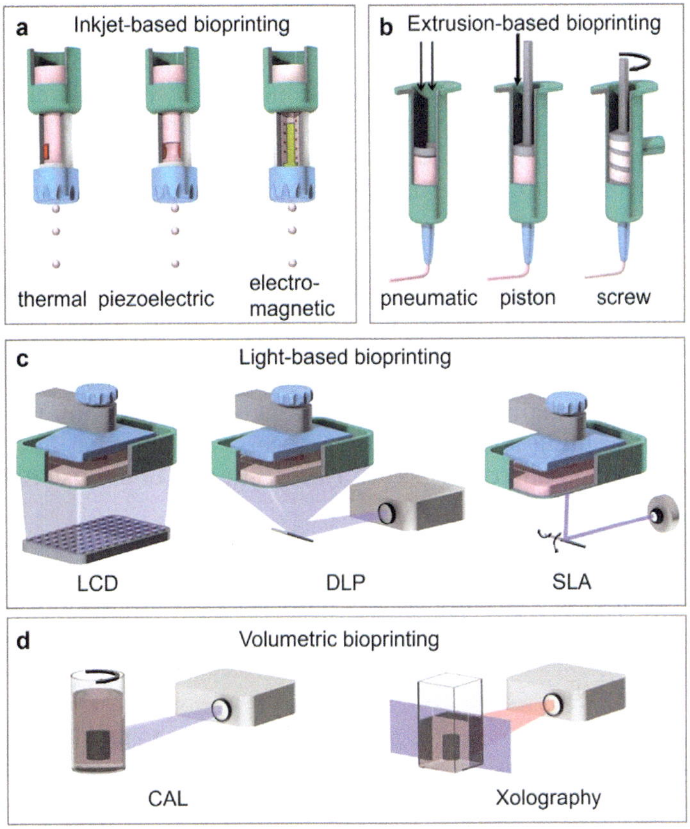

FIG. 1 Illustration of various 3D bioprinting technologies. Major features and mechanisms of (A) inkjet-based bioprinting (including thermal, piezoelectric, and electromagnetic) (B) extrusion-based bioprinting (including pneumatic, piston-driven, and screw-driven bioprinting) (C) light-based bioprinting technologies (liquid crystal display [LCD], digital light processing [DLP], stereolithography [SLA]), and (D) volumetric bioprinting (computed axial lithography (CAL) and Xolography) are depicted.

Other actuation techniques include electromagnetic, electrostatic, electrohydrodynamic, solenoid-valve, and acoustic inkjet printing but are less commonly used. A schematic description of different inkjet-based bioprinting techniques is shown in Fig. 1A.

Inkjet printing is particularly well suited for precise deposition of cells and biomolecules owing to its droplet-based dispensing mechanism. It has been successfully used to print various types of tissues, including cells dissociated from patient-derived tumor organoids, which are used to investigate intratumoral heterogeneity in bladder cancer,[25] and alveolar cells, which are used to develop a three-layered alveolar barrier model with unprecedented thickness.[26]

Inkjet bioprinting also has limitations. High-viscosity materials can clog the nozzles, limiting the range of materials that can be used and potentially affecting the structural stability of printed constructs. In addition, the rapid ejection of droplets can subject cells to shear stress, potentially compromising cell viability.[27] As a result, inkjet bioprinting is often used for applications that do not require complex constructs or for depositing thin layers of cells in applications such as skin printing.[28]

Extrusion-based bioprinting

Extrusion-based bioprinting is the most widely used technology in 3D bioprinting. In this method, bioinks are extruded through a nozzle using either a pneumatic or mechanical system, forming continuous strands of material that are deposited layer-by-layer. Extrusion-based bioprinting is adaptable to a wide range of bioinks, including those with high viscosities, making it suitable for printing tissues with varying mechanical properties, such as cartilage or bone.[29] A key advantage of extrusion-based bioprinting is its ability to print larger structures with multiple cell types and materials. Continuous extrusion of material allows the creation of complex architectures.[15] However, extrusion-based bioprinting comes with trade-offs. It is generally slower than inkjet bioprinting because it relies on continuous deposition of material rather than rapid droplet ejection. In addition, the extrusion process, especially when using smaller nozzle diameters[29] or higher-viscosity materials, can generate higher shear forces, potentially affecting cell viability. Careful optimization of extrusion parameters, such as printing speed, nozzle diameter, and extrusion force, is necessary to balance printing resolution and cell viability. Different types of extrusion-based bioprinting are shown in Fig. 1B.

Extrusion-based bioprinting can create a variety of tissues for different applications, including lung tumor cells embedded in human primary lung fibroblasts used for radiotherapy research,[30] as well as for developing liver models for infection and toxicity studies.[31,32]

Light-based bioprinting

Light-based bioprinting technologies rely on light energy to solidify photosensitive biomaterials, precisely building cell-laden structures or cell-free scaffolds. This approach enables the creation of highly detailed and complex architectures that mimic natural tissue environments, providing a high level of control over cell placement and structure formation.

In SLA bioprinting, a focused ultraviolet (UV) or visible-light laser selectively cures bioink layer-by-layer in a vat. The bioink contains living cells and biomaterials that solidify when exposed to UV light. This process relies on a photoinitiator, a chemical compound within the bioink that absorbs light energy and initiates a radical polymerization reaction, causing controlled solidification of the material. Following the completion of a layer, the build platform, which supports the structure being printed, moves slightly into or away from the vat to allow the deposition of the next layer of bioink. This movement is typically measured in micrometers, allowing the formation of intricate features without compromising structural integrity. Once the new layer is added, the UV laser selectively cures the bioink, repeating this process until the desired structure is fully formed. SLA's ability to produce detailed structures with smooth surfaces makes it ideal for generating tissue scaffolds or small, intricate biological structures, such as corneal stroma[21] and vascular networks.[33] However, the slow speed of SLA, due to its point-by-point laser approach, can limit its use in producing larger tissue constructs.

DLP bioprinting, also known as digital micro-mirror device-based SLA, is another light-based bioprinting technology. DLP relies on projecting entire layers of UV light onto the bioink at once, based on thousands of micromirrors that reflect light in the shape of each layer, curing the bioink in a single exposure. DLP's higher throughput enables the production of complex, patterned tissue structures,[34] including bone scaffolds,[35] skin grafts,[36] and perfusable cancer models,[37] while maintaining high resolution and cell integrity. Different types of light-based bioprinting are shown in Fig. 1C.

Volumetric bioprinting

Volumetric bioprinting is a newer technique that allows the creation of 3D biological structures in seconds by illuminating an entire volume of bioink at once in a vial. It differs from common 3D printing methods, which build tissues in a layer-by-layer manner. For example, computed axial lithography, the most popular volumetric bioprinting method, projects dynamic light patterns into a rotating vial of bioink, where polymerization occurs in regions that receive a cumulative light dose above a defined threshold. This volumetric approach significantly reduces the time needed for tissue fabrication, making it ideal for bioprinting larger or more complex constructs that require support for printing—such as a human auricle model[38] or a heterocellular bone model[39]—with minimal stress on the embedded cells and greatly improved cell viability.

Xolography is a variation of volumetric bioprinting that uses two intersecting light fields to initiate polymerization within a confined region, enabling high-speed fabrication with improved structural fidelity and uniform cell distribution.[40] This process involves a two-step mechanism, typically a laser light sheet to activate photoswitchable photoinitiators and a DLP-projected patterned light to selectively cure photopolymerizable bioinks layer-by-layer, offering high resolution at lower cost, though it is limited by light penetration depth.[41] Volumetric bioprinting technologies are shown in Fig. 1D.

Bioinks and biomaterials in 3D bioprinting

There are two primary components of bioinks: cells and biomaterials. The selection of appropriate bioinks is one of the most critical aspects of successful 3D bioprinting. A suitable bioink must meet several biological and physical criteria, such as biocompatibility, mechanical integrity, and printability. The bioink environment must also support cell proliferation, differentiation, and tissue formation over time.

The choice of bioink is crucial to achieving high biological performance and structural integrity and must be carefully adapted to the specific bioprinting technology, as each method imposes distinct physical and chemical requirements. Extrusion-based bioprinting requires bioinks with shear-thinning properties to facilitate smooth flow through the nozzle while maintaining shape fidelity after deposition. Inkjet bioprinting demands bioinks with low viscosity to enable precise droplet formation without clogging the nozzle. Light-based bioprinting relies on bioinks containing photopolymerizable components and compatible photoinitiators. Different categories of bioinks are summarized in Fig. 2.

Natural and synthetic bioinks

Bioinks can be broadly categorized according to their sources into natural and synthetic materials, each offering distinct advantages and limitations.

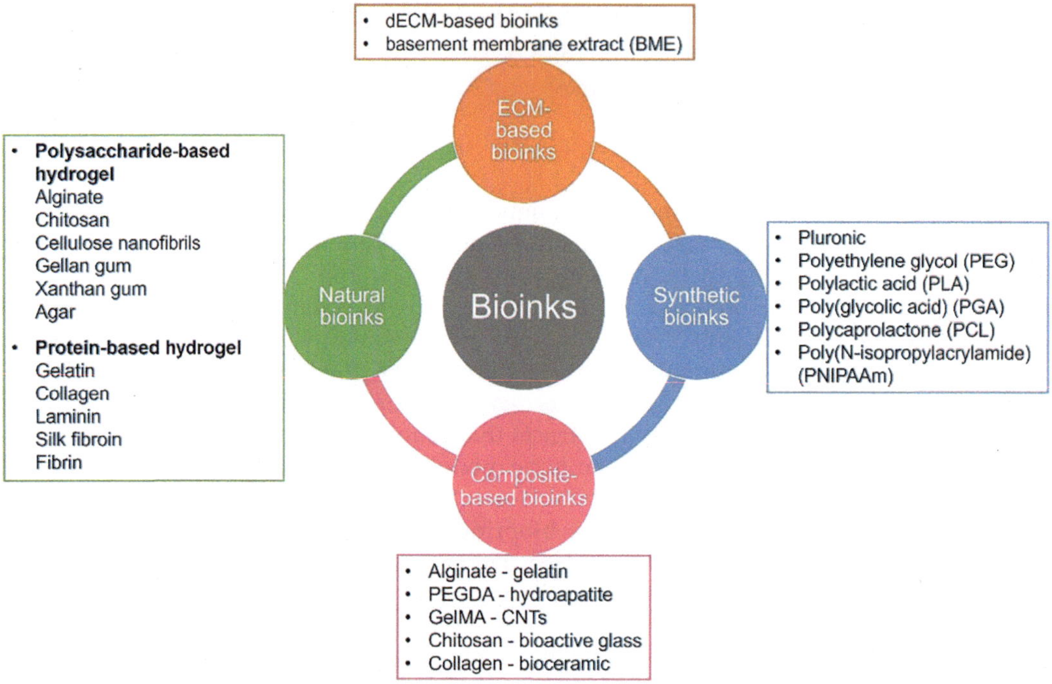

FIG. 2 Categories of Different Bioinks. Bioinks can be classified into four types according to their sources: natural bioinks, synthetic bioinks, ECM-based bioinks, and composite-based bioinks. Representative bioink components are also indicated in the figure.

Natural bioinks

Natural bioinks are derived from biological sources and typically mimic the extracellular matrix (ECM) of tissues. They are inherently biocompatible and possess biochemical cues that promote cell adhesion, proliferation, and differentiation. Common examples include alginate, collagen, gelatin, fibrin, and hyaluronic acid. These materials are often used to replicate the environment found in soft tissues, such as skin, liver, or cartilage.

Alginate, a polysaccharide derived from seaweed, is widely used as a natural bioink. Alginate can form hydrogels under mild conditions, making it ideal for encapsulating cells and bioactive molecules.[42] Its gelation typically occurs through ionic crosslinking with divalent cations, such as calcium ions, which allows rapid and reversible formation of a stable matrix. Despite these advantages, pure alginate lacks the cellular adhesion sites necessary for optimal cell attachment and interaction.[43] Consequently, alginate is often combined with other biomaterials, such as collagen,[44] gelatin,[45] or fibrin,[46] to enhance its biological properties. Alternatively, alginate can be functionalized with bioactive peptides, such as the arginine-glycine-aspartic acid (RGD) sequence, a well-known cell adhesion motif recognized by integrin receptors on cell surfaces.

Collagen, a key component of the decellularized extracellular matrix (dECM), is the most abundant protein in the body and is widely found in animal connective tissues. As a result, collagen has become one of the most extensively studied biomolecules in ECM research and is commonly applied in biofabrication. However, when used in 3D bioprinting, collagen often demonstrates poor mechanical strength and structural stability, similar to dECM.[47] Blending collagen with other polymeric materials, such as sodium alginate, has shown promise in improving its mechanical attributes and printability.[32]

Gelatin is a denatured form of collagen and is derived from animal sources, including bones, skin, and connective tissues.[48] It maintains important bioactive sequences, such as the RGD motif, which promote cell adhesion and proliferation. Gelatin undergoes thermally induced gelation, forming a gel at lower temperatures by associating into helical structures. However, at physiological temperatures around 37°C, gelatin gels liquefy, limiting their stability in many biological applications.[49] Despite its favorable biological properties, native gelatin has poor mechanical strength, which limits its use in bioprinting without blending it with other biomaterials, such as alginate.[45] To enhance gelatin's stability and mechanical properties while preserving its bioactivity, gelatin methacryloyl (GelMA) has been developed as a widely used derivative. GelMA is synthesized by modifying gelatin with methacrylate groups, allowing it to undergo photocrosslinking upon exposure to UV or visible light in the presence of a photoinitiator. GelMA is commonly used in light-based bioprinting technologies, including DLP and SLA bioprinters, as a standalone bioink or mixed with other biomaterials, such as alginate.[50]

Synthetic bioinks

Synthetic bioinks offer greater control over mechanical properties and degradation rates. They can be engineered to have specific physical and chemical characteristics, making them highly versatile for a variety of applications. Common synthetic bioinks include polyethylene glycol (PEG), a synthetic poly(ether) with excellent biocompatibility, affordability, and solubility in both organic and aqueous solutions. Modifications, such as incorporating cell-binding motifs like RGD peptides into PEG hydrogels, enhance cell adhesion. PEG-based microgels enable easy extrusion and exhibit high stability due to strong intraparticle adhesion.[51] Therapeutic proteins can also be embedded into nanoengineered biomaterial inks made from hydrolytically degradable polymers and 2D nanosilicates, creating shear-thinning inks with adjustable swelling, mechanical properties, and degradation rates.[52]

ECM-based bioinks

dECM can be prepared from native tissues by removing the cellular components using chemical, physical, and enzymatic processes, while preserving structural and biochemical components of the ECM, such as collagen, elastin, and glycosaminoglycans. This process leaves behind a matrix that mimics the natural environment of cells, providing ideal conditions for tissue growth and development. dECM bioinks have been successfully used to print a wide range of tissues, including skin,[53] liver,[54] heart,[55] and lung tissue.[56] Basement membrane extract (BME) is also widely used in bioinks. Matrigel is a BME that primarily consists of laminin, collagen, entactin, and heparan sulfate proteoglycans, along with various growth factors.[57,58]

Xeno-free bioinks

As new nonanimal methods like bioprinting continue to be developed, a paradox remains: animal-derived components are still widely used as key biomaterials. This reliance on animal-derived materials (ADMs) leads to the creation of chimeric models with human and animal biological elements, reducing their translatability to human physiology and perpetuating ethical concerns. To overcome these challenges, there has been a growing demand for the development of completely animal-free (xeno-free) models. Progress in bioprinting has been limited, though; a recent review of publications up to June 2023 found no fully xeno-free bioprinted liver models.[59]

A major challenge in developing xeno-free bioprinted models lies in replacing widely used ADMs such as BME, gelatin, dECM, and collagen, which closely mimic in vivo conditions. In addition, a majority of models are cultured in media supplemented with fetal bovine serum (FBS) and animal-derived growth factors, further complicating efforts to eliminate ADMs. However, human-derived and synthetic alternatives have the potential to replace ADMs. Human placental collagen provides a viable replacement for rat-tail collagen, while recombinant growth factors can substitute their animal-derived counterparts. Similarly, recombinant laminin and fibronectin offer functional replacements for animal-derived ECM proteins in bioinks. Other promising biomaterials include hyaluronic acid, chitosan, methylcellulose, silk fibroin, and sodium alginate, as well as a range of synthetic polymers that can be tailored to meet specific mechanical and biological requirements for bioinks.[60] In 2024, a fully xeno-free bioprinted liver model was developed using a collagen-alginate bioink with a chemically defined, FBS-free culture medium.[32]

Crosslinking mechanisms in 3D bioprinting

Crosslinking is a vital process in 3D bioprinting, as it solidifies the bioink into a stable, 3D structure. Without crosslinking, the printed tissue would lack the mechanical integrity needed to maintain its shape and function.[61] Crosslinking can occur through a variety of mechanisms, including physical, chemical, and enzymatic processes, each with advantages and limitations. A summary of widely used crosslinking mechanisms is shown in Fig. 3.

Physical crosslinking

Physical crosslinking is widely used in bioprinting as it avoids exogenous agents, thereby reducing contamination risks. It relies on environmental changes such as ion concentration, temperature, or pH.[61]

Ionic crosslinking occurs through electrostatic interactions between oppositely charged molecules. Ionic crosslinking can sometimes result in uneven or inconsistent gelation, leading to variations in the mechanical properties of the printed tissue. This can be overcome by using an automatic mixing machine to enhance the homogeneity and reproducibility of alginate-based bioinks.[62]

Thermal crosslinking involves heating or cooling polymer solutions to induce gelation. Agarose, an uncharged polysaccharide, exemplifies a thermally crosslinked bioink.

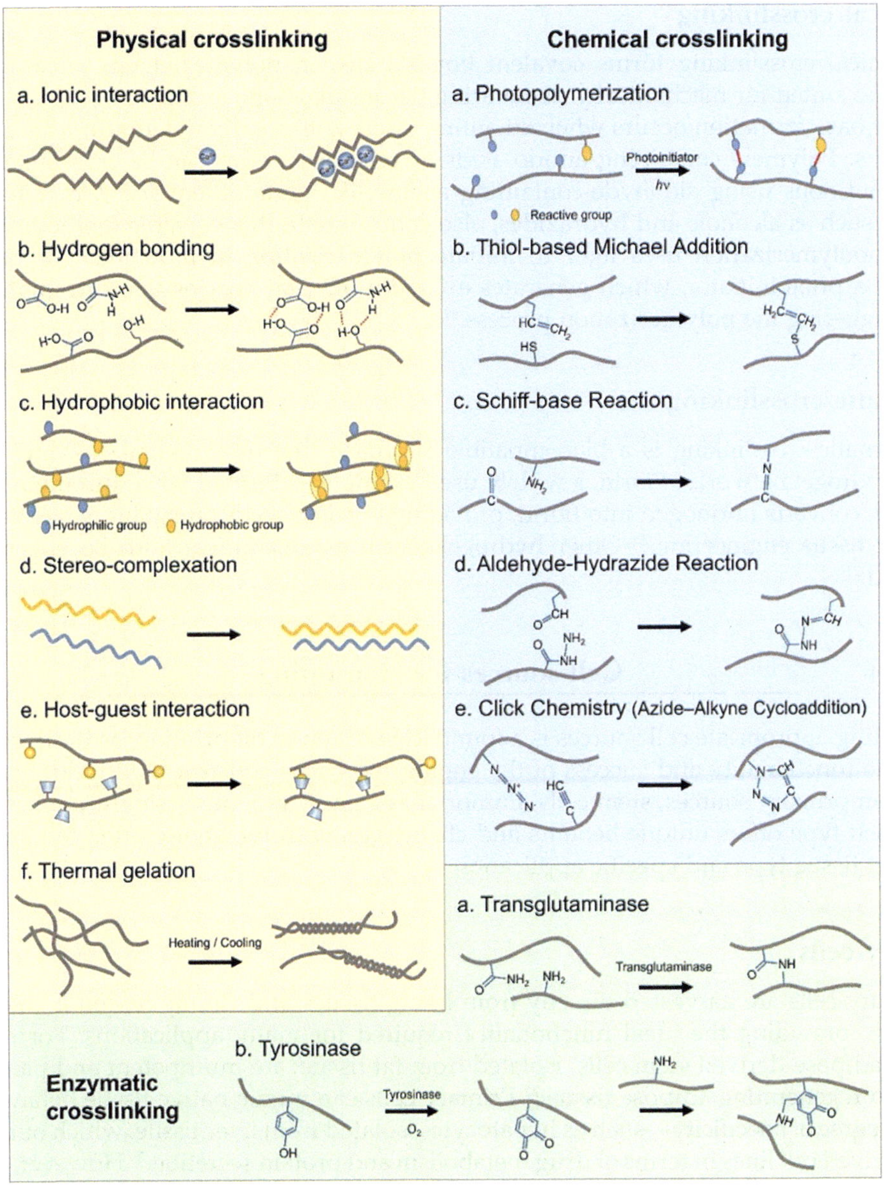

FIG. 3 Common crosslinking of bioinks in 3D bioprinting. Bioinks can be crosslinked and stabilized through various approaches, including physical crosslinking, chemical crosslinking, and enzymatic crosslinking.

At temperatures above 40°C, agarose exhibits a random coil conformation; upon cooling to approximately 32°C, it transitions to a helical structure, resulting in the formation of a 3D network.[63] Other thermocrosslinkable bioinks used for 3D bioprinting include collagen,[64] poloxamers,[65] and N-isopropylacrylamide copolymers.[66]

Chemical crosslinking

Chemical crosslinking forms covalent bonds between polymer chains, creating stable structures suited for mechanically demanding tissues like bone and cartilage.

Schiff base formation occurs when an amine reacts with a carbonyl group, forming stable crosslinks. Polymers containing amino acids or polysaccharides can be crosslinked under mild conditions using aldehyde-containing agents like glutaraldehyde.[67] Other functional groups, such as alcohols and hydrazides, also contribute to this versatile method.[68]

Photopolymerization uses light to initiate polymerization in monomers or oligomers through a photoinitiator, which generates excited molecular species upon irradiation with light, triggering the polymerization process.[69]

Enzymatic crosslinking

Enzymatic crosslinking is a biocompatible approach that uses natural enzymes to form stable hydrogel networks. Fibrin, a widely used example, is formed when thrombin, a serine protease, converts fibrinogen into fibrin, providing excellent cytocompatibility and cell adhesion for tissue engineering.[70] Other hydrogels, such as gelatin, can also be enzymatically modified.[71]

Cell sources for bioprinting

Selecting appropriate cell sources is a foundational step in bioprinting, as it directly influences the functionality and success of the engineered tissue constructs. Bioprinting can use cells from primary sources, stem cells, immortalized lines, and, increasingly, patient-specific cells. Each type offers unique benefits and challenges, with the choice often tailored to the intended tissue type and specific application.

Primary cells

Primary cells are harvested directly from living tissues and closely resemble their tissue of origin, providing the ideal functionality required for many applications. For example, human adipose-derived stem cells, isolated from fat tissue, are multipotent and particularly useful in regenerating adipose tissue.[72] Primary cells can mimic native tissue behavior with a high degree of specificity—such as hepatocytes isolated from liver tissue, which outperform liver-derived cell lines in terms of drug metabolism and protein secretion.[73] However, primary cells have limited proliferation in culture, are sensitive to in vitro environments, and are often scarce, restricting their scalability in large-scale applications.

Stem cells

Stem cells, particularly pluripotent stem cells, have garnered significant interest for bioprinting due to their ability to differentiate into various cell types. Embryonic stem cells and induced pluripotent stem cells (iPSCs) can be differentiated into specific cell lineages, enabling the production of diverse tissues. For example, iPSCs can be differentiated into hepatocyte-like cells for drug screening and toxicity bioprinting applications.[74] iPSCs can also be used to study inter-individual variation, such as sex, age, and ethnicity.[75] While stem cells offer tremendous potential, the differentiation protocols for stem cells can be complex and may not always yield pure populations of the desired cell type, which can impact tissue functionality.

Immortalized cell lines

Immortalized cell lines, derived from cancer cells or modified to proliferate indefinitely, offer a reproducible and robust cell source. They are valuable in bioprinting applications, such as viral infection studies and drug screening, due to their ease of culture and stable growth characteristics.[13,32] Immortalized cells are optimal for proof-of-concept work in bioprinting device development, as they do not require the stringent culture conditions of primary or stem cells. However, their tumorigenic origin limits their applicability in bioprinting functional tissues, where non-cancerous, highly specialized cells are required. Immortalized cells are generally more resilient to shear stress during 3D bioprinting compared to primary or iPSC-derived cells, which are more sensitive to mechanical forces, impacting their viability and functionality.[76–78]

Patient-specific cells

With advances in personalized medicine, using patient-derived cells in bioprinting is gaining traction. One method involves obtaining fibroblasts from a skin biopsy, which are reprogrammed into iPSCs. These iPSCs can then be differentiated into specific cell types required for bioprinting, such as cardiomyocytes for heart repair or pancreatic beta cells for diabetes treatment.[79,80] In addition to iPSCs, primary cells directly isolated from patient tissues are also increasingly used, as they retain donor-specific physiological and genetic characteristics.[81] The scalability of personalized cells for widespread therapeutic applications continues to be a key area of ongoing research and development. While challenges remain in optimizing cell differentiation, functional integration, and large-scale production, ongoing advancements in bioprinting technology and stem cell research are steadily bringing these innovations closer to clinical application.[82]

Applications of 3D bioprinting

3D bioprinting has a potential impact on biomedical research and clinical applications by enabling the precise fabrication of complex tissue models that closely mimic human physiology. In this section, we highlight its key areas of application, as illustrated in Fig. 4.

Cancer research

The heterogeneity of tumor cells, along with the interactions between tumor cells and ECM, is challenging to study using traditional 2D and 3D tumor models, as these models do not accurately replicate the complex 3D architecture of tumors. 3D bioprinting can precisely

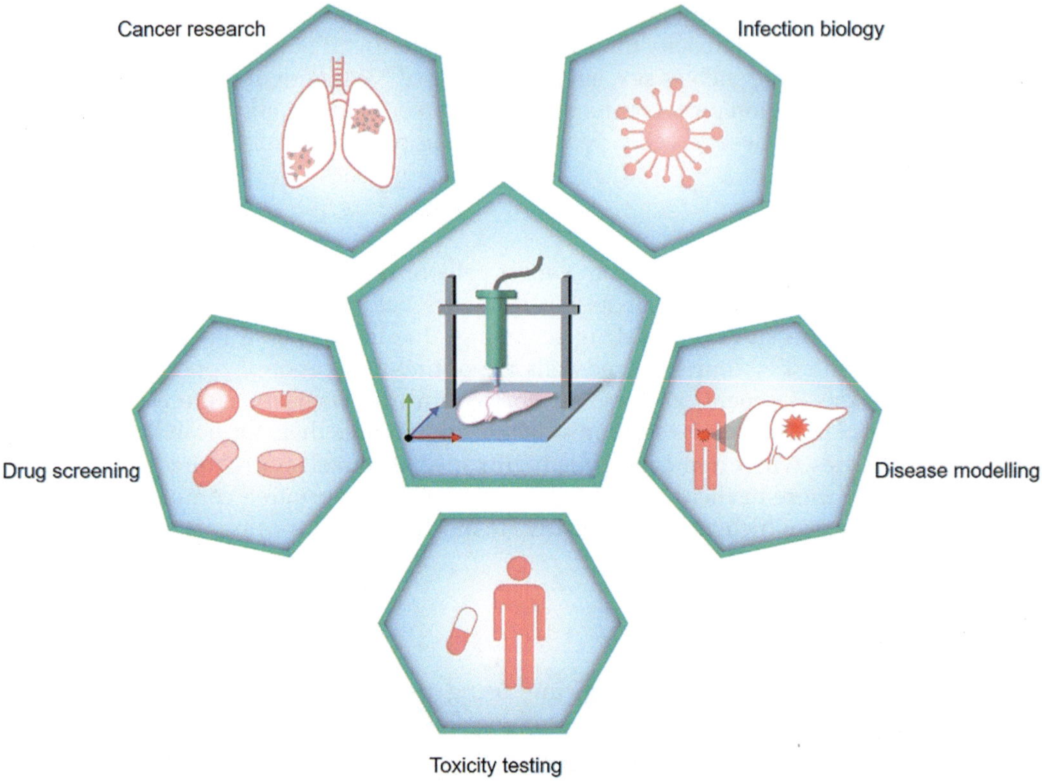

FIG. 4 Application fields of 3D bioprinting. 3D bioprinting can be used in various biomedical research areas, e.g., cancer research, drug screening, toxicity testing, disease modeling and infection biology.

position cells in a controlled spatial arrangement, enabling the creation of in vitro tumor models that more accurately replicate the tumor microenvironment. For example, a perfusable bioprinted cancer model within a hepatic microenvironment successfully demonstrated a treatment's inhibition of breast cancer growth (Fig. 5A).[15] In another approach, a bioprinted, microvascularized neuroblastoma tumor-environment model incorporating a gelatin-methacrylate/fibrin-based matrix that supports microvessel formation and mimics the tumor microenvironment enabled patient-specific drug testing.[83]

Drug screening

3D bioprinting has been instrumental in developing in vitro tissue models for drug screening. For instance, over 40 compounds were tested on a full-thickness bioprinted skin model using fibroblasts and keratinocytes.[84] In another study, a 3D bioprinted hepatocellular carcinoma model based on a composite bioink of gelatin, sodium alginate, and HepG2 cells was used to evaluate cancer drugs.[85] Endothelialized liver lobule-like structures were constructed for drug screening using dot extrusion printing with GelMA hydrogel microbeads combined with gelatin microbeads containing human umbilical vein endothelial cells to form a vascular

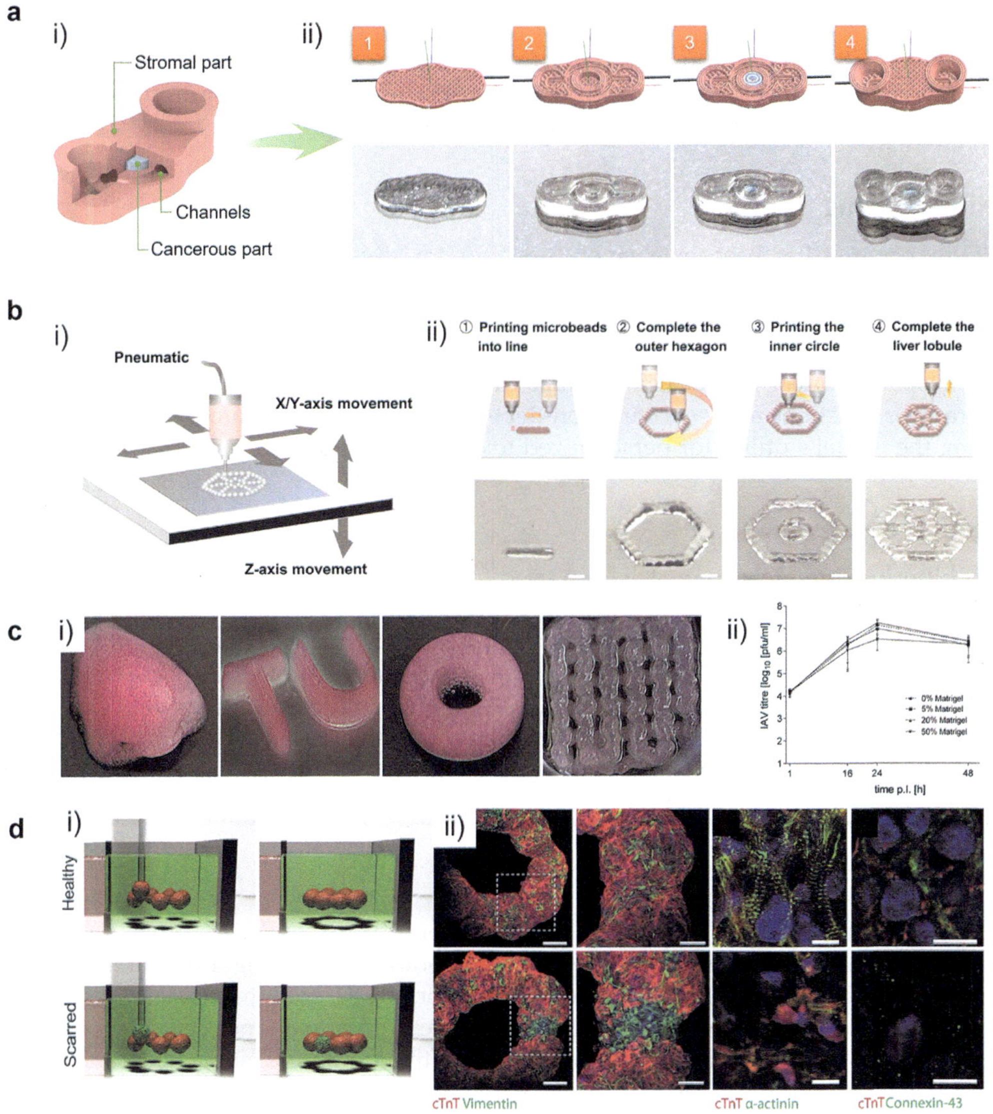

FIG. 5 Representative application cases of 3D bioprinting. (A) 3D bioprinted perfusable hepatic cancer models. (i) Schematic of the perfusable bioprinted cancer models; (ii) Schematic and images of the bioprinting process. (B) Bioprinted endothelialized liver lobule-like constructs for drug screening. (i) Schematic illustration of liver lobule-like structures using GelMA hydrogel beads; (ii) Fabrication steps for liver lobule-like structures and corresponding images. (C) 3D bioprinted human lung model for infection study. (i) Images of different constructs by 3D bioprinting with the developed bioink; (ii) Influenza A virus replication in 3D bioprinted A549 cell-laden constructs with varying Matrigel concentrations. (D) Bioprinted heterogeneous tissue models for cardiac disease modeling. (i) Schematic demonstrating the 3D bioprinting process of healthy and scarred cardiac microtissues; (ii) Characterization of the bioprinted healthy and scarred cardiac microtissues by immunofluorescence staining. *Panel A: Wu D, Pang S, Berg J, et al. Bioprinting of perfusable vascularized organ models for drug development via sacrificial-free direct ink writing. Adv FunctMater. 2024;34:2314171. Copyright, 2024 The Authors. Available under the terms of the Creative Commons CC BY license. Panel B: Fan Z, Wei X, Chen K, et al. 3D bioprinting of an endothelialized liver lobule-like construct as a tumor-scale drug screening platform. Micromachines. 2023;14:878. Copyright, 2023 The Authors. Available under the terms of the Creative Commons CC BY license. Panel C: Berg J, Hiller T, Kissner MS, et al. Optimization of cell-laden bioinks for 3D bioprinting and efficient infection with influenza A virus. Sci Rep. 2018;8:13877. Copyright, 2018 The Authors. Available under the terms of the Creative Commons CC BY license. Panel D: Daly AC, Davidson MD, Burdick JA. 3D bioprinting of high cell-density heterogeneous tissue models through spheroid fusion within self-healing hydrogels. Nat Commun. 2021;12:753. Copyright, 2021 The Authors. Available under the terms of the Creative Commons CC BY license.*

network (Fig. 5B).[86] Bioprinting is also useful for high-throughput drug screening, such as a melanoma model developed using pneumatic extrusion-based bioprinting of cell-laden droplets in a 96-well plate to enable efficient drug screening.[87]

Toxicity testing

3D bioprinting has been extensively used for toxicity testing. For example, bioprinted human liver tissues composed of patient-derived hepatocytes and nonparenchymal cells arranged in a scaffold-free, in vivo-like architecture successfully demonstrated hepatotoxic effects, highlighting the model's potential for assessing drug-induced liver injury in vitro.[88] A 3D bioprinted liver model using human-induced hepatocytes from fibroblasts showed mature liver functions and concentration-dependent toxicity of hepatotoxic agents, demonstrating the model's utility for personalized toxicity screening.[89]

Infection biology

3D bioprinted models have also been used for infectious disease research. A lung model created using extrusion-based bioprinting with a bioink formulated from alginate, gelatin, and Matrigel was employed to investigate influenza A virus infection, demonstrating that the bioink composition could sustain viral replication and promote the release of proinflammatory interferons in infected cells (Fig. 5C).[13] In a separate study, a model of bioprinted biofilms that accurately replicated native biofilms and retained their structure when printed onto human bronchial epithelial cells was used to investigate bacterial biofilm infection caused by *E. coli*.[90]

Disease modeling

3D bioprinting has been used for modeling several diseases. For example, a patient-derived model for studying Alzheimer's disease was developed using a microfluidic printer. The model utilized a bioink mixture containing fibrinogen, sodium alginate, genipin, and neural progenitor cells differentiated from patient-derived human induced pluripotent stem cells and exhibited key Alzheimer's markers.[91] Extrusion-based bioprinted cardiac tissue models were created using patient-derived cardiomyocytes embedded in a hydrogel formulated with a commercial laminin-based bioink. These constructs maintained consistent contraction over a 30-day period, and gene expression analysis indicated enhanced maturation compared to traditional 2D cell cultures.[79] Finally, a high cell-density bioprinted cardiac disease model was developed by patterning healthy and scarred spheroids within a self-healing support hydrogel and used to study cardiac regenerative therapies (Fig. 5D).[92]

Comparison of 3D modeling techniques

Each 3D modeling technique has unique advantages and limitations, which define its potential applications. Bioprinting offers distinct advantages over organoids and organ-on-a-chip technologies. For example, the precise spatial deposition of cells enables controlled organization of multiple cell types within a scaffold. In contrast, organoids rely on self-organization, often resulting in variability in tissue structure, while organ-on-a-chip devices incorporate microfluidic channels to simulate physiological conditions but struggle to recreate full tissue complexity on a larger scale. Another key advantage of bioprinting is its ability to introduce vascular channels within constructs that have been successfully perfused. This is a major challenge for organoids, which depend on passive diffusion for nutrient and oxygen transport. Techniques like sacrificial printing—where temporary materials are printed and later removed to form hollow channels—and coaxial extrusion—which enables the simultaneous deposition of core–shell structures—allow for the fabrication of intricate vascular networks, enhancing tissue viability and making it more suitable for applications requiring functional and perfusable tissues.

While bioprinting, organoids, and organ-on-chip technologies each have their strengths, they are not mutually exclusive. Integrating bioprinting with organoid and organ-on-chip microfluidics may offer a powerful strategy for engineering more complex and functional tissues.

Future directions and conclusions

The roadmap for 3D bioprinting encompasses key advancements, ongoing challenges, and future directions that will shape the trajectory of this transformative technology. As 3D bioprinting continues to evolve, researchers and industry stakeholders must navigate various scientific, technical, and regulatory hurdles to achieve its full potential in biomedical applications.

A primary objective in the near future is improving the resolution and precision of bioprinting techniques to better mimic the intricate microarchitecture of native tissues. Current technologies, such as extrusion-based, inkjet, and light-based bioprinting, have demonstrated promising results in fabricating tissue-like constructs. However, achieving the fine spatial control necessary for functional tissue regeneration remains a critical challenge. Future research should focus on optimizing printing resolution, developing novel bioinks with enhanced biocompatibility, and refining postprinting maturation processes to ensure structural and functional fidelity.

Another crucial challenge is integrating vascularization strategies into bioprinted tissues. The successful fabrication of large, complex tissues and organs requires an embedded vascular network capable of delivering nutrients and oxygen to sustain cellular viability. Various approaches, including sacrificial ink printing, coaxial extrusion, and bioprinting with endothelial cells, have been explored to create perfusable vascular channels. Advancements in this area will significantly enhance the feasibility of bioprinted constructs for transplantation and in vitro disease modeling.

The convergence of 3D bioprinting with organ-on-a-chip technologies and other new approach methodologies (NAMs) will further revolutionize biomedical research. Organoids and microfluidic-based organ chip systems have provided valuable platforms for studying tissue function, disease progression, and drug responses. (See Chapter 6 for a deeper discussion of these approaches.) By integrating bioprinting with these technologies, researchers can fabricate highly complex, patient-specific models that offer superior physiological relevance compared to traditional in vitro and animal models. A comparative analysis of bioprinting and these NAMs will help define their respective strengths and limitations, guiding their application in regenerative medicine, pharmaceutical testing, and precision medicine.

Looking ahead, artificial intelligence (AI) and machine learning are expected to play an instrumental role in advancing bioprinting methodologies. AI-driven design optimization, automated quality control, and predictive modeling of tissue development can enhance the efficiency and reproducibility of the bioprinting process. By leveraging AI, researchers can refine bioink compositions, optimize printing parameters, and predict tissue maturation outcomes, ultimately accelerating the path to clinical translation.[93] (See Chapter 9 for a deeper discussion about the role of AI in nonanimal research and testing.)

In addition to technological advancements, addressing regulatory and ethical considerations is essential for the widespread adoption of bioprinting in clinical applications. Standardizing bioink formulations, ensuring reproducibility of printed constructs, and establishing guidelines for clinical translation are key areas that require collaborative efforts among researchers, regulatory agencies, and policymakers. Ethical concerns, such as the use of human-derived cells and the long-term safety of bioprinted tissues, must be carefully considered to foster public trust and acceptance of this emerging technology.

The establishment of bioprinting facilities within hospitals and research institutions will be a pivotal step in translating bioprinted tissues from the laboratory to clinical settings. On-site bioprinting capabilities could allow for real-time fabrication of patient-specific grafts and implants, reducing waiting times for organ transplants and personalized therapeutic applications. The integration of bioprinting into clinical workflows will require extensive validation, regulatory approvals, and training for healthcare professionals, ensuring seamless implementation into standard medical practice.

In summary, this chapter has outlined the foundational principles, key methodologies, and applications of 3D bioprinting. By enabling the precise fabrication of complex multicellular structures, bioprinting offers unprecedented opportunities to replicate human tissue architecture in vitro. Ultimately, its continued evolution holds the promise of reducing and potentially replacing animal testing, aligning scientific innovation with ethical imperatives.

References

[1] Zhou L, Miller J, Vezza J, et al. Additive manufacturing: a comprehensive review. *Sensors (Basel)*. 2024;24:2668.

[2] Bishop ES, Mostafa S, Pakvasa M, et al. 3-D bioprinting technologies in tissue engineering and regenerative medicine: current and future trends. *Genes Dis*. 2017;4:185–195.

[3] Mironov V, Boland T, Trust T, et al. Organ printing: computer-aided jet-based 3D tissue engineering. *Trends Biotech*. 2003;21:157–161.

[4] Murphy SV, Atala A. 3D bioprinting of tissues and organs. *Nat Biotechnol*. 2014;32:773–785.

[5] Gu Z, Fu J, Lin H, He Y. Development of 3D bioprinting: from printing methods to biomedical applications. *Asian J Pharm Sci*. 2020;15:529–557.

[6] Cui X, Boland T, D'Lima DD, Lotz MK. Thermal inkjet printing in tissue engineering and regenerative medicine. *Recent PAT Drug DelivFormul*. 2012;6:149–155.

[7] Li X, Liu B, Pei B, et al. Inkjet bioprinting of biomaterials. *Chem Rev*. 2020;120:10793–10833.

[8] Suntornnond R, Ng WL, Shkolnikov S, Yeong WY. A facile method to fabricate cell-laden hydrogel microparticles of tunable sizes using thermal inkjet bioprinting. *Droplet*. 2024;3:e144.

[9] Cui X, Breitenkamp K, Finn MG, et al. Direct human cartilage repair using three-dimensional bioprinting technology. *Tissue Eng Part A*. 2012;18:1304–1312.

[10] Hewes S, Wong AD, Searson PC. Bioprinting microvessels using an inkjet printer. *Bioprinting*. 2017;7:14–18.

[11] Takagi D, Lin W, Matsumoto T, et al. High-precision three-dimensional inkjet technology for live cell bioprinting. *Int J Bioprint*. 2019;5:208.

[12] Nothdurfter D, Ploner C, Coraca-Huber DC, et al. 3D bioprinted, vascularized neuroblastoma tumor environment in fluidic chip devices for precision medicine drug testing. *Biofabrication*. 2022;14:035002.

[13] Berg J, Hiller T, Kissner MS, et al. Optimization of cell-laden bioinks for 3D bioprinting and efficient infection with influenza A virus. *Sci Rep*. 2018;8:13877.

[14] Yan Y, Li X, Gao Y, et al. 3D bioprinting of human neural tissues with functional connectivity. *Cell Stem Cell*. 2024;31:260–274.e267.

[15] Wu D, Pang S, Berg J, et al. Bioprinting of perfusable vascularized organ models for drug development via sacrificial-free direct ink writing. *Adv FunctMater*. 2024;34:2314171.

[16] Liu K, Yang C, Li J, et al. Development of a piston-actuated 3D bio-printer with performance prediction system for the reliable printing of tubular structures. *J Micromech Microeng*. 2023;33:115010.

[17] Bhattacharyya A, Janarthanan G, Tran HN, et al. Bioink homogeneity control during 3D bioprinting of multicomponent micro/nanocomposite hydrogel for even tissue regeneration using novel twin screw extrusion system. *Chem Engin J*. 2021;415:128971.

[18] Wang B, Ye X, Chen G, et al. Fabrication and properties of PLA/β-TCP scaffolds using liquid crystal display (LCD) photocuring 3D printing for bone tissue engineering. *Front Bioeng Biotechnol*. 2024;12:1273541.

[19] Grigoryan B, Pause SJ, Corbett DC, et al. Multivascular networks and functional intravascular topologies within biocompatible hydrogels. *Science*. 2019;364:458–464.

[20] You S, Xiang Y, Hwang HH, et al. High cell density and high-resolution 3D bioprinting for fabricating vascularized tissues. *Sci Adv*. 2023;9:eade7923.

[21] Mahdavi SS, Abdekhodaie MJ, Kumar H, et al. Stereolithography 3D bioprinting method for fabrication of human corneal stroma equivalent. *Ann Biomed Eng*. 2020;48:1955–1970.

[22] Lian L, Xie M, Luo L, et al. Rapid volumetric bioprinting of decellularized extracellular matrix bioinks. *Adv Mater*. 2024;36:2304846.

[23] Sgarminato V, Madrid-Wolff J, Boniface A, et al. 3D in vitro modeling of the exocrine pancreatic unit using tomographic volumetric bioprinting. *Biofabrication*. 2024;16:045034.

[24] Stoecker L, Cedillo-Servin G, König NF, et al. Xolography for biomedical applications: dual-color light-sheet printing of hydrogels with local control over shape and stiffness. *Adv Mater*. 2025;37:2410292.

[25] Yoon WH, Lee H, Kim S, et al. Use of inkjet-printed single cells to quantify intratumoral heterogeneity. Biofabrication. 2020;12(3):035030.

[26] Kang D, Park JA, Kim W, et al. All-Inkjet-printed 3D alveolar barrier model with physiologically relevant microarchitecture. *Adv Sci (Weinh)*. 2021;8:2004990.

[27] Tripathi S, Mandal SS, Bauri S, et al. 3D bioprinting and its innovative approach for biomedical applications. *MedComm*. 2023;4:e194.

[28] Michael S, Sorg H, Peck CT, et al., Tissue engineered skin substitutes created by laser-assisted bioprinting form skin-like structures in the dorsal skin fold chamber in mice. *PLoS One*. 2013;8:e57741.

[29] Ramesh S, Harrysson OLA, Rao PK, et al. Extrusion bioprinting: recent progress, challenges, and future opportunities. *Bioprinting*. 2021;21:e00116.

[30] Mei Y, Lakotsenina E, Wegner M, et al. Three-dimensional-bioprinted non-small cell lung cancer models in a mouse phantom for radiotherapy research. *Int J Mol Sci*. 2024;25:10268.

[31] Hiller T, Berg J, Elomaa L, et al. Generation of a 3D liver model comprising human extracellular matrix in an alginate/gelatin-based bioink by extrusion bioprinting for infection and transduction studies. *Int J Mol Sci*. 2018;19:3129.

[32] Ali ASM, Berg J, Roehrs V, et al. Xeno-free 3D bioprinted liver model for hepatotoxicity assessment. *Int J Mol Sci*. 2024;25:1811.

[33] Cui H, Esworthy T, Zhou X, et al. Engineering a novel 3D printed vascularized tissue model for investigating breast cancer metastasis to bone. *Adv Healthc Mater*. 2020;9:1900924.

[34] Lu Y, Mapili G, Suhali G, et al. A digital micro-mirror device-based system for the microfabrication of complex, spatially patterned tissue engineering scaffolds. *J Biomed Mater Res A*. 2006;77:396–405.

[35] Rajput M, Mondal P, Yadav P, et al. Light-based 3D bioprinting of bone tissue scaffolds with tunable mechanical properties and architecture from photocurable silk fibroin. *Int J Biol Macromol*. 2022;202:644–656.

[36] Zhou F, Hong Y, Lian R, et al. Rapid printing of bio-inspired 3D tissue constructs for skin regeneration. *Biomaterials*. 2020;258:120287.

[37] Mei Y, Wu D, Berg J, et al. Generation of a perfusable 3D lung cancer model by digital light processing. *Int J Mol Sci*. 2023;24:6071.

[38] Bernal PN, Delrot P, Loterie D, et al. Volumetric bioprinting of complex living-tissue constructs within seconds. *Adv Mater*. 2019;31:1904209.

[39] Gehlen J, Qiu W, Schädli GN, et al. Tomographic volumetric bioprinting of heterocellular bone-like tissues in seconds. *Acta Biomater*. 2023;156:49–60.

[40] König NF, Reuter M, Reuß, et al. Xolography for 3D printing in microgravity. *Adv Mater*. 2025;37:2413391.

[41] Shafique H, Karamzadeh V, Kim G, et al. High-resolution low-cost LCD 3D printing of microfluidics and organ-on-chip devices. *Lab Chip*. 2024;24:2774–2790.

[42] Ahmad RR, Nawawi WMFW, Nasarujddin RR. Alginate and alginate composites for biomedical applications. *Asian J Pharm Sci*. 2021;16:280–306.

[43] Lee KY, Mooney DJ. Alginate: properties and biomedical applications. *Prog Polym Sci*. 2012;37:106–126.

[44] Zimmerling A, Zhou Y, Chen X. Synthesis of alginate/collagen bioink for bioprinting respiratory tissue models. *J Funct Biomater*. 2024;15:90.

[45] Kim J, Choi YJ, Gal CW, et al. Development of an alginate-gelatin bioink enhancing osteogenic differentiation by gelatin release. *Int J Bioprinting*. 2023;9:660.

[46] Cavallo A, Kayal TA, Mero A, et al. Fibrinogen-based bioink for application in skin equivalent 3D bioprinting. *J Funct Biomate*. 2023;14:459.

[47] Shim JH, Kim JY, Park M, et al. Development of a hybrid scaffold with synthetic biomaterials and hydrogel using solid freeform fabrication technology. *Biofabrication*. 2011;3:034102.

[48] Rather JA, Akhter N, Ashraf QS, et al. A comprehensive review on gelatin: understanding impact of the sources, extraction methods, and modifications on potential packaging applications. *Food Pack Shelf Life*. 2022;34:100945.

[49] Asim S, Tabish TA, Liaqat U, et al. Advances in gelatin bioinks to optimize bioprinted cell functions. *Adv Healthc Mater*. 2023;12:2203148.

[50] Shukla P, Mitruka M, Pati V, et al. The effect of the synthetic route on the biophysiochemical properties of methacrylated gelatin (GelMA) based hydrogel for development of GelMA-based bioinks for 3D bioprinting applications. *Materialia*. 2022;25:101542.

[51] Xin S, Chimene D, Garza JE, et al. Clickable PEG hydrogel microspheres as building blocks for 3D bioprinting. *Biomater Sci*. 2019;7:1179–1187.

[52] Skardal A, Zhang J, Prestwich GD. Bioprinting vessel-like constructs using hyaluronan hydrogels crosslinked with tetrahedral polyethylene glycol tetracrylates. *Biomaterials*. 2010;31:6173–6181.

[53] Won JY, Lee MH, Kim MJ, et al. A potential dermal substitute using decellularized dermis extracellular matrix derived bio-ink. *Artif Cells Nanomed Biotechnol*. 2019;47:644–649.

[54] Kim MK, Jeong W, Jeon S, et al. 3D bioprinting of dECM-incorporated hepatocyte spheroid for simultaneous promotion of cell-cell and -ECM interactions. *Font Bioeng Biotechnol*. 2023;11:1305023.

[55] Basara G, Ozcebe SG, Ellis BW, et al. Tunable human myocardium derived decellularized extracellular matrix for 3D bioprinting and cardiac tissue engineering. *Gels*. 2021;7:70.

[56] Dabaghi M, Saraei N, Carpio MB, et al. A robust protocol for decellularized human lung bioink generation amenable to 2D and 3D lung cell culture. *Cells*. 2021;10:1538.

[57] Rijal G. Bioinks of natural biomaterials for printing tissues. *Bioeng (Basel)*. 2023;10:705.

[58] Benton G, Arnaoutova I, George J, et al. Matrigel: from discovery and ECM mimicry to assays and models for cancer research. *Adv Drug Deliv Rev*. 2014;79–80:3–18.

[59] Ali ASM, Wu D, Bannach-Brown A, et al. 3D bioprinting of liver models: a systematic scoping review of methods, bioinks, and reporting quality. *Mater Today Bio*. 2024;26:100991.

[60] Khoeini R, Nosrati H, Akbarzadeh A, et al. Natural and synthetic bioinks for 3D bioprinting. *Adv Nanobio Res*. 2021;1:2000097.

[61] GhavamiNejad A, Ashammakhi N, Wu XY, et al. Crosslinking strategies for 3D bioprinting of polymeric hydrogels. *Small*. 2020;16:e2002931.

[62] Wu D, Pang S, Röhrs V, et al. Man vs. machine: automated bioink mixing device improves reliability and reproducibility of bioprinting results compared to human operators. *Int J Bioprinting*. 2024;10:1974.

[63] Wenger L, Radtke CP, Gerisch E, et al. Systematic evaluation of agarose- and agar-based bioinks for extrusion-based bioprinting of enzymatically active hydrogels. *Front Bioeng Biotechnol*. 2022;10:928878.

[64] Li Z, Ruan C, Niu X. Collagen-based bioinks for regenerative medicine: fabrication, application and prospective. *Med/Novel Technol Devices*. 2023;17:100211.
[65] Russo E, Villa C. Poloxamer hydrogels for biomedical applications. *Pharmaceutics*. 2019;11:671.
[66] Navara AM, Kim YS, Xu Y, et al. A dual-gelling poly(N-isopropylacrylamide)-based ink and thermoreversible poloxamer support bath for high-resolution bioprinting. *Bioact Mater*. 2022;14:302–312.
[67] Alavarse AC, Frachini ECG, da Silva RLCG, et al. Crosslinkers for polysaccharides and proteins: synthesis conditions, mechanisms, and crosslinking efficiency, a review. *Int J Biol Macromol*. 2022;202:558–596.
[68] Kohlan TB, Wen Y, Mini C, et al. Schiff base crosslinked hyaluronic acid hydrogels with tunable and cell instructive time-dependent mechanical properties. *Carbohydr Polym*. 2024;338:122173.
[69] Elkhoury K, Zuazola J, Vijayavenkataraman S. Bioprinting the future using light: a review on photocrosslinking reactions, photoreactive groups, and photoinitiators. *SLAS Technol*. 2023;28:142–151.
[70] Sanz-Horta R, Matesanz A, Gallardo A, et al. Technological advances in fibrin for tissue engineering. *J Tissue Eng*. 2023;14:20417314231190288.
[71] Song K, Ren B, Zhai Y, et al. Effects of transglutaminase cross-linking process on printability of gelatin microgel-gelatin solution composite bioink. *Biofabrication*. 2021;14:015014.
[72] Albrecht FB, Schmidt FF, Volz AC, et al. Bioprinting of 3D adipose tissue models using a GelMA-bioink with human mature adipocytes or human adipose-derived stem cells. *Gels*. 2022;8:611.
[73] Nyberg SL, Remmel RP, Mann HG, et al. Primary hepatocytes outperform Hep G2 cells as the source of biotransformation functions in a bioartificial liver. *Ann Surg*. 1994;220:59–67.
[74] He J, Wang J, Pang Y, et al. Bioprinting of a hepatic tissue model using human-induced pluripotent stem cell-derived hepatocytes for drug-induced hepatotoxicity evaluation. *Int J Bioprint*. 2022;8:581.
[75] Volpato V, Webber C. Addressing variability in iPSC-derived models of human disease: guidelines to promote reproducibility. *Dis Model Mech*. 2020;13:042317.
[76] Müller SJ, Fabry B, Gekle S. Predicting cell stress and strain during extrusion bioprinting. *Phys Rev Applied*. 2023;19:064061.
[77] Blaeser A, Campos DFD, Puster U, et al. Controlling shear stress in 3D bioprinting is a key factor to balance printing resolution and stem cell integrity. *Adv Healthc Mater*. 2016;5:326–333.
[78] Xu HQ, Liu JC, Zhang ZY, et al. A review on cell damage, viability, and functionality during 3D bioprinting. *Mil Med Res*. 2022;9:70.
[79] Wolfe JT, He W, Kim MS, et al. 3D-bioprinting of patient-derived cardiac tissue models for studying congenital heart disease. *Front Cardiovasc Med*. 2023;10:1162731.
[80] Veres A, Faust AL, Bushnell HL, et al. Charting cellular identity during human in vitro β-cell differentiation. *Nature*. 2019;569:368–373.
[81] Mao S, He J, Zhao Y, et al. Bioprinting of patient-derived in vitro intrahepatic cholangiocarcinoma tumor model: establishment, evaluation and anti-cancer drug testing. *Biofabrication*. 2020;12:045014.
[82] Sharma A, Sances S, Workman M, et al. Multi-lineage human iPSC-derived platforms for disease modeling and drug discovery. *Cell Stem Cell*. 2020;26:309–329.
[83] Delgrosso E, Scocozza F, Cansolino L, et al. 3D bioprinted osteosarcoma model for experimental boron neutron capture therapy (BNCT) applications: preliminary assessment. *J Biomed Mater Res*. 2023;111:1571–1580.
[84] Wei Z, Liu X, Ooka M, et al. Two-dimensional cellular and three-dimensional bio-printed skin models to screen topical-use compounds for irritation potential. Front Bioeng Biotechnol. 2020;8:109.
[85] Sun L, Yang H, Wang Y, et al. Application of a 3D bioprinted hepatocellular carcinoma cell model in antitumor drug research. *Front Oncol*. 2020;10:878.
[86] Fan Z, Wei X, Chen K, et al. 3D bioprinting of an endothelialized liver lobule-like construct as a tumor-scale drug screening platform. *Micromachines*. 2023;14:878.
[87] de Villiers M, Kotze AF, du Plessis LH. Pneumatic extrusion bioprinting-based high throughput fabrication of a melanoma 3D cell culture model for anti-cancer drug screening. *Biomed Mater*. 2024;19:055034.
[88] Nguyen DG, Funk J, Robbins JB, et al. Bioprinted 3D primary liver tissues allow assessment of organ-level response to clinical drug induced toxicity in vitro. *PLoS One*. 2016;11:e0158674.
[89] Ma Y, He R, Deng B, et al. Advanced 3D bioprinted liver models with human-induced hepatocytes for personalized toxicity screening. *J Tissue Eng*. 2025;16:20417314241313341.
[90] Aliyazdi S, Frisch S, Hidalgo A, et al. 3D bioprinting of *E. coli* MG1655 biofilms on human lung epithelial cells for building complex *in vitro* infection models. *Biofabrication*. 2023;15:035019.

[91] Benwood C, Walters-Shumka J, Scheck K, et al. 3D bioprinting patient-derived induced pluripotent stem cell models of Alzheimer's disease using a smart bioink. *Bioelectron Med*. 2023;9:10.
[92] Daly AC, Davidson MD, Burdick JA. 3D bioprinting of high cell-density heterogeneous tissue models through spheroid fusion within self-healing hydrogels. *Nat Commun*. 2021;12:753.
[93] Chen H, Zhang B, Huang J. Recent advances and applications of artificial intelligence in 3D bioprinting. *Biophys Rev*. 2024;5:031301.

CHAPTER

9

The role of artificial intelligence in overcoming animal testing

Thomas Hartung[a,b,c], *Thomas Luechtefeld*[a,d], *and Alexandra Maertens*[a]

[a]Center for Alternatives to Animal Testing (CAAT), Bloomberg School of Public Health, Johns Hopkins University, Baltimore, MD, United States
[b]Doerenkamp-Zbinden-Chair for Evidence-Based Toxicology, Baltimore, MD, United States
[c]CAAT-Europe, University of Konstanz, Konstanz, Germany
[d]Insilica LLC and ToxTrack Inc., Bethesda, MD, United States

Abstract

The rapid advancement of artificial intelligence (AI) is transforming science. We explore the role of AI in advancing nonanimal approaches for chemical safety assessment and drug discovery. We discuss the limitations of animal models and then examine how AI can serve as a key enabler for nonanimal methods, leveraging the growing availability of big data, natural language processing for literature mining, machine learning for predictive modeling, and explainable AI for model interpretation. The synergies between AI and other innovative technologies, such as microphysiological systems, adverse outcome pathways, and integrated testing strategies, are highlighted for creating human-relevant, predictive toxicology. We also address the challenges and future directions of AI, including data quality and bias issues, validation and regulatory acceptance, the need for responsible and human-centered AI, and the importance of multidisciplinary collaboration. We conclude with a forward-looking perspective on the transformative potential of AI in toxicology and pharmacology, envisioning AI as a toxicology copilot, the development of tox chatbots, and the acceleration of drug discovery. Ultimately, we call for embracing these new technologies while upholding scientific rigor and ethical principles, and for a future integration of AI with nonanimal approaches that revolutionize chemical safety assessment and drug discovery and reduce reliance on animal testing.

Keywords: Artificial intelligence; Machine learning; Regulatory science; 3Rs; New approach methods

Introduction

Animal testing has long been a cornerstone of medical research and drug development. For decades, animals have served as the primary models for understanding disease mechanisms, testing the safety and efficacy of new treatments, and predicting potential side effects. However,

Moving Toward Nonanimal Approaches in Medical Research and Testing
https://doi.org/10.1016/B978-0-443-26514-3.00007-2

the reliance on animal models has come under increasing scrutiny in recent years due to ethical concerns, economic needs,[1] and scientific limitations.[2] There is growing recognition that animal models often fail to accurately predict human responses to drugs and diseases.[3] Species differences in physiology, metabolism, and genetics can cause misleading results and costly failures when translating findings from animals to humans. The high attrition rates in clinical trials, where many promising drug candidates fail due to lack of efficacy or unexpected toxicities, underscore the limitations of animal-based testing. Moreover, the use of animals in research raises significant ethical concerns about animal welfare and the moral justification for their suffering.

In light of these challenges, there is a pressing need for nonanimal alternatives that can provide more human-relevant and predictive information for medical research and testing. Advances in fields such as stem cell biology, tissue engineering, and computational modeling have opened up new possibilities for developing innovative nonanimal approaches. Among these emerging technologies, artificial intelligence (AI) stands out as a potential game-changer with the capacity to transform the landscape of medical research and testing.

The application of AI in medicine and safety sciences is not entirely new, but recent years have witnessed a surge of interest and progress in this field. Early AI systems in medicine focused on expert systems and rule-based approaches for diagnosis and decision support. In toxicology, quantitative structure-activity relationship models were developed to predict chemical toxicity based on structural features. However, these early efforts were limited by the availability of large-scale data, computational power, and advanced algorithms.

The convergence of several factors has now created a perfect storm for AI to make a transformative impact. The explosion of biomedical data from sources such as electronic health records, omics technologies, and high-throughput screening has provided the fuel for data-hungry AI algorithms. Open-access publishing makes more scientific literature machine-readable each year. Concurrent advances in computing power, storage, and cloud computing have made it feasible to process and analyze massive datasets. Finally, the development of powerful machine learning techniques, particularly deep learning, has enabled the creation of AI models that can learn from complex, high-dimensional data and make accurate predictions.

The stage is now set for AI to revolutionize medical research and testing by enabling faster, cheaper, and more accurate nonanimal approaches. AI-based methods have the potential to integrate and analyze diverse types of data—from molecular and cellular information to clinical records and imaging—to generate predictive models of disease and drug response. Machine learning can help identify novel drug targets, optimize lead compounds, and predict toxicity and side effects. Natural language processing (NLP) can extract valuable information from scientific literature and generate new hypotheses. AI can also be coupled with other nonanimal technologies, such as organ-on-chip (OOC) systems and 3D tissue models, to create more comprehensive and human-relevant testing platforms. As we stand at the cusp of this exciting era, it is important to explore the role of AI in advancing nonanimal approaches and its potential to transform the future of medical research and testing.

Limitations of animal models in drug development and safety sciences

Animal models have long been the bedrock of drug development and safety testing, but while animal studies have contributed to many medical advances, they often fail to predict human responses accurately. This section discusses the shortcomings of animal models through anecdotal examples, systematic evaluations, and the challenges posed by the reproducibility crisis and the prevalence problem. (See Chapter 3 for more on the poor translatability of animal research, Chapter 4 for more on the cost of animal research on animals, and Chapter 5 for more on the financial and human costs of animal research.)

One of the most striking examples of the limitations of animal models is the case of thalidomide. This drug, prescribed to pregnant women for morning sickness in the 1950s and 1960s, caused severe birth defects in thousands of children. No routine testing of reproductive and developmental toxicity was in place when the drug was developed, but the problem may not have been identified even if there had been: thalidomide is teratogenic in humans and certain animal species, such as the New Zealand white rabbit, but not in others, such as mice and rats. Historical control data reveal considerable variability in the sensitivity and outcomes of rabbit prenatal developmental toxicity studies across laboratories and study designs, which can impact the consistent detection of teratogenic effects.[4–7] Retrospectively tested on animals, the teratogenic effects of thalidomide were not detected in most species, highlighting the potential for species differences to mask critical safety issues. Similarly, the arthritis drug Vioxx, which was withdrawn from the market due to an increased risk of heart attacks and strokes, showed no such adverse effects in animal studies. These cases underscore the need for caution when extrapolating animal findings to humans.

Systematic evaluations of animal studies have further revealed their limitations. A 2014 study by Pound and Bracken[8] found that only 37% of animal studies translated to successful human trials, indicating a high failure rate. The Collaborative Approach to Meta-Analysis and Review of Animal Data from Experimental Studies (CAMARADES[9]) initiative has also identified significant methodological flaws and biases in many animal studies, casting doubt on their reliability. These evaluations suggest that relying heavily on animal models may lead to inefficient and misleading drug development and safety testing.

In addition to scientific concerns, animal testing raises significant ethical and economic issues. The use of animals in research is increasingly seen as morally problematic, with growing public opposition to animal experimentation.[10–12] From an economic perspective, animal studies are costly and time-consuming, contributing to the high failure rates and expenses in drug development.[1] The European Union's ban on animal testing for cosmetics resonates with the growing recognition of these ethical and economic considerations.

The reproducibility crisis in biomedical research further compounds the limitations of animal models. Many animal studies have been found to be poorly designed, underpowered, and subject to bias, leading to irreproducible results.[13] This lack of reproducibility undermines the reliability of animal-based findings and contributes to the high attrition rates in drug development. (See Chapter 3 for more information on the lack of reliability and validity of animal research.) Efforts to improve the quality and transparency of animal research, such as the ARRIVE guidelines[14] and the TOP guidelines,[15] aim to address this crisis but have yet to show significant impact.

The prevalence problem poses another challenge for animal models in drug development. The vast majority (about 92%) of drug candidates fail in clinical trials,[16] with only a small fraction making it to market. This "needle in a haystack" problem means that even highly accurate animal models may struggle to identify the rare effective drug candidates among the many failures.[2] Similarly, the low prevalence of rare toxic effects can make them difficult to detect in animal studies.[17] To illustrate the difficulty of finding rare events, consider a hypothetical animal model that is 90% accurate in predicting drug efficacy. If only one in 10,000 drug candidates is truly effective, then out of 10,000 compounds tested, the model would correctly identify the one effective drug but also falsely flag around 999 ineffective compounds as promising. Thus even a highly accurate animal model can be overwhelmed by the sheer rarity of successful drug candidates.

The limitations of animal models in drug development and safety testing are substantial and multifaceted. From ethical concerns to scientific shortcomings, the reliance on animal studies is increasingly being questioned.[13] The reproducibility crisis and the prevalence problem further highlight the challenges of translating animal findings to humans. As we seek to improve the efficiency and human relevance of drug development and safety assessment, it is crucial to explore and embrace alternative approaches, such as those offered by AI and other advanced technologies.

AI as a key enabler for nonanimal approaches

The rapid advancement of AI has opened up new possibilities for developing nonanimal approaches in toxicology and pharmacology. AI's ability to process vast amounts of data, identify patterns, and make predictions has the potential to revolutionize the way we assess chemical safety and discover new drugs.[18,19] This section explores how AI is enabling a shift away from animal testing by leveraging the data revolution, enhancing literature mining, improving predictive modeling, and increasing model interpretability.[20]

The data revolution in toxicology and pharmacology has been a significant driver for the adoption of AI. The increasing availability of large-scale datasets, such as high-throughput screening assays, omics data, and electronic health records, has provided a rich resource for training AI models. Initiatives like the Tox21[21] and ToxCast[22] programs have generated vast amounts of in vitro toxicity data, while efforts like the Illuminating the Druggable Genome Pharos project[23] have curated information on drug targets and pathways. These datasets have enabled the development of more sophisticated and predictive AI models for toxicity assessment and drug discovery.

AI can also play a crucial role in avoiding duplication of animal experiments by facilitating the retrieval of earlier work. NLP techniques can be applied to mine the scientific literature and legacy data, extracting relevant information on previous animal studies and their outcomes. By identifying similar experiments that have already been conducted, researchers can avoid unnecessary repetition and reduce the use of animals. NLP tools can also assist in the curation and annotation of large toxicology databases, making it easier for researchers to access and utilize existing knowledge.

Machine learning, a subset of AI, has shown great promise in predictive toxicology and drug discovery. Techniques such as deep learning can detect complex patterns and relationships from large datasets, enabling the development of predictive models for various toxicity endpoints[24,25] and drug properties.[26–28] For example, convolutional neural networks have been used to predict the toxicity of chemicals based on their structural features,[29] while recurrent neural networks have been applied to model the pharmacokinetics and pharmacodynamics of drugs.[30,31] These AI-based approaches can help prioritize compounds for further testing, identify potential safety issues early in the development process,[32] and guide the design of safer and more effective drugs.

Explainable AI (xAI) is another important aspect of AI that is particularly relevant for nonanimal approaches. Many AI models, especially deep learning models, are considered "black boxes" due to their complex inner workings. This lack of transparency can hinder the acceptance and trust in AI-based predictions, particularly in the context of regulatory decision-making. xAI techniques aim to provide interpretable explanations for AI model outputs, allowing researchers to understand the underlying reasoning and identify potential biases or limitations. By making AI models more transparent and interpretable, xAI can facilitate the validation and adoption of nonanimal methods in toxicology and pharmacology.

AI is making significant contributions to drug discovery through advances in protein structure prediction and molecular docking. DeepMind's AlphaFold, an AI system that predicts 3D protein structures with unprecedented accuracy, has the potential to revolutionize drug design and development.[33,34] By accurately predicting the structure of drug targets, AlphaFold enables researchers to perform virtual screening and molecular docking studies, identifying potential drug candidates without the need for animal testing. Its latest iteration, AlphaFold 3.0, extends from proteins to other biological molecules,[35] further expanding these opportunities. This AI-driven approach can accelerate the drug discovery process, reduce costs, and minimize the use of animals in the early stages of drug development.

Finally, AI promises to identify alternative approaches for research projects that rely on animal testing.[36] By analyzing the scientific literature, research proposals, and funding applications, AI algorithms can suggest nonanimal methods that may be suitable for addressing specific research questions. For example, AI could recommend the use of human-relevant in vitro models, such as OOC systems or 3D cell cultures, based on the biological processes or diseases being studied. By providing researchers with information on available alternative methods and their applicability, AI can help drive the transition toward more human-relevant and ethical research practices.

AI is a key enabler for nonanimal approaches in toxicology and pharmacology. The data revolution, combined with advances in NLP, machine learning, and explainable AI, has created new opportunities for predicting toxicity, discovering drugs, and replacing animal experiments. By harnessing the power of AI responsibly and integrating it with other innovative technologies, we can accelerate the development of more effective, efficient, and ethical approaches to chemical safety assessment and drug development.

Synergies of AI with other new approach methodologies

The potential of AI to transform toxicology and pharmacology is further enhanced by its synergies with other emerging technologies and approaches. By integrating AI with microphysiological systems (MPS), adverse outcome pathways (AOPs), integrated testing strategies, and evidence-based approaches, we can create powerful platforms for predicting toxicity, understanding biological mechanisms, and supporting decision-making. This section will explore these synergies and their implications for advancing nonanimal methods.

MPS, including OOC technologies, are innovative in vitro models that aim to recreate the complex structure and function of human tissues and organs[37–39] (see Chapter 6 for more on this topic). By combining advanced microfluidics, 3D cell culture, and biomaterials, these systems can mimic the physiological microenvironment and capture the interactions between different cell types. The field is progressing and has formed an International Society[40] and a series of annual World Summits.[41] AI can significantly enhance the development and application of MPS and OOC systems by optimizing their design, analyzing the complex data they generate, and predicting the human relevance of the results. For example, machine learning algorithms can be used to model the pharmacokinetics and pharmacodynamics of drugs in these systems, enabling more accurate predictions of in vivo responses.[42,43]

AOPs, also known as pathways of toxicity,[44] provide a mechanistic framework for understanding the biological processes that lead from a molecular initiating event to an adverse outcome at the organism or population level.[45] AOPs aim to formally describe the pathways perturbed by classes of chemicals leading to hazard manifestation, thereby forming a peer-reviewed basis for mechanistic discussions in toxicology and also providing guidance for the design of testing strategies comprehensively mapping relevant mechanisms. AOPs are increasingly being used to guide the development of predictive toxicity models and inform regulatory decision-making. AI can support the development and application of AOPs in several ways. NLP can be used to mine the scientific literature and identify relevant knowledge for constructing AOPs. Machine learning algorithms can help identify patterns and relationships in omics data, supporting the elucidation of key events and their connections.[46–49] AI-based models can also be used to quantitatively predict the outcomes of AOPs, enabling a more predictive and quantitative approach to risk assessment.

Integrated approaches to testing and assessment (IATA) are flexible frameworks that combine multiple sources of information, including in silico models, in vitro assays, and existing data, to inform regulatory decision-making.[50] IATA aim to optimize the use of available data and minimize the reliance on animal testing.[51] AI can play a crucial role in IATA by integrating and analyzing diverse types of data, predicting missing information, and supporting weight-of-evidence assessments. Machine learning models can be used to predict the results of in vitro assays based on chemical structures, reducing the need for actual testing. AI-based systems can also assist in the selection of the most informative assays for a given chemical or endpoint, optimizing the use of resources and minimizing animal use.

Evidence-based approaches, such as systematic reviews and metaanalyses, are increasingly being applied in toxicology and pharmacology to synthesize and critically appraise the available evidence on a specific topic.[52–55] These approaches help to identify gaps in knowledge, assess the reliability of existing studies, and support evidence-based decision-making. AI can streamline and enhance evidence-based approaches by automating the literature search and screening process, extracting relevant data from studies,[56,57] and assisting in the quality assessment of the evidence. NLP tools can be used to identify and extract key information from scientific publications, reducing the time and effort required for manual review.[58] Machine learning algorithms can also be applied to detect biases and inconsistencies in the evidence, supporting a more rigorous and transparent evaluation process.

The synergies between AI and these new approach methodologies offer exciting opportunities for advancing nonanimal methods and improving the efficiency and effectiveness of toxicology and pharmacology research. By leveraging the strengths of each approach and integrating them into a cohesive framework, we can create powerful platforms for predicting toxicity, understanding biological mechanisms, and supporting decision-making. For example, AI-based models trained on data from MPS such as OOC systems can provide more accurate and human-relevant predictions of drug safety and efficacy. An example is our work on organoid intelligence,[59,60] where the combination of AI with brain organoids allows simple forms of learning and memory. AOPs can guide the development of mechanistically relevant AI models and help interpret their predictions in a biological context. IATA can incorporate AI-based predictions alongside other sources of evidence to support a more comprehensive and reliable assessment of chemical safety.

To fully realize the potential of these synergies, it is essential to foster multidisciplinary collaboration and data sharing among researchers, industry, and regulatory agencies. The development of standardized data formats, ontologies, and reporting guidelines can facilitate the integration and reuse of data across different platforms and domains (see Chapter 17 for more on this topic). Investment in infrastructure and training programs can help build the necessary capacity and expertise to effectively implement these new approaches.[61]

The synergies between AI and other new approach methodologies, such as MPS,[7] AOPs,[44] integrated testing strategies,[62,63] and evidence-based approaches,[64] hold immense promise for advancing nonanimal methods in toxicology and pharmacology. By harnessing these synergies and promoting multidisciplinary collaboration, we can create a more predictive, mechanistically informed, and evidence-based framework for assessing chemical safety and developing new drugs. This paradigm shift toward integrating AI with innovative nonanimal approaches has the potential to revolutionize the field, reducing the reliance on animal testing while improving the efficiency, effectiveness, and human relevance of toxicology and pharmacology research.[13,18]

AI-based alternative methods for Green Toxicology and front-loading toxicology in drug development

The integration of AI with nonanimal approaches not only reduces the reliance on animal testing but also enables the implementation of Green Toxicology principles and the front-loading of toxicology in drug development.[32] Green Toxicology is an emerging field that focuses on the design and application of safer, more sustainable chemicals and products.[65,66] It aims to minimize the environmental and health impacts of chemicals throughout their life cycle, from production to use and disposal.[67] AI-based alternative methods can support Green Toxicology by enabling the early prediction of toxicity and the prioritization of safer chemical design.

One of the key aspects of Green Toxicology[68] is the shift toward a proactive, rather than reactive, approach to chemical safety assessment. Traditional toxicology often relies on animal testing to identify potential hazards after a chemical has been developed, leading to a "test and fix" approach. In contrast, Green Toxicology emphasizes the importance of designing safer chemicals from the start, using predictive tools and alternative methods to guide the development process. AI can play a crucial role in this paradigm shift by providing rapid, cost-effective, and human-relevant predictions of toxicity early in the chemical design phase.

Machine learning models trained on large datasets of chemical structures and toxicity endpoints can be used to screen virtual libraries of compounds and identify those with the most favorable safety profiles. This in silico approach allows researchers to prioritize the synthesis and testing of chemicals that are predicted to have lower toxicity, reducing the need for animal testing and minimizing the potential for adverse effects. AI-based models can also be used to optimize chemical structures for desired properties, such as improved efficacy and reduced toxicity, guiding the design of safer and more sustainable chemicals.

In the context of drug development, AI-based alternative methods can enable the front-loading of toxicology, where safety assessment is performed earlier in the discovery and preclinical stages. Traditionally, toxicology studies are conducted late in the drug development process, often after significant investments have been made in a particular compound. This approach can lead to costly failures and delays when toxicity issues are identified at later stages. By using AI-based methods to predict toxicity early on, drug developers can make informed decisions about which compounds to advance and which to deprioritize, saving time and resources.

Front-loading toxicology with AI can involve the integration of multiple alternative approaches, such as in silico models, in vitro assays, and read-across methods.[69] For example, machine learning models can be used to predict the results of in vitro assays based on chemical structure, reducing the need for actual testing. NLP can be applied to mine literature and databases for relevant toxicity information, supporting read-across and weight-of-evidence assessments.[70]

The front-loading of toxicology with AI also enables a more iterative and adaptive approach to drug development. By generating toxicity predictions early and often, drug developers can make data-driven decisions and adjust their strategies as needed. This agile approach can help identify and address potential safety issues before they become significant problems, reducing the risk of late-stage failures and improving the overall success rate of drug development.

AI-based alternative methods enable the implementation of Green Toxicology principles and the front-loading of toxicology in drug development. By providing rapid, cost-effective, and human-relevant predictions of toxicity early in the chemical design and drug discovery process, AI can guide the development of safer and more sustainable chemicals and medicines. The integration of AI with other alternative approaches, such as in vitro assays and OOC systems, can create a comprehensive and efficient testing strategy that minimizes the use of animals and supports data-driven decision-making. As we move toward a more proactive and sustainable approach to chemical safety assessment and drug development, the role of AI in enabling Green Toxicology and front-loading toxicology will become increasingly important, driving innovation and progress in these fields.

Challenges and future directions

AI offers tremendous potential for advancing nonanimal approaches in toxicology and pharmacology; however, it also faces challenges related to responsible and effective implementation, data quality, bias, and availability, validation and regulatory acceptance, responsible and human-centered AI development, and the need for multidisciplinary collaboration.

Most progress in recent years has been made by increasing data use and computational capacity, resulting in enormous energy needs, water use, and carbon footprint. Progress has already been made with more efficient algorithms, smaller models, and the prospect of quantum computing, but it will be important to monitor the ecological consequences of these technologies and provide incentives for more energy-efficient approaches.

One of the most significant challenges in applying AI to toxicology and pharmacology is the quality, bias, and availability of data. AI models are only as good as the data they are trained on, and the lack of high-quality, diverse, and representative datasets can limit their performance and generalizability. Many existing toxicology databases suffer from data quality issues, such as inconsistent reporting, missing values, and a lack of standardization. Bias in data collection and labeling can also lead to AI models that perpetuate or amplify existing biases, resulting in unfair or inaccurate predictions. Addressing these challenges requires the development of robust data quality standards, data curation and harmonization efforts, and the use of techniques such as data augmentation and transfer learning to leverage limited data.

Validation and regulatory acceptance of AI models pose another significant challenge. Regulators and decision-makers need to have confidence in the reliability, reproducibility, and interpretability of AI-based predictions.[71] However, the complexity and opaqueness of many AI models can make it difficult to assess their validity and suitability for regulatory purposes. There is a need for clear validation frameworks and guidelines that assess the performance, robustness, and domain of applicability of AI models. Collaborative efforts between researchers, industry, and regulatory agencies are essential to establish best practices and build trust in AI-based approaches. The development of explainable AI techniques and the use of mechanistic information, such as AOPs, can also help to increase the interpretability and acceptability of AI models.

Responsible and human-centered AI development is another critical challenge that needs to be addressed. AI systems should be designed and deployed in a way that respects human values, rights, and welfare.[72] This includes ensuring the transparency, accountability, and fairness of AI models, and considering the ethical implications of their use. Engaging diverse stakeholders, including toxicologists, pharmacologists, computer scientists, ethicists, and the public, in the development and governance of AI systems is essential to ensure that they align with societal values and priorities.[73] The development of AI literacy programs and the integration of AI ethics into toxicology and pharmacology curricula can help to foster a culture of responsible innovation.[61]

Multidisciplinary collaboration is key to realizing the full potential of AI in advancing nonanimal approaches. The development and application of AI in toxicology and pharmacology require expertise from multiple domains, including computer science, data science, toxicology, pharmacology, and systems biology. Fostering collaboration and knowledge exchange among these disciplines is essential to ensure that AI models are based on sound scientific principles and are fit for purpose. The establishment of interdisciplinary research centers, consortia, and networks can facilitate the sharing of data, methods, and best practices, and promote the development of innovative AI-based solutions. Collaborative efforts can also help to identify and prioritize research needs and to develop coordinated strategies for advancing nonanimal approaches.

Looking to the future, there are several promising directions for the application of AI in toxicology and pharmacology. One area of active research is the development of federated learning approaches,[74,75] which allow AI models to be trained on decentralized datasets without the need for data sharing. This can help to overcome data privacy and confidentiality concerns and enable the development of more robust and generalizable models. The use of AI for drug repurposing[76] and personalized medicine[77,78] is also an area of increasing interest, leveraging large-scale omics data and electronic health records to identify new therapeutic indications and optimize treatment strategies. While not replacing animal testing directly, these areas demonstrate that the new approach methods are enabling technologies, allowing more than can be done with animals and thereby building trust in these methods.

To fully realize the potential of AI in advancing nonanimal approaches, sustained investment in research, infrastructure, and training is essential. This includes funding for the development and maintenance of high-quality toxicology and pharmacology databases, the establishment of data standards and ontologies, and the creation of AI-ready data pipelines. Investment in computational infrastructure, such as high-performance computing systems and cloud computing platforms, is also necessary to support the training and deployment of large-scale AI models. The development of educational programs and training opportunities in AI and data science for toxicologists and pharmacologists is crucial to building the necessary skills and expertise to effectively leverage these technologies.

While there are significant challenges to be addressed, the future of AI in advancing nonanimal approaches in toxicology and pharmacology is bright. By addressing issues related to data quality, bias, and availability; validation and regulatory acceptance; responsible and human-centered AI development; and multidisciplinary collaboration, we can create a more robust and reliable framework for the application of AI in these fields. Sustained investment in research, infrastructure, and training, coupled with a commitment to responsible innovation and stakeholder engagement, will be key to realizing the full potential of AI in transforming toxicology and pharmacology and reducing the reliance on animal testing. As we move forward, it is essential to remain focused on the ultimate goal of developing more predictive, efficient, and human-relevant approaches to assess the safety and efficacy of chemicals and drugs, while ensuring the protection of human health and the environment.

Conclusion

The rapid advancements in AI have opened up new avenues for transforming toxicology and pharmacology, particularly in the development of nonanimal approaches. AI's ability to process vast amounts of data, identify complex patterns, and generate predictive models has the potential to revolutionize the way we assess chemical safety and develop new drugs. By leveraging the power of machine learning, NLP, and other AI techniques, we can create more efficient, predictive, and human-relevant testing strategies that minimize the reliance on animal experimentation.

However, it is important to recognize that AI is not a standalone solution, but rather an enabling tool that must be used alongside other advancements in the field. The integration of AI with emerging technologies, such as MPS and advanced in vitro assays, can create powerful testing platforms that better recapitulate human biology and provide more reliable predictions of toxicity and efficacy. AI can also be used to enhance the development and application of AOPs, providing a mechanistic framework for understanding the biological processes underlying toxic effects.

The vision of an AI-based toxicology copilot is an exciting prospect, where AI systems can assist researchers and decision-makers in navigating the complex landscape of chemical safety assessment. Tox chatbots, powered by NLP and knowledge graphs, could provide instant access to relevant toxicology information and guidance, streamlining the research process and facilitating the adoption of nonanimal methods. AI-driven drug discovery, as demonstrated by companies like Insilico Medicine[79]—the first company to bring an AI-first drug candidate into human phase II trials—has the potential to accelerate the development of new therapies by identifying promising drug candidates and optimizing their properties using machine learning algorithms.

As we embrace the potential of AI in advancing nonanimal approaches, it is crucial to recognize the importance of continued improvement in primary evidence generation. The quality, reliability, and relevance of the data used to train and validate AI models are critical factors in determining their performance and applicability. Efforts to enhance data quality, standardization, and reporting, such as the implementation of FAIR (Findable, Accessible, Interoperable, and Reusable) data principles, are essential to ensure the robustness and reproducibility of AI-based approaches. The integration of AI with evidence-based approaches, such as systematic reviews and metaanalyses, can help identify and prioritize the most relevant and reliable evidence for informing decision-making.

Open science and data sharing are also key enablers for the successful application of AI in toxicology and pharmacology. The development of large-scale, diverse, and high-quality datasets is essential for training and validating AI models that can generalize across different chemical spaces and biological contexts. Collaborative efforts to create shared data resources, such as the Tox21 and ToxCast programs, can accelerate the development and adoption of AI-based approaches. The promotion of data sharing and transparency can also foster trust and reproducibility, ensuring that the benefits of AI are realized in a responsible and equitable manner.

The merging of AI and evidence-based approaches represents a powerful paradigm for advancing toxicology and pharmacology. By combining the predictive power of AI with the rigor and transparency of systematic review and metaanalysis, we can create a more comprehensive and reliable framework for assessing chemical safety and developing new drugs. This integration can help identify gaps in knowledge, prioritize research needs, and support evidence-based decision-making, ultimately leading to more effective and efficient nonanimal testing strategies.

As we move forward, it is essential to embrace new technologies while upholding scientific rigor and ethical principles. The responsible development and deployment of AI in toxicology and pharmacology requires ongoing collaboration among researchers, industry, regulators, and other stakeholders. The establishment of best practices, validation frameworks, and governance structures is necessary to ensure the transparency, accountability, and fairness of AI-based approaches. Engaging diverse perspectives, including those of toxicologists, computer scientists, ethicists, and the public, can help align the development of AI with societal values and priorities.

The integration of AI with nonanimal approaches represents a transformative opportunity for toxicology and pharmacology. By leveraging the power of machine learning, NLP, and other AI techniques, we can create more predictive, efficient, and human-relevant testing strategies that reduce reliance on animal experimentation. The vision of an AI-based toxicology co-pilot, tox chatbots, and accelerated drug discovery is within reach, but it requires continued investment in research, infrastructure, and training, as well as a commitment to open science, data sharing, and evidence-based approaches. As we embrace these new technologies, we must do so with scientific rigor, ethical responsibility, and a focus on the ultimate goal of protecting human health and the environment.

Acknowledgment

This work has received funding from the European Union's Horizon 2020 research and innovation program under grant agreement No. 963845 (ONTOX).

Data availability

No datasets were generated in this study.

Conflict of interest

Thomas Luechtefeld is the founder and owner of Tox-Track Inc. and Insilica LLC. He, along with Thomas Hartung, serves as a consultant for computational toxicology for Underwriters Laboratories (UL) and receives shares from their respective sales. Thomas Hartung is a member of Apple's Green Chemistry Advisory Board. He also holds stock options in, and consults for, Tox-Track Inc. and Insilica LLC.

References

[1] Meigs L, Smirnova L, Rovida C, Leist M, Hartung T. Animal testing and its alternatives—the most important omics is economics. *ALTEX*. 2018;35:275–305. doi:10.14573/altex.1807041.
[2] Hartung T. The (misleading) role of animal models in drug development. *Front Drug Discov*. 2024;4:1355044. doi:10.3389/fddsv.2024.1355044.
[3] Schmeisser S, Miccoli A, von Bergen M, et al. New approach methodologies in human regulatory toxicology—not if, but how and when! *Environ Int*. 2023;178:108082. doi:10.1016/j.envint.2023.108082.
[4] Feussner EL, Lightkep GE, Hennesy RA, Hoberman AM, Christian MS. A decade of rabbit fertility data: study of historical control animals. *Teratology*. 1992;46:349–365. doi:10.1002/tera.1420460406.
[5] Ema M, Aoyama H, Arima A, et al. Historical control data on prenatal developmental toxicity studies in rabbits. *Congenit Anom (Kyoto)*. 2012;52:155–161. doi:10.1111/j.1741-4520.2012.00365.x.
[6] Basketter DA, Clewell H, Kimber I, et al. A roadmap for the development of alternative (nonanimal) methods for systemic toxicity testing. *ALTEX*. 2012;29:3–91. doi:10.14573/altex.2012.1.003.
[7] Smirnova L, Kleinstreuer N, Corvi R, et al. 3S—systematic, systemic, and systems biology and toxicology. *ALTEX*. 2018;35:139–162. doi:10.14573/altex.1804051.
[8] Pound P, Bracken MB. Is animal research sufficiently evidence based to be a cornerstone of biomedical research? *BMJ*. 2014;348:g3387. doi:10.1136/bmj.g3387.
[9] Collaborative approach to meta analysis and review of animal data from experimental studies (CAMARADES). University of Edinburgh. n.d. Available at: https://www.research.ed.ac.uk/en/organisations/collaborative-approach-to-meta-analysis-and-review-of-animal-data/. Accessed 9 August 2026.
[10] Physicians Committee for Social Responsibility. U.S. gen pop beliefs on animal testing; n.d. Available at: https://pcrm.widen.net/s/qzfxtfh7bw/animal-testing-survey. Accessed 23 May 2025.
[11] IPSOS.com. Public atttitudes to animal research 2018. London UK; n.d. Available at: https://www.ipsos.com/en-uk/public-attitudes-animal-research-2018. Accessed 9 August 2026.
[12] Funk C, Rainie L. Chapter 7: Opinion about the use of animals in research. Pew Research Center; 2015. Available at: https://www.pewresearch.org/science/2015/07/01/chapter-7-opinion-about-the-use-of-animals-in-research/ Accessed 23 May 2025.
[13] Hartung T. Look back in anger—what clinical studies tell us about preclinical work. *ALTEX*. 2013;30:275–291. doi:10.14573/altex.2013.3.275.
[14] The ARRIVE guidelines. n.d. Available at: https://arriveguidelines.org. Accessed 23 May 2025.
[15] Center for Open Science (COS). The TOP guidelines were created by journals, funders and societies to align scientific ideals with practices; 2025. Available at: https://www.cos.io/initiatives/top-guidelines. Accessed 23 May 2025.
[16] PharmaIntelligence Informa: qualitative life sciences. Clinical development success rates and contributing factors 2011-2020; n.d. Available at: https://go.bio.org/rs/490-EHZ-999/images/ClinicalDevelopmentSuccessRates2011_2020.pdf. Accessed 23 May 2025.

[17] Hoffmann S, Hartung T. Diagnosis: toxic!—trying to apply approaches of clinical diagnostics and prevalence in toxicology considerations. *Toxicol Sci.* 2005;85:422–428. doi:10.1093/toxsci/kfi099.
[18] Hartung T. ToxAIcology—the evolving role of artificial intelligence in advancing toxicology and modernizing regulatory science. *ALTEX.* 2023;40:559–570. doi:10.14573/altex.2309191.
[19] Hartung T. AI as the new frontier in chemical risk assessment. *Front Artif Intell.* 2023a;40:559–570. doi:10.3389/frai.2023.1269932.
[20] Kleinstreuer N, Hartung T. Artificial intelligence (AI) – it's the end of the tox as we know it (and I feel fine)—AI for predictive toxicology. *Arch Toxicol.* 2024;98:735–754. doi:10.1007/s00204-023-03666-2.
[21] Tox21: toxicology in the 21st century. n.d. Available at: https://tox21.gov/data-and-tools/. Accessed 23 May 2025.
[22] United States Environmental Protection Agency. Toxicity forecasting (ToxCast); n.d. Available at: https://www.epa.gov/comptox-tools/toxicity-forecasting-toxcast. Accessed 23 May 2025.
[23] National Institutes of Health. Illuminating the druggable genome (IDG); n.d. Available at: https://ncats.nih.gov/research/research-activities/idg. Accessed 9 August 2026.
[24] Hartung T. Making big sense from big data in toxicology by read-across. *ALTEX.* 2016;33:83–93. doi:10.14573/altex.1603091.
[25] Luechtefeld T, Hartung T. Computational approaches to chemical hazard assessment. *ALTEX.* 2017;34:459–478. doi:10.14573/altex.1710141.
[26] Fleming N. How artificial intelligence is changing drug discovery. *Nature.* 2018;557(7707):S55–S57. doi:10.1038/d41586-018-05267-x.
[27] Blanco-González A, Cabezón A, Seco-González A, et al. The role of AI in drug discovery: challenges, opportunities, and strategies. *Pharmaceuticals.* 2023;16:891. doi:10.3390/ph16060891.
[28] Hasselgren C, Oprea TI. Artificial intelligence for drug discovery: are we there yet? *Annu Rev Pharmacol Toxicol.* 2024;64:527–550. doi:10.1146/annurev-pharmtox-040323-040828.
[29] Luechtefeld T, Marsh D, Rowlands C, Hartung T. Machine learning of toxicological big data enables read-across structure activity relationships (RASAR) outperforming animal test reproducibility. *Toxicol Sci.* 2018;165:198–212. doi:10.1093/toxsci/kfy152.
[30] Liu X, Liu C, Huang R, et al. Long short-term memory recurrent neural network for pharmacokinetic-pharmacodynamic modeling. *Int J Clin Pharmacol Ther.* 2021;59:138–146. doi:10.5414/CP203800.
[31] Tang A. Machine learning for pharmacokinetic/pharmacodynamic modeling. *J Pharm Sci.* 2023;112:1460–1475. doi:10.1016/j.xphs.2023.01.010.
[32] Maertens A, Luechtefeld T, Hartung T. Alternative methods go green! Green toxicology as a sustainable approach for assessing chemical safety and designing safer chemicals. *ALTEX.* 2024;41:3–19. doi:10.14573/altex.2312291.
[33] Karelina M, Noh JJ, Dror RO. How accurately can one predict drug binding modes using AlphaFold models? *ELife.* 2023;12:RP89386.
[34] Arnold C. AlphaFold touted as next big thing for drug discovery—but is it? *Nature.* 2023;622(7981):15–17. doi:10.1038/d41586-023-02984-w.
[35] Abramson J, Adler J, Dunger J, et al. Accurate structure prediction of biomolecular interactions with AlphaFold 3. *Nature.* 2024;630(8016):493–500. doi:10.1038/s41586-024-07487-w.
[36] National Institutes of Health. Complement animal research in experimentatation (complement-ARIE) program (team Insilica); n.d. Available at: https://commonfund.nih.gov/complementarie/challengewinnersummaries#namkg Accessed 23 May 2025.
[37] Marx U, Andersson TB, Bahinski A, et al. Biology-inspired microphysiological system approaches to solve the prediction dilemma of substance testing using animals. *ALTEX.* 2016;33:272–321. doi:10.14573/altex.1603161.
[38] Marx U, Akabane T, Andersson TB, et al. Biology-inspired microphysiological systems to advance medicines for patient benefit and animal welfare. *ALTEX.* 2020;37:364–394. doi:10.14573/altex.2001241.
[39] Roth A, Berlin MPS-WS. Human microphysiological systems for drug development. *Science.* 2021;373:1304–1306. doi:10.1126/science.abc3734.
[40] International MPS Society. Connect, exchange, educate. n.d. https://impss.org. Accessed 23 May 2025.
[41] Microphysiological Systems World Summit. 2025. https://mpsworldsummit.com. Accessed 23 May 2025.
[42] Young AT, Rivera KR, Erb PD, Daniele MA. Monitoring of microphysiological systems: integrating sensors and real-time data analysis toward autonomous decision-making. *ACS Sens.* 2019;4:1454–1464.

[43] Tronolone JJ, Mathur T, Chaftari CP, et al. Evaluation of the morphological and biological functions of vascularized microphysiological systems with supervised machine learning. *Ann Biomed Eng*. 2023;51:1723–1737. doi:10.1007/s10439-023-03177-2.

[44] Kleensang A, Maertens A, Rosenberg M, et al. Pathways of toxicity. *ALTEX*. 2014;31:53–61. doi:10.14573/altex.1309261.

[45] Leist M, Ghallab A, Graepel R, et al. Adverse outcome pathways: opportunities, limitations and open questions. *Arch Toxicol*. 2017;91(11):3477–3505. doi:10.1007/s00204-017-2045-3.

[46] Nicora G, Vitali F, Dagliati A, Geifman N, Bellazzi R. Integrated multi-omics analyses in oncology: a review of machine learning methods and tools. *Front Oncol*. 2020;10:1030. doi:10.3389/fonc.2020.01030.

[47] Reel PS, Reel S, Pearson E, Trucco E, Jefferson E. Using machine learning approaches for multi-omics data analysis: a review. *Biotechnol Adv*. 2021;49:107739. doi:10.1016/j.biotechadv.2021.107739.

[48] Li R, Li L, Xu Y, Yang J. Machine learning meets omics: applications and perspectives. *Brief Bioinform*. 2022;23:bbab460.

[49] Arjmand B, Hamidpour SK, Tayanloo-Beik A, et al. Machine learning: a new prospect in multi-omics data analysis of cancer. *Front Genet*. 2022;13:824451. doi:10.3389/fgene.2022.824451.

[50] Caloni F, De Angelis I, Hartung T. Replacement of animal testing by integrated approaches to testing and assessment (IATA): a call for in vivitrosi. *Arch Toxicol*. 2022;96:1935–1950. doi:10.1007/s00204-022-03299-x.

[51] Tollefsen KE, Scholz S, Cronin MT, et al. Applying adverse outcome pathways (AOPs) to support integrated approaches to testing and assessment (IATA). *Reg Toxicol Pharmacol*. 2014;70:629–640. doi:10.1016/j.yrtph.2014.09.009.

[52] Bozada T, Borden J, Workman J, et al. Sysrev: a FAIR platform for data curation and systematic evidence review. *Front Artif Intell*. 2021;4:685298. doi:10.3389/frai.2021.685298.

[53] Farhat N, Tsaioun K, Saunders-Hastings P, et al. Systematic review in evidence-based risk assessment. *ALTEX*. 2022;39:463–479. doi:10.14573/altex.2004111.

[54] Krewski D, Saunders-Hastings P, Baan R, et al. Workshop report: development of an evidence-based risk assessment framework. *ALTEX*. 2022;39:667–693. doi:10.14573/altex.2004041.

[55] Hartung T, Tsaioun K. Evidence-based approaches in toxicology: their origins, challenges, and future directions. *Evid Based Toxicol*. 2024;2(1):2421187. doi:10.1080/2833373X.2024.2421187.

[56] Fabiano N, Gupta A, Bhambra N, et al. How to optimize the systematic review process using AI tools. *JCPP Adv*. 2024;4:e12234. doi:10.1002/jcv2.12234.

[57] Ofori-Boateng R, Aceves-Martins M, Wiratunga N, et al. Towards the automation of systematic reviews using natural language processing, machine learning, and deep learning: a comprehensive review. *Artif Intell Rev*. 2024;57:200. doi:10.1007/s10462-024-10844-w.

[58] van Dijk SHB, Brusse-Keizer MGJ, Bucsán CC, et al. Artificial intelligence in systematic reviews: promising when appropriately used. *BMJ Open*. 2023;13:e072254. doi:10.1136/bmjopen-2023-072254.

[59] Smirnova L, Caffo BS, Gracias DH, et al. Organoid intelligence (OI): the new frontier in biocomputing and intelligence-in-a-dish. *Front Sci*. 2023a;(1):1017235. doi:10.3389/fsci.2023.1017235.

[60] Smirnova L, Morales Pantoja IE, Hartung T. Organoid Intelligence (OI)—the ultimate functionality of a brain microphysiological system. *ALTEX*. 2023b;40:191–203. doi:10.14573/altex.2303261.

[61] Hartung T, Navas-Acien A, Chiu WA. Future directions workshop: advancing the next scientific revolution in toxicology; 2023. Available at: https://basicresearch.defense.gov/Portals/61/Documents/future-directions/Future%20Directions%20Workshop%20-%20Advancing%20the%20Next%20Scientific%20Revolution%20in%20Toxicology.pdf?ver=q0_CyJCAT-aj4HVv_W0a9Q%3d%3d. Accessed 23 May 2025.

[62] Hartung T, Luechtefeld T, Maertens A, Kleensang A. Integrated testing strategies for safety assessments. *ALTEX*. 2013;30:3–18. doi:10.14573/altex.2013.1.003.

[63] Rovida C, Alépée N, Api AM, et al. Integrated testing strategies (ITS) for safety assessment. *ALTEX*. 2015;32:171–181. doi:10.14573/altex.1506201.

[64] Hoffmann S, Aiassa E, Angrish M, et al. Application of evidence-based methods to construct mechanism-driven chemical assessment frameworks. *ALTEX*. 2022;39:499–518. doi:10.14573/altex.2202141.

[65] Maertens A, Anastas N, Spencer PJ, et al. Green toxicology. *ALTEX*. 2014;31:243–249.

[66] Crawford SE, Hartung T, Hollert H, et al. Green toxicology: a strategy for sustainable chemical and material development. *Environ Sci Eur*. 2017;29:16. doi:10.1186/s12302-017-0115-z.

[67] Maertens A, Hartung T. Green toxicology—know early about and avoid toxic product liabilities. *Toxicol Sci*. 2018;161:285–289. doi:10.1093/toxsci/kfx243.

[68] Maertens A. Green toxicology: making chemicals benign by design. *R Soc Chem*;2022. doi:10.1039/9781839164392.
[69] Ball N, Cronin MTD, Shen J, et al. Toward good read-across practice (GRAP) guidance. *ALTEX*. 2016;33:149–166. doi:10.14573/altex.1601251.
[70] Linkov I, Massey O, Keisler J, Rusyn I, Hartung T. From "weight of evidence" to quantitative data integration using multicriteria decision analysis and Bayesian methods. *ALTEX*. 2015;32:3–8. doi:10.14573/altex.1412231.
[71] Leist M, Hasiwa M, Daneshian M, Hartung T. Validation and quality control of replacement alternatives—current status and future challenges. *Toxicol Res*. 2012;1(8). doi:10.1039/C2TX20011B.
[72] Singer P, Fai TY AI ethics: beyond the species boundary. A lecture by Professor Peter Singer; n.d. Available at: https://www.youtube.com/watch?v=_TqiRDONSiY. Accessed 13 October 2024.
[73] Ayinla BS, Amoo OO, Atadoga A, et al. Ethical AI in practice: balancing technological advancements with human values. *Int J Sci Res Arch*. 2024;11(01):1311–1326. doi:10.30574/ijsra.2024.11.1.0218.
[74] Rapicavoli RV, Alaimo S, Ferro A, Pulvirenti A. Computational methods for drug repurposing. *Adv Exp Med Biol*. 2022;1361:119–141. doi:10.1007/978-3-030-91836-1_7.
[75] Bassani D, Brigo A, Andrews-Morger A. Federated learning in computational toxicology: an industrial perspective on the Effiris hackathon. *Chem Res Toxicol*. 2023;36:1503–1517. doi:10.1021/acs.chemrestox.3c00137.
[76] Ghandikota SK, Jegga AG. Application of artificial intelligence and machine learning in drug repurposing. *Prog Mol Biol Transl Sci*. 2024;205:171–211. doi:10.1016/bs.pmbts.2024.03.030.
[77] Schork NJ. Artificial intelligence and personalized medicine. *Cancer Treat Res*. 2019;178:265–283. doi:10.1007/978-3-030-16391-4_11.
[78] Singh AV, Chandrasekar V, Paudel N, et al. Integrative toxicogenomics: advancing precision medicine and toxicology through artificial intelligence and OMICs technology. *Biomed Pharmacother*. 2023;163:114784. doi:10.1016/j.biopha.2023.114784.
[79] Insilico medicine. n.d. Available at: https://insilico.com.

CHAPTER

10

Biobanks: Yielding extraordinary returns for human health and disease

Victoria Corbit, Azra Hasan, Thomas J. Bell, and Melissa W. VonDran

National Disease Research Interchange (NDRI), Philadelphia, PA, United States

Abstract

Human tissues serve as an ideal model system for expanding our understanding of human health and disease because they allow researchers to connect mechanistic findings from cell models with relevant applications in humans. Access to human biospecimens is made possible by biobanks that procure, store, and distribute samples for research. The types of samples available in biorepositories around the world vary widely in sample type, donor demographics, and preservation methods, enabling research across a range of areas. This chapter describes key information for researchers interested in conducting research with human biospecimens. We describe the following: (1) the options for obtaining human tissue for research and the different types of bioresources that exist; (2) ethical and legal requirements for human biospecimen collections and implications for biobank governance, including informed consent, and (3) innovations within biospecimen and biorepository science that are enhancing rigor, reproducibility, and the increased use of human biospecimens for research.

Keywords: Human tissue; Biobank; Informed consent; Biobank governance; Ethical, Legal, and Social Implications (ELSI)

Introduction

The overarching goal of most biomedical research efforts is to advance our understanding of human health and disease. To that end, using human biospecimens to validate the underpinnings of research hypotheses continues to prevail as the gold standard in translational research. Not only are they used in experimental models, but they also connect investigational findings from other experimental models to well-defined biological mechanisms in human physiology and pathology. Ongoing innovations in molecular techniques and genetic data analysis have propelled single-cell and spatial analyses on human tissues from all body systems to evolve into standard laboratory procedures, thereby accelerating the ability of investigators to shed light on genetic variations in human disease and etiology. Biobanks—entities that

https://doi.org/10.1016/B978-0-443-26514-3.00003-5

collect, store, process, and distribute human biospecimens—are a vital resource for investigators who require human tissue for their studies. Policies and procedures in biobanking, from consent to biospecimen storage, are progressing in tandem with these groundbreaking experimental advances to appropriately align ethical and technical practices for donor families and scientists. Collectively, experimental and ethical advancements in studies utilizing human biospecimens have thrust the field of biobanking to an inflection point. This chapter reviews the foundational elements and challenges for bioresources to meet the emerging needs of biomedical research by procuring, processing, storing, and managing the distribution of suitable biospecimens to scientists studying human biology, physiology, diseases, and disorders.

The use of human anatomical samples to understand health and disease dates back to 300 BC in Greece. In the United States, human biospecimens have been stored for research purposes for over 100 years, initially in hospital pathology departments.[1] One of the first large-scale initiatives to procure biospecimens exclusively for research was the Framingham Heart Study, which has been collecting samples and clinical data from over 5000 participants in the Framingham, Massachusetts area since 1948.[2] Since then, numerous biobanks around the world have been established, providing a wide range of sample types for scientists to advance their translational research. Biobanks, also called biorepositories, maintain human biospecimens and associated data that vary widely in sample type, donor demographics, and preservation methods, offering many opportunities for scientists to use human samples to advance their translational work.

The use of human tissue in experimental models, or as an experimental model system itself, eliminates a caveat ubiquitous in animal research—that experimental results from animals may only apply to that species. Findings from experiments conducted on human samples are inherently more relevant to human health and disease than results from animal models. Uniquely human traits are present across biological systems and may contribute to a widespread disconnect between animal experiments and clinical trials in humans,[3] as described at length in Chapter 3. One consequence of these species-specific differences is that many human diseases do not have a fully valid animal correlation, and partially valid models may approximate certain symptoms but do not encompass an entire disorder. Thus the conclusions that can be drawn about human health and disease from animal experiments are limited.

While using human biospecimens greatly increases the translational relevance of research, one of the biggest hurdles to this approach is access to samples. The public generally supports the donation of biospecimens in the interest of scientific progress.[4] Surveys have found that patient cohorts are motivated to donate specimens and believe that biospecimens are needed to develop new drugs or treatments that could benefit them or their families in the future.[5] Nevertheless, it can be challenging for investigators to find suitable samples to fit their purpose. Concerns about experimental capabilities with human tissue models may be part of the reason animal use remains central to biomedical research.[6] However, many scientists likely overestimate the barriers to working with human biospecimens, and some investigators may not even know that such resources exist. As the use of human biospecimens for research increases globally, education and resources for these approaches will continue to grow, as will the understanding of the benefits of approaches involving them.

In this chapter, we describe key information for those interested in research using human biospecimens, including: (1) options for obtaining human tissue for research and the different types of bioresources that exist; (2) ethical and legal requirements for human biospecimen collections and implications for biobank governance, including informed consent; and (3) innovations within biospecimen and biorepository science that are enhancing rigor, reproducibility, and the increased use of human biospecimens for research.

Biobanks are sources of human biospecimens for biomedical research

The mission of a biobank is to provide high-quality biospecimens for research, which requires biobanks to follow standards in sample and data collection, processing, storage, and distribution. Several organizations have developed standards, including the International Society for Biological and Environmental Repositories (ISBER), the Office of Biorepositories and Biospecimen Research, and the International Organization for Standardization (ISO), to ensure best practices for managing biospecimen storage and distribution to support rigor in researchers' downstream analyses.[7–9] When scientists use human biospecimens, it is critical to consider behavioral, genetic, and experiential factors of the donor that may affect experimental results. The amount and types of data associated with biospecimens vary with the biobank and the intended use of biospecimens, but can include basic demographics, information from health records or behavioral reports, and tissue recovery and processing details.[10] Efforts to standardize and enhance donor metadata collection and management for biobanks have significantly increased in recent years.[11] The following sections will explain the processes for procuring biospecimens from living and deceased donors and summarize the different types of bioresources that are created for different purposes (Fig. 1).

Living donor specimen collections

Biospecimens collected from living donors fall into two categories: 1) bodily fluids or biopsies collected specifically for research (e.g., blood draws, buccal swabs, skin biopsies) or 2) residual tissues following routine clinical care procedures. Biofluids (e.g., blood, saliva,

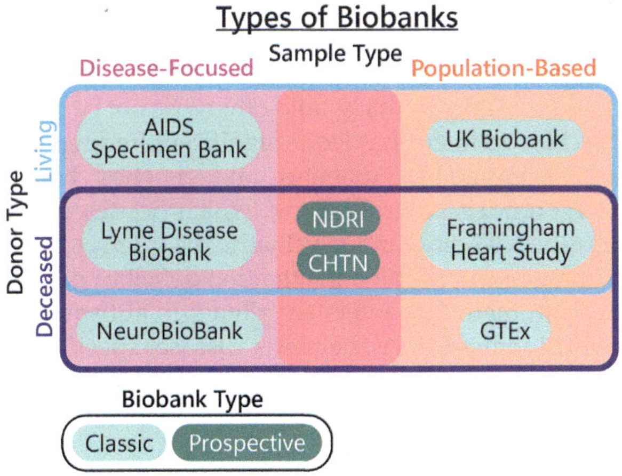

FIG. 1 **Summary of types of biobanks by biospecimen and donor type.** Biobanks can be organized by the type of specimen they contain, including living vs. deceased donor specimens and disease-focused vs. population-based cohorts, as well as by classic vs. prospective collection methods. Several biobanks offer a combination of specimen or donor cohort types, which are outlined in a central position in the graphic. *CHTN: Cooperative Human Tissue Network; NDRI: National Disease Research Interchange; GTEx: Genotype Tissue Expression Project.*

and urine) are ideal for studying biomarkers for diagnostics or monitoring disease and treatment effects. They are generally low-risk to collect and have high translational value. For instance, the UK Biobank, one of the largest biobanks of living donor fluid samples,[12] has contributed to numerous translational findings. Analyses of blood samples, physiological data, donor cognitive tests, and other donor medical or experiential information have identified risk factors and biomarkers for early-onset dementia and Alzheimer's disease, linked COVID-19 infection to future gastrointestinal diseases, and assessed the interaction between air pollution and the risk of chronic kidney disease.[13–16]

The second type of living donor specimens—residual clinical care samples—are obtained from preventative or elective surgery from consenting patients. After clinical samples are used in diagnosis and care, remaining tissue may be stored in a biobank (which is often located within the hospital) to be used in future research. Examples include tissue from common surgical procedures like resections (e.g., cancer, epilepsy) or replacements (e.g., joint replacements, organ explants). Residual tumor samples have improved our understanding of cancer, allowing researchers to accurately model tumor microenvironments and their impact on disease progression and treatment.[17,18] Explanted human lungs, for example, were used by scientists at Vertex Pharmaceuticals to design and assess drugs that target a cystic fibrosis-specific protein abnormality, leading to revolutionary treatments for the disease.[19]

Deceased donor specimen collections

Biospecimens from deceased donors provide a broad range of tissue types and donor histories. Donation opportunities may arise from rapid autopsy programs at hospitals, private donation programs for researching specific disorders, or organizations that provide organs and tissue for transplant (**Box 1**). Postmortem biospecimens can be recovered from any biological system (unlike specimens acquired from living donors), making it possible to study multiple body sites in the same donor and allowing powerful linked analyses regarding health and disease. Furthermore, postmortem specimens can be recovered from a diverse range of donors with variations in medical history, demographics, and experiences. Postmortem samples can include tissue involving diseases that do not require surgical intervention, often providing nondiseased or healthy control samples that would otherwise not be possible to obtain. Research with postmortem human biospecimens has led to significant findings in many fields, including neuroscience, cardiology, immunology, diabetes, and genetics. For example, the National Institutes of Health (NIH)-funded NeuroBioBank, which houses postmortem brain samples from donors with various neurological and psychiatric disorders,[20] has played a critical role in identifying specific brain regions implicated in attention deficit hyperactivity disorder,[21] understanding how genes in brain memory centers are altered in autism,[22] and discovering inflammatory factors in the development of Alzheimer's disease, specifically in those with Down syndrome.[23]

Types of biobanks

Some biobanks focus on a specific disease or disease group, while others aim to represent the "normal" population by collecting and distributing samples from a large, inclusive group of donors. Logistically, biobank models can either operate as a classic "predicted-need" model or a "prospective need-specific" model.

BOX 1

Deceased donor transplant organizations advance medical research by integrating research initiatives into their core missions

Transplant organizations serve as a bridge between the scientific community and the public, ensuring that research is both well supported and well understood. There are several types of transplant organizations that can support deceased donor gifts for research. **Organ procurement organizations** are federally designated in the United States to screen all acute death referrals from hospitals within their donor service areas and recover whole organs for transplantation. **Tissue and eye banks** recover tissues, such as bones, tendons, ligaments, skin, cardiovascular tissue, and corneas, for transplantation, including allografts. When these organizations approach families for organ and tissue donation for transplant, they also offer the opportunity for donation for research and education. By integrating research and education donation into their mission and vision, these organizations unite doctors, scientists, patients, and donors around the common goal of saving and improving lives through innovation. This dual focus not only enhances patient care but also accelerates the development of new therapies. Human tissues obtained through transplantation and donation programs serve as invaluable resources for a wide range of scientific disciplines, enabling groundbreaking discoveries that shape modern medicine. Their commitment to research has accelerated the development of therapies that were once only dreams—from drugs that prevent organ rejection to techniques that may one day grow new organs in the lab. By championing research at every level, transplant organizations advance scientific progress for the patients and families they support.

Disease-focused biobanks

Many biorepositories are funded by government agencies or patient advocacy foundations to study specific diseases. Examples include biobanks that have been developed following infectious disease epidemics. The UCSF AIDS Specimen Bank, which started in 1982 at the beginning of the AIDS epidemic,[24] has been instrumental in understanding and treating HIV/AIDS. A key challenge for disease-specific biobanks is finding donors, and these biobanks are often started in collaboration with clinical care centers or patient advocacy groups.[25] The Lyme Disease Biobank (LDB), funded by the Bay Area Lyme Foundation, collects biofluids from patients with Lyme disease and other tick-borne infections across the country, and their Tissue Collection Program preserves surgical remnant specimens and postmortem samples to distribute for future research use.[26] These samples are well-characterized with Lyme disease-specific health, diagnostic, and behavioral information, allowing numerous advances in understanding Lyme pathology and diagnosis since the inception of the bank in 2014.[27,28]

Population-based biobanks

Population-based collections help model the general population by collecting a large number of samples from living or deceased donors with few exclusion or inclusion criteria. The UK Biobank has a long-term commitment from over 500,000 participants, aged 40–69, to regularly provide blood, urine, and saliva samples, along with detailed information about their lifestyle.[12] Paired with additional data from cognitive testing, genetic sequencing, and physiological measurements, the UK Biobank serves as a resource for researchers studying diseases that often develop in middle- and older-aged adults.[29,30] The Genotype-Tissue Expression (GTEx) project is a population-based biobank focused on collecting a large number of postmortem specimens from healthy, nondiseased donors. This model allowed for the analysis of multiple tissue and cell types across all body systems from the same individual. This ground-breaking effort yielded a distinctive research resource that provided comprehensive molecular and genetic insights regarding how human genome expression varies under nonpathological conditions across numerous human tissues to support their respective biological functions.[31]

Classic biobank model

In a classic biobank, biospecimens are recovered based on predicted researcher needs and stored long-term in either a frozen or fixed preparation. Depending on the goal of the biobank, samples may be disease-specific or population-based and may come from living or deceased donors. Classic biobanks offer some key advantages: because the samples are already collected and stored, researchers can usually obtain tissue quickly once they receive approval. In addition, because classic biobanks are often created based on specific initiatives or cohorts, the samples and accompanying data can be standardized to allow for rigorous experiments. Classic biobanks may also be the only option to access samples from rare diseases. Depending on research needs, it can be challenging for scientists to find a biobank containing the specimens they require. While classic biobanks exist for some diseases, such as neurological/psychiatric conditions (NeuroBioBank), HIV/AIDS (AIDS Specimen Bank), or cancer (hospital pathology departments), many diseases do not have a dedicated biobank. Similarly, researchers who need samples from healthy, nondiseased control donors or samples that meet specific collection parameters (e.g., preservation method, tissue type, or location) may not find what they need in existing classic biobanks.

Prospective biobank model

In a prospective biobank model, the distribution of specimens is prioritized, and the biospecimens recovered are directly tied to researcher needs. An investigator will typically approach a prospective bioresource with a specific request that includes details on donor inclusion and exclusion criteria, as well as any necessary recovery and preservation parameters. The biobank determines whether those requests are reasonable and within their capabilities before specimens are collected for the researcher.[32] For example, investigators conducting experiments that require tissue samples for functional assays would not be able to use fixed or frozen samples that are available at classic biobanks. The National Disease Research Interchange (NDRI) follows a prospective model by utilizing a diverse network of tissue collection partners to provide either disease-specific or population-based samples from both living and deceased donors.[33] The Cooperative Human Tissue Network (CHTN) is a disease-specific prospective biobank that works with several US hospitals to provide cancer resections and other living-donor samples, as well as population-based samples from deceased donors, with investigator-specific criteria.[34]

Accessing biobanks

Distribution of samples is a key goal of biobanks, but they must also consider the ethical and legal implications of sharing biospecimens and data from human donors. Most organizations have a request or application process before samples are accessible, and most will require information on proposed research uses, credentials of the users, acknowledgment of ethical considerations, and a material transfer or data use agreement. Internal dedicated committees are often employed to evaluate requests on a case-by-case basis. For investigators interested in using human biospecimens, some resources to locate biobanks include the Biobank Resource Centre Biobank Locator[35] and ISBER's International Repository Locator.[36]

Biobank governance and ethical considerations

Biobank governance ensures ethical, legal, and operational integrity in managing biological specimens and data and establishes principles that guide responsible decision-making and operations. Effective biobank governance strategies consist of three key components.[37] *Transparency* requires the biobank to clearly communicate its goals, operations, and policies to stakeholders, including donors, researchers, and the public. *Accountability* encompasses the ways a biobank addresses stakeholder inquiries regarding its operations, including how it manages biospecimen distribution, upholds fair practices, safeguards donor rights, and ensures informed consent and confidentiality. *Oversight* relates to the policies and procedures the biobank employs to manage its responsibilities and infrastructure, including ensuring financial stability, operational efficiency, and supporting long-term sustainability.

Models of biobank governance

Biobanks foster partnerships between patients who donate biospecimens and researchers, and they must establish a governance model that engages internal and external stakeholders throughout their lifecycle to ensure biospecimen collections comply with regulations and ethical principles and meet the quality standards of end users. Internal stakeholders are individuals within the biobank involved in its operations, while external stakeholders are those outside the organization who are affected by or interested in its activities, such as regulatory agencies, advisors, contractors, or donors. Governance models vary with the size and mission of the biobank, as well as its operational and business needs, but their goal should be to protect the interests of all stakeholders involved in the biobank. Ultimately, this establishes good stewardship of biospecimens, reassures participants about their contributions, and helps encourage broader participation.

A governance plan outlines specific policies and procedures that a biobank utilizes to ensure it operates ethically and legally. The plan should include a clear structure with defined oversight responsibilities from all stakeholders, representing the expertise needed to provide feedback on scientific direction, legal and financial responsibilities, and donor interests and privacy (including scientific, ethical, legal, health, safety, data, finance, business, and regulatory expertise; Fig. 2). Region-specific regulations and accreditations govern the policies and procedures required for biobank operations. Further governance support may be provided

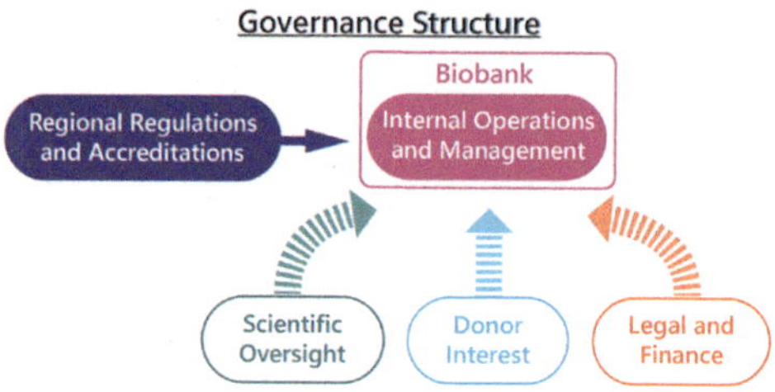

FIG. 2 **Biobank governance structure.** Day-to-day oversight of the biobank is the responsibility of the biobank management (pink). External regulatory agencies and accrediting bodies provide input on policies and procedures required for the operations of the biobank (purple). Biobank governance should include oversight and input on critical aspects of biobank support: legal and finance, scientific use and direction, and the interests of donors (including ethics and privacy). The type and number of individuals or committees that provide this support will vary based on the size, mission, and operational needs of the biobank and may include an independent board of directors, external committees, and advisory groups.

by a board of directors, other external committees, or advisory boards. The National Cancer Institute Best Practices for Biospecimen Resources[8] recommends that a governance plan include policies and procedures to address how the biobank will: (1) ensure the physical integrity of the biospecimens and data, (2) approve the distribution of biospecimens to investigators, and (3) define the roles and responsibilities of the institution and biobank director. Governance plans should include transparent biospecimen and data access policies, as well as retention policies that specify criteria for storage duration, including assessing the continued research value and quality of biospecimens and plans for transfer to prevent valuable biospecimens from being wasted.[38] Distinct approaches to governance by two biobanks are described in more detail below.

The UK biobank

UK Biobank governance is overseen by internal and external bodies to ensure effective management and accountability. Day-to-day operations are handled by the Principal Investigator and the Executive Management Team, who possess relevant experience in biobank management, epidemiology, and clinical research. Several external boards and committees provide guidance to the Principal Investigator and are composed of experts in various fields. The Board of Directors includes members from several areas of scientific research, technology, data management, and ethical backgrounds. It holds overall responsibility for the strategic direction and management of the biobank. Additional scientific oversight is provided by the International Scientific Advisory Board, which consists of scientists representing key users of the biobank and offers advisory support on the strategic direction of the biobank, and the Access Committee, which is responsible for guiding scientific access to samples in the UK Biobank, particularly those that are in short supply or will be used for potentially contentious research. The Audit and Risk Committee provides guidance on the biobank's approach to legal and financial risk management. The Ethics Advisory Committee supplies independent oversight of the biobank's Ethics and Governance Framework and advises on ethical, legal, and social issues, ensuring that the biobank aligns with ethical standards and the interests of participants and the public. The UK Biobank also receives input from participants via the Participant Advisory Group, which represents biobank donors and provides guidance on communication with the public and participants, as well as issues related to specimens and data.

In addition to seeking feedback from experts, the UK Biobank follows regulatory guidance provided by local authorities. Protocols and informed consent are approved by the Human Tissue Authority, the regulatory authority for the UK Biobank's collection of human biospecimens. It also maintains accreditation with ISO standards that ensure the biobank operates according to best practices for biobanking.[30,39]

The National Disease Research Interchange (NDRI)

NDRI provides investigators with specimens from both living and deceased donors via a prospective procurement model.[33] Similar to the UK Biobank, day-to-day operations are handled by the CEO and the Executive Management Team. Its governance is overseen by a Board of Directors composed of experts from diverse fields, including scientists from various research disciplines, professionals in finance, procurement organizations, ethics, biobanking, and data management. The board provides strategic guidance to the CEO and the Executive Management Team regarding the scientific, data management, ethical, financial, and technology directions of NDRI's services and programs. Additional scientific oversight is provided by committees, including: (1) an External Review Committee that evaluates new tissue request applications, (2) a Pathology Consultant Committee comprised of board-certified pathologists and biobank and bioethics experts who provide input on pathology services for the program, and (3) NIH Steering Committees that provide feedback on relevant NIH services. The Finance and Audit Committee oversees the organization's budget, ensuring financial stability and accountability.

Regulatory and accrediting oversight ensure that NDRI is aligned with best practices for biobanking. NDRI maintains approval for its tissue procurement program via an Institutional Review Board protocol to ensure compliance with ethical standards for human tissue donation according to the Common Rule. In addition, NDRI maintains accreditations from the College of American Pathologists (CAP) as a Biorepository and licensure from New York State as a nontransplant anatomic bank. These governance structures enable NDRI to maintain rigorous operational standards and support innovative biomedical research.

Ethical, legal, and social considerations for biobanks

Ethical, legal, and social aspects of biobank governance focus on protecting donor welfare and rights. Along with the consent of the participant for tissue donation, biobanks must consider issues related to the return of research findings and incidental discoveries. A biobank may have the capability to provide results to patients, but it must balance the duty to disclose actionable health information with the risk of causing unnecessary anxiety if results lack clinical validation.[40,41] Governance should thus balance beneficence with autonomy and consider data privacy, consent, equitable access to follow-up care, the participant's interest in knowing their results, and transparency.[42,43] These issues are of critical importance to public trust, especially from underrepresented groups. Ethical oversight may also address public concerns, promote community engagement, ensure diverse representation in biobank samples, and mitigate skepticism, particularly in commercial contexts.[44]

Informed consent

Ethical standards and regulatory guidance on informed consent

Informed consent is based on universal ethical principles guiding all human subjects research, including research involving biobank specimens and data: respect for persons, beneficence, and justice (Fig. 3). Informed consent is essential for biorepository operations, driving the overall use of biospecimens for research and building trust between biobanks and the donor communities they serve. Best practices for informed consent require that it: (1) discloses information needed to make an informed decision, (2) facilitates the understanding of what has been disclosed, and (3) assures that the decision of whether or not to participate in the research is voluntary. Ethical standards for informed consent vary internationally, and while most countries do not have specific regulations governing biobanks,[45] guidance for the collection of biospecimens is provided in human subject research and privacy regulations. In the United States, the Common Rule outlines regulations for the protection of human subjects[46] and requires that informed consent be obtained for human biospecimen donation from living individuals. The Uniform Anatomical Gift Act[47] allows research on biospecimens from deceased donors if the donor previously authorized it. In practice, many US biobanks follow NIH recommendations to obtain consent from a deceased donor's next-of-kin for the use of biospecimens for research.[48] Similarly, the Human Tissue Act provides guidance on activities related to human tissue use in England, Wales, and Northern Ireland, including the need for informed consent, and applies to both living and deceased individuals.

Approaches to informed consent for biobanks

Regulatory guidance requires the review of consent practices by an Institutional Review Board or other applicable ethical review committee. Seeking feedback from community members and other external stakeholders, particularly on enhancing understanding for patients, can strengthen the approach to informed consent.[50,51] Biobanks must consider the

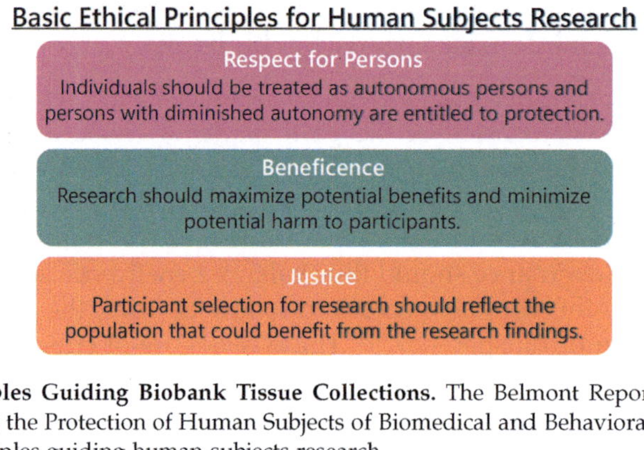

FIG. 3 **Ethical Principles Guiding Biobank Tissue Collections.** The Belmont Report, developed by the U.S. National Commission for the Protection of Human Subjects of Biomedical and Behavioral Research in 1978,[49] summarizes the ethical principles guiding human subjects research.

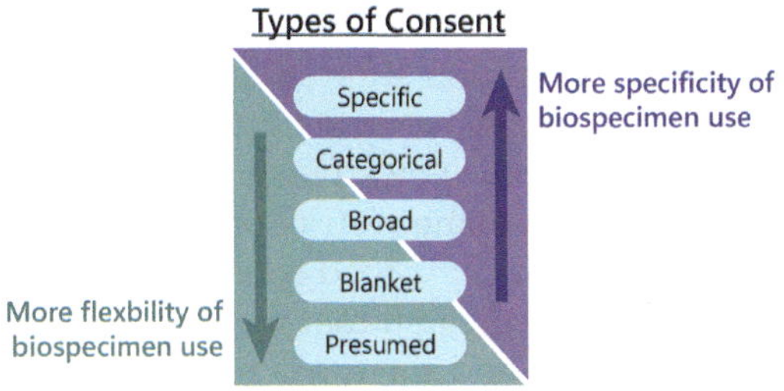

FIG. 4 **Types of consent for biospecimen donation.** Five common approaches to informed consent are shown. The type of consent where the donor receives the most detail on biospecimen use and the biobank has the least flexibility with biospecimen use is shown at the top. The type of consent where the donor receives the least detail on biospecimen use and the biobank has the most flexibility with biospecimen use is shown at the bottom.

balance between providing details on research uses to donors and not overly restricting future utilization of the donated biospecimens and data (Fig. 4). Five common approaches to informed consent used by biobanks are defined below.

Specific consent

In specific consent, the details of the research project for which the biospecimens will be utilized are clearly described.[52] This approach provides the donor with the most information regarding the use of the biospecimen, which is strictly defined and usually supports a single scientific study. If additional uses are desired in the future, the donor must be recontacted for their consent for the additional use, which can be burdensome on the biobank and the donor.[53] Specific consent contains all elements of informed consent as outlined in US regulations 45 CFR 46.116(a), the gold standard for ensuring patients are informed about research participation. These elements include: (1) an explanation of the research, including who will have access and for how long; (2) how specimens and associated information will be collected; (3) any risks associated with donation; (4) benefits to the donor or others associated with the research; (5) processes to protect confidentiality; (6) whether the donor will incur costs from participation or profits from commercial products generated; (7) whom to contact for study details; and (8) a statement that participation is voluntary.

Categorical consent

Categorical consent, also called "tiered consent,"[54] provides patients with several options for potential research uses of biospecimens.[52] Donors are able to choose the types of research uses that align with their interests and decline those that do not.[54] This approach allows biospecimens to be utilized for multiple studies. The biobank must have a system in place to track the options selected in the consent for each biospecimen to ensure the donor's preferences are upheld. Dynamic consent is a version of categorical consent in which participants can opt in to research studies and manage their consent options over time.[55] Dynamic consent allows participants to use a personalized, web-based interface to review additional requests for the use of biospecimens and remain engaged with the biobank. This approach ensures participants' preferences are maintained and enables the biobank to expand the utility of biospecimens as needed.

Broad consent

Following revisions to the US Common Rule in 2017, broad consent became a recommended approach for biobank collections because of its focus on the secondary research use of specimens and data.[56] Broad consent allows biospecimens and data to be utilized for a range of future research studies, with some restrictions based on the original consent, but without requiring recontact with donors.[56,57] Broad consent requires that donors are provided with: (1) a description of the types of research that may be conducted, (2) the biospecimens and associated data that will be used and the types of institutions or researchers that may have access to the biospecimens and data, (3) the length of time the biospecimens and data will be stored and used (which could be indefinitely), and (4) details on whether or not they will be informed of future research results. The biobank is responsible for ensuring oversight of biospecimen and data distribution that aligns with the details of the broad consent but does not need to manage individual patient choices as in the categorical consent approach.

Blanket consent

Blanket consent indicates that biospecimens and associated data can be used for future use with no limitations.[52] This approach does not provide donors with significant information regarding the types of research that may be conducted on biospecimens and provides the biobank with flexibility for the distribution and use of biospecimens. Patients uncomfortable with the range of use implied by this approach can choose not to participate.

Presumed consent

Presumed consent assumes that participants who consent to the collection of biospecimens and associated information for clinical care also consent to future research uses of the samples and data.[52,53] Donors can opt out of having their biospecimens and data included in future use. This approach increases the availability of specimens for secondary research use but provides patients with the least detail on research use.

Donor privacy

A key responsibility of biobanks is to protect patients' confidentiality and privacy. The US Privacy Rule, under the Health Insurance Portability and Accountability Act (HIPAA), limits the disclosure of protected health information, which includes medical records and other identifiable information, by covered entities, including healthcare providers.[58] In the EU and UK, data-sharing laws such as the Data Protection Act and the European Data Protection Directive regulate how organizations can collect, store, and use personal data.[45] Biobanks must have policies regarding what is considered protected health information and personal data and how to protect their release to external parties. Most bioresources protect donor data from reidentification by coding it.[59] Biobanks may also receive biospecimens and data without any identifying information, making the biospecimens and data anonymized and not requiring additional deidentification.

Genomic information can be considered identifiable information and must be protected. Sharing DNA sequences among investigators on public platforms increases the risk of identification of donors and their relatives and may reveal information regarding risks for disease, which heightens the risk of genetic discrimination.[60,61] The risks of reidentification through genomic sequencing are likely low in most cases, as it is only possible if the publicly available deidentified data is linked to identifiable genomic data accessible via another source.[62] Nevertheless, biobanks must protect donors from these risks through their governance, access policies, and distribution policies, as well as material transfer and data use agreements.

Innovations in biobanking

Advances in biospecimen processing and operations

As scientific methodologies have advanced, the biospecimen and biorepository science field has evolved to address the needs of the biomedical research and patient communities.[63] Advancements in experimental methodologies, driven by innovations in genomic and multiomic approaches, are transforming sample collection and processing techniques to support data accuracy, experimental rigor, and reproducibility.[64] Increasingly, biobank collections must consider the effects of preanalytic variables that impact downstream molecular analyses and institute quality control measures. The GTEx study evaluated the effects of preanalytic variables on specimen quality, analyzing over 30,000 tissue specimens from deceased donors[31] and identifying variables such as autolysis and postmortem interval that impact tissue-specific RNA integrity.[65] To support spatial genomic techniques, which map high-resolution genomic features within a tissue, biobank collections must consider a common coordinate framework for tissue recovery. Such frameworks can include metadata associated with tissue recovery locations that allow consistency across experimental replicates, as well as microdissection and precision aliquoting to ensure accurate distribution of specimens, reduced variability, and optimized downstream analyses.

Advances in technology for biobanks

Computerized inventory systems allow automated sample storage, enhancing sample tracking, reducing human error, and ensuring optimal preservation conditions.[66] Data analysis utilizing artificial intelligence (AI) enables rapid identification of patterns, enhancing quality control by detecting anomalies in sample integrity and improving sample classification for research and clinical applications.[67] Integrating automation and AI-driven solutions will further streamline sample processing, reducing errors and enhancing efficiency in biobanking for biomedical research and clinical applications. The emergence of virtual biobanks, which integrate decentralized data from multiple sources, allows researchers worldwide to access standardized, high-quality datasets without physical transfers.[68] These digital platforms enhance access and collaboration, streamline regulatory compliance, and easily support large-scale studies, ultimately accelerating discoveries in biomedical research.

Large-scale tissue and data collection efforts, which also serve to create genetic databases for future downstream analyses, provide a significant contribution to genomic studies. The impact of these large-scale projects is enhanced when their datasets are linked with a common coordinate framework that drives the development and implementation of identical or very similar collection processes. This enables scientists to study biospecimens that have been collected in a standardized manner, thereby elevating the rigor and reproducibility of the datasets by reducing experimental artifacts or data analysis errors that occur due to variations in tissue collection, processing, and preservation procedures. An example of such an effort is the Somatic Mosaicism across Human Tissues (SMaHT) project, which is developing the first large-scale, systematic catalog of somatic variation across normal adult tissues and supporting the advancement of new approaches to diagnosing, treating, and curing human diseases driven by somatic genetic variations.[69] The SMaHT biobank is initiating the collection of approximately 20 tissue types from deceased donors representing all three developmental layers using a common coordinate framework that parallels the collection methodologies for overlapping tissues when possible. In addition, a precision aliquoting approach is being utilized for sample distributions to maintain spatial detail. Finally, SMaHT is standardizing its collection approach with other similar resources to strengthen its utility and enable more impactful genomic analysis across large-scale efforts.

Using biospecimens in experimental models

Biobanks are ideal sources for the standardized, quality-controlled primary tissues or cell lines needed to create innovative experimental culture models. Advanced techniques now allow investigators to isolate distinct cell types from biospecimens to establish organs-on-a-chip, where cell types within an organ are grown in a device engineered to mimic the function of the whole organ.[70] This system is ideal for examining disease pathophysiology or drug effects on cells within their relevant microenvironment. Similarly, another burgeoning model system of great interest to translational researchers is organoids—3D structures made up of organ-specific cell types that are grown from stem cells via their intrinsic developmental cell programs. As these new approach methodologies (NAMs) using human tissue become more developed and widely available, the speed at which investigators are able to translate preclinical findings to the clinic will accelerate, benefiting patients across a spectrum of diseases. For a deeper discussion of these experimental models, see Chapter 6.

Conclusions

Biobanks are essential research resources that frequently support high-priority advancements in translational research. As experimental methodologies continue their progression to more multifaceted spatial sequencing and mapping echelons, the demand for access to biobank resources with suitable samples will continue to grow; this will include the need for a broader range of tissues, anatomical structures, cell types, and donor cohorts. In addition, the donor metadata repositories in biobanks hold the promise of contributing to the successful progression of big data and personalized medicine. By adopting new innovations in tissue collection, data management, and ethical processes, biobanks will ensure their function as state-of-the-art research resources for donors and scientists, contributing to the future development of cutting-edge research methodologies and pioneering translational research outcomes.

References

[1] Eiseman E, Haga SB *Handbook of Human Tissue Sources. A National Resource of Human Tissue Samples.*; 1999. Accessed January 31, 2025. https://apps.dtic.mil/sti/citations/ADA373679

[2] Mahmood SS, Levy D, Vasan RS, Wang TJ. The Framingham Heart Study and the Epidemiology of Cardiovascular Diseases: a Historical Perspective. *Lancet.* 2014;383(9921):999–1008. doi:10.1016/S0140-6736(13)61752-3.

[3] Pound P, Ritskes-Hoitinga M. Is it possible to overcome issues of external validity in preclinical animal research? Why most animal models are bound to fail. *J Transl Med.* 2018;16(1):304. doi:10.1186/s12967-018-1678-1.

[4] Lewis C, Clotworthy M, Hilton S, et al. Public views on the donation and use of human biological samples in biomedical research: a mixed methods study. *BMJ Open.* 2013;3(8):e003056. doi:10.1136/bmjopen-2013-003056.

[5] Brothers KB, Morrison DR, Clayton EW. Two large-scale surveys on community attitudes toward an opt-out biobank. *Am J Med Genet A.* 2011;155(12):2982–2990. doi:10.1002/ajmg.a.34304.

[6] Robinson NB, Krieger K, Khan FM, et al. The current state of animal models in research: a review. *Int J Surg.* 2019;72:9–13. doi:10.1016/j.ijsu.2019.10.015.

[7] International Organization for Standardization. ISO. 2018. Accessed February 26, 2025. https://www.iso.org/standard/67888.html

[8] National Cancer Institute. NCI best practices for biospecimen resources. Published online March 2016. https://biospecimens.cancer.gov/bestpractices/2016-NCIBestPractices.pdf

[9] Snapes E, Astrin JJ, Bertheussen Krüger N, et al. Updating international society for biological and environmental repositories best practices, fifth edition: a new process for relevance in an evolving landscape. *Biopreserv Biobank.* 2023;21(6):537–546. doi:10.1089/bio.2023.0140.

[10] Robb JA, Gulley ML, Fitzgibbons PL, et al. A call to standardize preanalytic data elements for biospecimens. *Arch Pathol Lab Med.* 2014;138(4):526–537. doi:10.5858/arpa.2013-0250-CP.

[11] Alkhatib R, Gaede KI. Data management in biobanking: strategies, challenges, and future directions. *BioTech.* 2024;13(3):34. doi:10.3390/biotech13030034.

[12] UK Biobank. 2025. Accessed February 7, 2025. https://www.ukbiobank.ac.uk

[13] Hendriks S, Ranson JM, Peetoom K, et al. Risk factors for young-onset dementia in the UK biobank. *JAMA Neurol.* 2023;81(2):134–142. doi:10.1001/jamaneurol.2023.4929.

[14] Ma Y, Zhang L, Wei R, et al. Risks of digestive diseases in long COVID: evidence from a population-based cohort study. *BMC Med.* 2024;22(1):14. doi:10.1186/s12916-023-03236-4.

[15] Wang J, Li D, Sun Y, Tian Y. Air pollutants, genetic factors, and risk of chronic kidney disease: findings from the UK Biobank. *Ecotoxicol Environ Saf* 2022;247:114219. doi:10.1016/j.ecoenv.2022.114219.

[16] Wang X, Shi Z, Qiu Y, Sun D, Zhou H. Peripheral GFAP and NfL as early biomarkers for dementia: longitudinal insights from the UK Biobank. *BMC Med.* 2024;22(1):192. doi:10.1186/s12916-024-03418-8.

[17] He J, Zhang C, Ozkan A, et al. Patient-derived tumor models and their distinctive applications in personalized drug therapy. *Mechanobiol Med.* 2023;1(2):100014. doi:10.1016/j.mbm.2023.100014.

[18] Miserocchi G, Mercatali L, Liverani C, et al. Management and potentialities of primary cancer cultures in preclinical and translational studies. *J Transl Med.* 2017;15(1):229. doi:10.1186/s12967-017-1328-z.

[19] Herper M, Feuerstein A. "A game-changer": how Vertex delivered on cystic fibrosis. STAT. October 23, 2019. Accessed February 12, 2025. https://www.statnews.com/2019/10/23/we-conquered-a-disease-how-vertex-delivered-a-transformative-medicine-for-cystic-fibrosis/

[20] Freund M, Taylor A, Ng C, Little AR. The NIH NeuroBioBank: creating opportunities for human brain research. *Handb Clin Neurol.* 2018;150:41–48. doi:10.1016/B978-0-444-63639-3.00004-9.

[21] Sudre G, Gildea DE, Shastri GG, et al. Mapping the cortico-striatal transcriptome in attention deficit hyperactivity disorder. *Mol Psychiatry.* 2023;28(2):792–800. doi:10.1038/s41380-022-01844-9.

[22] Rexrode LE, Hartley J, Showmaker KC, et al. Molecular profiling of the hippocampus of children with autism spectrum disorder. *Mol Psychiatry.* 2024;29(7):1968–1979. doi:10.1038/s41380-024-02441-8.

[23] Wilcock DM, Hurban J, Helman AM, et al. Down syndrome individuals with Alzheimer's disease have a distinct neuroinflammatory phenotype compared to sporadic Alzheimer's disease. *Neurobiol Aging.* 2015;36(9):2468–2474. doi:10.1016/j.neurobiolaging.2015.05.016.

[24] De Souza YG, Greenspan JS. Biobanking past, present and future: responsibilities and benefits. *AIDS.* 2013;27(3):303. doi:10.1097/QAD.0b013e32835c1244.

[25] VonDran M, Thomas JA, Freund MP, et al. *The National Disease Research Interchange and Collaborators On: What are the Major Hurdles to the Recovery of Human Tissue to Advance Research?* Biopres Biobanking; 2016; doi:10.1089/bio.2016.29014.mvd.

[26] Lyme Disease Biobank. Bay Area Lyme Foundation. Accessed February 5, 2025. https://www.bayarealyme.org/biobank/

[27] Ghosh R, Joung HA, Goncharov A, et al. Rapid single-tier serodiagnosis of Lyme disease. *Nat Commun.* 2024;15(1):7124. doi:10.1038/s41467-024-51067-5.

[28] Kelbauskas L, Legutki JB, Woodbury NW. Highly heterogenous humoral immune response in Lyme disease patients revealed by broad machine learning-assisted antibody binding profiling with random peptide arrays. *Front Immunol.* 2024;15. doi:10.3389/fimmu.2024.1335446.

[29] Littlejohns TJ, Sudlow C, Allen NE, Collins R. UK Biobank: opportunities for cardiovascular research. *Eur Heart J.* 2019;40(14):1158–1166. doi:10.1093/eurheartj/ehx254.

[30] Sudlow C, Gallacher J, Allen N, et al. UK Biobank: an open access resource for identifying the causes of a wide range of complex diseases of middle and old age. *PLOS Med.* 2015;12(3):e1001779. doi:10.1371/journal.pmed.1001779.

[31] Lonsdale J, Thomas J, Salvatore M, et al. The genotype-tissue expression (GTEx) project. *Nat Genet.* 2013;45(6):580–585. doi:10.1038/ng.2653.

[32] Grizzle WE, Bledsoe MJ, Al Diffalha S, Otali D, Sexton KC. The utilization of biospecimens: impact of the choice of biobanking model. *Biopreserv Biobank.* 2019;17(3):230–242. doi:10.1089/bio.2019.0008.

[33] Bell TJ, Leinweber B. Times are changing: 35 years of human biospecimen procurements for the national disease research interchange. *Biopreserv Biobank.* 2015;13(5):309–310. doi:10.1089/bio.2015.29032.tjb.

[34] Cooperative Human Tissue Network (CHTN). Cooperative human tissue network (CHTN). Accessed February 12, 2025. https://chtn.cancer.gov/

[35] The University of British Columbia (UBC) Office of biobank education and research (OBER), CTRNet. Biobank Resource Centre: Biobank Locator. 2025. Accessed February 26, 2025. https://biobanking.org/locator

[36] ISBER. ISBER International Repository Locator. 2025. Accessed February 26, 2025. https://irlocator.isber.org/search/

[37] Gille F, Vayena E, Blasimme A. Future-proofing biobanks' governance. *Eur J Hum Genet.* 2020;28(8):989–996. doi:10.1038/s41431-020-0646-4.

[38] Matzke LAM, Fombonne B, Watson PH, Moore HM. Fundamental Considerations for Biobank Legacy Planning. *Biopreserv Biobank.* 2016;14(2):99–106. doi:10.1089/bio.2015.0073.

[39] Biobank UK. UK Biobank ethics and governance framework: version 3.0. Published online 2007. https://www.ukbiobank.ac.uk/media/0xsbmfmw/egf.pdf

[40] De Clercq E, Kaye J, Wolf SM, Koenig BA, Elger BS. Returning results in biobank research: global trends and solutions. *Genet Test Mol Biomark.* 2017;21(3):128–131. doi:10.1089/gtmb.2016.0394.

[41] Wolf SM. Return of results in genomic biobank research: ethics matters. *Genet Med.* 2013;15(2):157–159. doi:10.1038/gim.2012.162.

[42] Ramoni RB, McGuire AL, Robinson JO, et al. Experiences and attitudes of genome investigators regarding return of individual genetic test results. *Genet Med.* 2013;15(11):882–887. doi:10.1038/gim.2013.58.

[43] Stein CM, Ponsaran R, Trapl ES, Goldenberg AJ. Experiences and perspectives on the return of secondary findings among genetic epidemiologists. *Genet Med.* 2019;21(7):1541–1547. doi:10.1038/s41436-018-0369-x.

[44] Adsit-Morris C, NaDejda Collins R, Goering S, et al. UnboundingELSI: the ongoing work of centering equity and justice. *Am J Bioeth.* 2023;23(7):103–105. doi:10.1080/15265161.2023.2214055.

[45] Rothstein MA, Knoppers BM, Harrell HL. Comparative approaches to biobanks and privacy. *J Law Med Ethics.* 2016;44(1):161–172. doi:10.1177/1073110516644207.

[46] Protections (OHRP) O for HR. 45 CFR 46. February 16, 2016. Accessed February 27, 2025. https://www.hhs.gov/ohrp/regulations-and-policy/regulations/45-cfr-46/index.html

[47] Glazier AK. The principles of gift law and the regulation of organ donation. *Transpl Int.* 2011;24(4):368–372. doi:10.1111/j.1432-2277.2011.01226.x.

[48] Wilson-Genderson M, Barker KL, Gardiner HM, et al. Authorization of tissues from deceased patients for genetic research. *Hum Genet.* 2017;137(1):63–71. doi:10.1007/s00439-017-1855-y.

[49] National Commission for the Protection of Human Subjects of Biomedical and Behavioral Research. The Belmont Report: ethical Principles and Guidelines for the Protection of Human Subjects of Research. Published online 1978.

[50] O'Doherty K, Hawkins A. Structuring public engagement for effective input in policy development on human tissue biobanking. *Public Health Genomics*. 2010;13(4):197–206. doi:10.1159/000279621.

[51] Rotimi C, Leppert M, Matsuda I, et al. Community engagement and informed consent in the International HapMap project. *Community Genet*. 2007;10(3):186–198. doi:10.1159/000101761.

[52] Greenberg B, Christian J, Henry LM, Leavy M, Moore H. *Biorepositories: Addendum to Registries for Evaluating Patient Outcomes: A User's Guide*, 3rd ed. Agency for Healthcare Research and Quality (US); 2018. https://www.ncbi.nlm.nih.gov/books/NBK493632/.

[53] Rosenbloom ST, Madison JL, Brothers KB, et al. Ethical and practical challenges to studying patients who opt out of large-scale biorepository research. *J Am Med Inform Assoc*. 2013;20(e2):e221–225. doi:10.1136/amiajnl-2013-001937.

[54] Vaught J, Lockhart NC. The evolution of biobanking best practices. *Clin Chim Acta*. 2012;413(19-20):1569–1575. doi:10.1016/j.cca.2012.04.030.

[55] Kaye J, Whitley EA, Lund D, et al. Dynamic consent: a patient interface for 21st century research networks. *Eur J Hum Genet*. 2015;23(2):141–146. doi:10.1038/ejhg.2014.71.

[56] Maloy JW, Bass PF. Understanding broad consent. *Ochsner J*. 2020;20(1):81–86. doi:10.31486/toj.19.0088.

[57] Grady C, Eckstein L, Berkman B, et al. Broad consent for research with biological samples: workshop conclusions. *Am J Bioeth*. 2015;15(9):34–42. doi:10.1080/15265161.2015.1062162.

[58] CDC. Health Insurance Portability and Accountability Act of 1996 (HIPAA). Public Health Law. September 10, 2024. Accessed February 27, 2025. https://www.cdc.gov/phlp/php/resources/health-insurance-portability-and-accountability-act-of-1996-hipaa.html

[59] Alkhatib R, Gaede KI. Data management in biobanking: strategies, challenges, and future directions. *BioTech*. 2024;13(3):34. doi:10.3390/biotech13030034.

[60] McGuire AL, Caulfield T, Cho MK. Research ethics and the challenge of whole-genome sequencing. *Nat Rev Genet*. 2008;9(2):152–156. doi:10.1038/nrg2302.

[61] Tabor HK, Berkman BE, Hull SC, Bamshad MJ. Genomics really gets personal: how exome and whole genome sequencing challenge the ethical framework of human genetics research. *Am J Med Genet A*. 2011;155A(12):2916–2924. doi:10.1002/ajmg.a.34357.

[62] Malin B, Karp D, Scheuermann RH. Technical and policy approaches to balancing patient privacy and data sharing in clinical and translational research. *J Investig Med*. 2010;58(1):11–18. doi:10.231/JIM.0b013e3181c9b2ea.

[63] Vaught J. Biobanking comes of age: the transition to biospecimen science. *Annu Rev Pharmacol Toxicol*. 2016;56:211–228. doi:10.1146/annurev-pharmtox-010715-103246.

[64] Molla G, Bitew M. Revolutionizing personalized medicine: synergy with multi-omics data generation, main hurdles, and future perspectives. *Biomedicines*. 2024;12(12):2750. doi:10.3390/biomedicines12122750.

[65] Sobin L, Barcus M, Branton PA, et al. Histologic and quality assessment of genotype-tissue expression (GTEx) research samples: a large postmortem tissue collection. *Arch Pathol Lab Med*. 2025;149(3):217–232. doi:10.5858/arpa.2023-0467-OA.

[66] Linsen L, Van Landuyt K, Ectors N. Automated sample storage in biobanking to enhance translational research: the bumpy road to implementation. *Front Med*. 2020;6:309. doi:10.3389/fmed.2019.00309.

[67] Akyüz K, Cano Abadía M, Goisauf M, Mayrhofer, MTh. Unlocking the potential of big data and AI in medicine: insights from biobanking. *Front Med* 2024;11:1336588. doi:10.3389/fmed.2024.1336588.

[68] Watson JW, Clark G, Williams DT. The value of virtual biobanks for transparency purposes with respect to reagents and samples used during test development and validation. *Rev Sci Tech*. 2021;40(1):253–259. doi:10.20506/rst.40.1.3222.

[69] Somatic Mosaicism Across Human Tissues (SMaHT). Accessed February 27, 2025. https://smaht.org/

[70] Park SE, Georgescu A, Huh D. Organoids-on-a-chip. *Science*. 2019;364(6444):960–965. doi:10.1126/science.aaw7894.

CHAPTER

11

Emphasizing ancestry in human-based biomedical research

Erika Moore[a], *Alison Veintimilla*[a], *Nana Owusu-Boaitey*[a], *and Connie J. Mulligan*[b]

[a]Fischell Department of Bioengineering, University of Maryland, College Park, MD, United States

[b]Department of Anthropology, Genetics Institute, University of Florida, Gainesville, FL, United States

Abstract

Social and genetic ancestry remain important determinants of human health. Preclinical research and wider society use population descriptors to characterize ancestry. The successful application of these descriptors requires clear, meaningful definitions and awareness of how those performing and evaluating research may use these terms in a biased manner. Ancestry descriptors can be employed together with effective human-based preclinical modeling methods. This encompasses ex vivo studies using cells and other components isolated from patients, in vitro analyses conducted in laboratory settings, and computer-based in silico studies. These preclinical modeling methods improve understanding of the role of ancestry in disease mechanisms, therapeutics, and diagnosis, providing important advantages over animal-based methods. Maximizing the strengths of preclinical modeling and mitigating its limitations may require collaboration between stakeholders such as researchers, research participants, and advocacy groups. Collaboration facilitates preclinical modeling that produces scientifically rigorous results on ancestry in a manner respectful of vulnerable communities.

Keywords: Ancestry; Population descriptors; Preclinical nonanimal studies

Introduction

Social and genetic aspects of ancestry shape individual and population health, including differences in disease outcomes for patients. This chapter evaluates the unique contributions provided by preclinical nonanimal studies, including the use of ex vivo, in vitro, and in silico models. We begin by defining population descriptors important for identifying and responsibly discussing social and genetic facets of ancestry. The utility of these descriptors is illustrated, for example, by applying them to selected populations and previously identified health disparities. Preclinical models are then defined and differentiated by characteristics such as the cell type employed, the extracellular environment, and the methods used for assessing cellular function.

Moving Toward Nonanimal Approaches in Medical Research and Testing
https://doi.org/10.1016/B978-0-443-26514-3.00004-7

Given the wide variety of preclinical models, researchers can identify or develop a model best suited for the disease or treatment of interest. We examine examples of these models, elucidating how ancestry impacts disease mechanisms, ranging from diseases of the immune system to cancer and detrimental responses to allergens. The application of preclinical models to diagnostics and therapeutics is also explored, with an emphasis on how ancestry can cause differing patient responses to these treatments. We close with suggested considerations regarding the strengths, weaknesses, and feasibility of preclinical modeling. The considerations discussed include methods for recruiting cell donors, medical distrust resulting from published preclinical modeling results on ancestry, and limitations of population descriptors. This chapter will also discuss examples regarding the use of genetic ancestry, which reflects differences amongst populations in the frequency of genetic variants that increase or decrease disease risk. It is important to note that genetic ancestry does not cause differences in disease prevalence or mechanism, but rather reflects differences in frequency of important genetic variants and can be used as a rough proxy for these variant frequencies.

Context of demographic descriptors and genetic techniques

Researchers have long understood that combined social and genetic factors impact disease and therapeutic outcomes. The inclusion of diverse human populations to capture the heterogeneity of disease and therapeutic response is critical to ensuring the safety and translatability of biomedical innovations for all patients. Responsible and relevant in vitro work with human primary samples should fully define their origins to engage with such work. This can be achieved through the use of ancestral population descriptors.

Population descriptors are terms that are widely utilized in society and healthcare settings to distinguish human demographics both at the individual and collective levels.[1] Genetic ancestry, genealogical ancestry, race, ethnicity, nationality, and indigeneity are examples of ancestral population descriptors. Population descriptors are relative, meaning they are subject to both setting and time. For example, concepts of race within the United States differ from concepts of race in Latin America and have evolved over time within the United States. Therefore it is crucial that ancestral population descriptors be used with intentionality and full context.

Social ancestry-focused population descriptors, such as race, nationality, and ethnicity, are often grouped together, but there is important nuance between these terms. Race is a human-derived classification scheme based on perceived physical and cultural attributes (e.g., hair texture). Race has historically and continues to be misused to establish hierarchies of power and propagate health inequities. Ethnicity refers to the grouping of populations based on shared language, customs, religious practices, and history. Ethnicity can also refer to an indigenous group from a given area (e.g., the Hmong people of China). Nationality is the grouping of individuals under a national identity defined by a present-day country. All of these terms can overlap with one another and are highly variable depending on geopolitical determinants.[1]

Race and ethnicity have been topics of discourse in the United States due to their impact on propagating health inequities. Sociocultural trauma, such as systemic racism, has been speculated by multiple disciplines to induce or maintain modifications in the genome that lead to altered downstream genetic and phenotypic expression.[2] While the impact of these power structures has been studied for their effects on healthcare outcomes, their impact on biological systems has yet to be elucidated. In addition, race has been leveraged as a proxy for environmental factors.[3] The US legacy of systemic racism has led to differential healthcare and housing opportunities, which may be determinants in environmental exposures that influence biological processes. This can be seen in diminished fertility rates for those living near major highways,[4] increased heart disease rates for those living near food deserts,[5] and increased cancer rates for those living near polluting infrastructures.[6] Established biomedical dogmas do not consider social population descriptors, although integration of these descriptors into biomedical research offers greater translatability beyond the laboratory benchtop. While potentially useful, it should also be noted that there have historically been limitations to the implementation of social population descriptors in biomedical research.[7]

A less controversial and possibly more objective ancestral population descriptor is genetic ancestry. In biology-related studies, the use of the term "ancestry" is often in reference to genetic ancestry. Genetic ancestry describes the inheritance of DNA from common ancestors based on geographical origins and human migrations.[1,8] It is based on statistical analysis of DNA similarity. Genetic ancestry can differ from genealogical ancestry in that only 50% of each parent's DNA is passed to an offspring. Genetic ancestry examines DNA similarities between individuals and ancestral populations, while genealogical ancestry focuses on individual and familial descendancy or pedigree.[8] Direct-to-consumer genetic ancestry testing (e.g., 23andMe) has become increasingly popular with a general public eager to know more about their ancestry.

With globalization and the integration of human populations, genetic admixture as a tool to describe genetic ancestry has become increasingly popular. Genetic admixture refers to the interbreeding of two genetically distinct populations over time to create a distinct new population, which is referred to as admixed.[9] For example, African American individuals in the United States are considered to be admixed individuals with major contributions from West/West-Central Africa and varying contributions from Europe.[10] Given the numerous migrations that have occurred throughout human evolution, virtually all human populations are admixed to varying degrees. There are several computational tools and databases available to researchers to investigate genetic ancestry and genetic admixture. While genetic ancestry and admixture have widespread and varied uses in biology, standardization of how these studies are conducted has not been established.

Ancestral population descriptors, while insightful, also have some limitations, which will be detailed at the end of this chapter. Therefore they should be used with intentionality and with disclosure of these limitations. When leveraged appropriately, the integration of ancestral population descriptors with preclinical modeling techniques can lead to the development of powerful hypotheses and experimental designs, and can promote the translation of research findings to all patient populations.

Human-based preclinical modeling techniques

The human-based preclinical modeling discussed in this chapter includes ex vivo investigations using tissue isolated from patients and in vitro experiments employing cells and extracellular environments under laboratory conditions. Other modeling methods also exist, such as in silico simulations involving computer-based analysis of datasets.[11] These in vitro and ex vivo models differ from in vivo research, which involves human subjects or uses nonhuman animal models. Human-based preclinical models are further distinguished by the cell type employed. Models may also differ in their extracellular environment, along with methods for assessing cellular and extracellular functions within the model. For a deeper discussion of ex vivo models, see Chapter 10; for in vitro models, see Chapter 6; and for in silico models, see Chapter 7.

Models vary in the type and source of cells they incorporate. Primary cells are isolated from individuals during medical assessments or tissue donation.[12] Immortalized cells, in contrast, are modified to survive long-term under laboratory conditions, with genetic characterization and use of the same cell line across research groups.[13] Primary cell donors vary across characteristics, including sex, health status, race, indigeneity, and genetic and genealogical ancestry. Cell types isolated also vary depending on the target questions for the model, such as immune cells for studying response to infection or cells lining the lung airway for studying asthma.[14,15] Cells isolated from the same donor capture interactions between cell types within an individual. This contrasts with modeling that uses cell lines from different donors. Analysis of primary cells allows modeling of correlations between in vitro cell phenotype and in vivo clinical response. Human-based preclinical models may also compare primary cells from healthy individuals with those from individuals with a particular condition. This analysis can reveal how the disease changes cell phenotype.

Human-based preclinical models differ not only in the cell populations they employ but also in terms of the extracellular environment. Modeled environments differ in complexity and structure based on the question under consideration. Two-dimensional systems capture interactions such as those at cell-to-cell interfaces, including epithelial layers exposed to soluble targets or to wounding in scratch assays.[16] Three-dimensional systems model additional processes such as cell migration through an extracellular matrix. The complexity of three-dimensional systems can vary from transwell systems containing a layer of cells on a porous membrane separating a lower and upper chamber to hydrogels in which cells are encapsulated in a polymer matrix.[17,18] Other three-dimensional models recapitulate further physiologically relevant processes. These include organ chips, where fluid can flow between multicellular structures in tissue grown on a chip, and cells self-organizing in vitro into complex three-dimensional structures known as organoids.[18,19]

Both two-dimensional and three-dimensional systems contain structural components that remain largely in place in space within the model, and soluble components that may move more freely.[20] These soluble and structural components can be modified to model changes correlated with genetic and social ancestry, such as changes in skin stiffness and keloid formation.[21] Changes in stiffness can be modeled in hydrogels by promoting different binding reactions between polymers within the hydrogel. Cells in two-dimensional systems can be plated on surfaces with varying stiffness and porosity.[17] Soluble factors may also be modified. These modifiable components include small membrane-bound compartments known as exosomes that transport their contents between cells, and heavy metal concentrations that may vary due to environmental exposure.[20,22,23] Preclinical modeling can therefore encompass the interplay between cells, soluble portions of the extracellular environment, and structural components. This modeling can capture, for instance, differences in cellular response to soluble endocrine-disrupting chemicals to which Black women are disproportionately exposed.[24]

Assessment of cellular and extracellular function in preclinical modeling is tailored to the research question. Assessment can occur at different levels, including determining the expression of individual genes or groups of genes. Transcriptional analysis determines gene expression by estimating the transcription of genomic DNA to RNA. Genome-wide association studies (GWAS) test for associations between variants in genes and in noncoding regions of the genome versus a phenotype of interest (e.g., a particular disease).[15] GWAS can analyze single-nucleotide polymorphisms (SNPs), which are variations in the nucleotide present at a specific position in the genome. GWAS of patient databases and transcriptional analysis of primary cells in vitro together suggest that changes in gene expression observed in vitro may shape clinical outcomes.

Analysis of cellular and extracellular function encompasses protein production and phenotypes impacted by proteins, among other factors. Proteomics investigates whether changes observed at the gene level translate to changes in protein structure and function. Broader cellular and extracellular functions can be assessed using methods such as in vitro assays of cell proliferation and apoptosis to inform preclinical assessments of toxicity, or biomechanical investigations of changes in extracellular matrix stiffness.[22,25] In silico, computer-based modeling may be used to inform biomechanical assessments.[25] In silico analysis can also characterize binding between soluble molecules and the cell surface, along with other assessments. Preclinical modeling may focus on evaluating cellular and extracellular functions that correlate with functions observed in disease, diagnostics, and treatment. In the following subsections, we explore examples of preclinical models to elucidate disease mechanisms and therapeutic outcomes, with consideration of strengths, limitations, and feasibility concerns.

Investigating disease mechanisms

The diversity of human-based preclinical modeling systems facilitates the development of models suited for investigating a variety of conditions. Importantly, human-based preclinical models can elucidate how race, ethnicity, and genetic ancestry shape disease mechanisms in ways that are less feasible with animal models. Human diseases vary widely in characteristics, such as whether the condition develops over time in response to environmental factors or whether clinical presentation depends on similar mechanisms in humans versus nonhuman animals. Conditions may also differ in the conservation of mechanisms across cell types and processes, as well as in disease prevalence across races.[15,19,26,27] The characteristics of a given condition shape how preclinical modeling can be used to assess the role of genetic and social ancestry in disease development.

As described in greater detail in Chapter 3, preclinical modeling elucidates species-specific mechanisms. Nonhuman animals differ from humans in their response to medication, which may hamper animal testing of medications intended for humans. The antiviral medication fialuridine, for example, can cause steatosis, a medication side effect involving the accumulation of fats in the liver, known as lipids. Animal testing, however, did not predict the steatosis observed with the administration of fialuridine in human clinical trials. Steatosis risk was instead explored using a liver-chip model. A liver-chip system using human cells showed lipid accumulation upon fialuridine administration, while a liver-chip using rat cells did not show accumulation. This result was consistent with prior research on species-specific differences in liver lipid accumulation.[19] A human liver-chip system could therefore inform risk assessment before clinical trials and allow for investigation of mechanisms underlying drug-induced lipid accumulation.

While some human-based disease models capture mechanisms that correspond to those observed in animal models, others reveal contrasting mechanisms. Mouse models, for instance, capture some of the human hormonal mechanisms underlying the narrowing of the aortic valve, a section of tissue that regulates the flow of blood out of the heart.[26,28] Human-based modeling can reveal disease processes missing in some animal models, such as steatosis following drug administration in humans but not in rats.[19] Similarly, some mouse models of Coronavirus Disease 2019 (COVID-19) inadequately modeled human symptoms of the disease.[29,30]

Research that distinguishes different human genetic and social ancestry groups can characterize important mechanisms that would otherwise be missed. Distinguishing racial groups facilitated research on the genetic reprogramming of differentiated cells into induced pluripotent stem cells (iPSCs), which can be used to generate personalized disease models. Bisogno and colleagues reprogrammed iPSCs using fibroblasts from self-reported African American or White American donors and used transcriptomic data to assess reprogramming efficiency. The expression of hundreds of genes involved in reprogramming efficiency was specific to African Americans or White Americans, with very few genes showing significance in the combined cohort when ancestry was not taken into account.[27] These findings would not have been identified without a racially diverse set of samples or if the different racial groups did not have identified genetic ancestry, revealing that race can impact cell reprogramming and is thus important to account for in human-based cellular disease models. Additional research is needed to determine the factors that account for the ancestry-dependent altered gene expression associated with reprogramming efficiency.

Human-based preclinical models can uncover cellular processes that depend on cell type and the extracellular environment to promote human health or disease. In the study described above, ancestry-dependent gene expression that correlated with lower reprogramming efficiency also appeared in pathways promoting wound healing and cancer cell development. Specifically, gene expression linked to lower reprogramming efficiency was predictive of lower five-year breast cancer survival in both African American and White American donors. Similar processes, therefore, shape the fate of cancer cells, iPSCs, and impact tissue repair, suggesting that associations with genetic ancestry have differing consequences for human health.[27] Another human cell-based model of triple-negative breast cancer identified ancestry-dependent DNA regions with an open chromatin structure that made the DNA more accessible. These DNA regions in African American-derived cell lines exhibited increased accessibility for dozens of transcription factors associated with phenotypes such as cancer cell proliferation and chemotherapeutic resistance. Analysis of identified genes revealed enrichment in several biological pathways, including hypoxia, and culturing of cells under hypoxia demonstrated ancestry-specific stress responses. The authors concluded that the open chromatin structure of African American triple-negative breast cancer may contribute to a more lethal disease.[31]

Studies of human social and genetic ancestry can also provide disease mechanism information across groups of different sizes. Social and genetic ancestry groups can be large and broadly defined, as in studies of ancestry-dependent infection and immune responses in individuals of European versus African genetic ancestry.[14] Groups can also be smaller and more narrowly defined, as in asthma research on Hutterite versus Amish communal religious groups of European ancestry. Despite Hutterite and Amish children sharing similar genetic ancestry, asthma prevalence was greater among Hutterite children than among Amish children. This increased prevalence was linked to farming practices that differ between Amish and Hutterite social ancestry groups.[32] Revealing this role of social ancestry in asthma risk does not negate the need for studies of genetic ancestry.

Analyses of social ancestry complement analyses of genetic ancestry. A GWAS of a Hutterite population, for instance, identified asthma-associated genes that may shape the behavior of smooth muscle cells. Smooth muscle cells contribute to asthma symptoms by contracting to constrict lung airways. Contractility-associated genes act as a surrogate for the airway constriction that occurs in asthma patients. Smooth muscle cells in a preclinical model were exposed to the asthma-promoting immune signaling molecules interleukin-13 (IL-13) and interleukin-17A (IL-17A). The cells then increased the expression of asthma-associated genes identified in the GWAS of Hutterite individuals. Smooth muscle cells also increased the expression of contractility-associated genes in response to IL-17A, but not in response to IL-13 treatment.[15] Preclinical investigation of human genetic ancestry in a smaller, well-defined population therefore facilitated further understanding of asthma disease processes.

Analyzing therapeutics and diagnostics

Pharmacogenetics and precision medicine are emerging fields of study as therapeutics trend toward addressing the heterogeneity of disease to personalize treatment regimens. These fields aim to answer or provide context for varied therapeutic responses among different populations. However, these studies are predominantly conducted in non-Hispanic Whites of European descent.[33] Some studies that integrate descriptors of ethnicity, race, and genetic ancestry as facets of study are described as follows. These studies serve as potential examples for leveraging diverse human populations for greater physiological relevance and translatability compared to animal model counterparts.

Organic anion transporting polypeptide (OAT1B1) is an important transporter responsible for the initial liver uptake and clearance of drugs and blood compounds. There is particular interest in this polypeptide with regard to the use of statins, a drug class commonly used to treat high cholesterol. The pharmacokinetics of this transporter exhibit an unexplained ethnic variability that leads to differences in drug disposition and response.[33,34] Li and colleagues leveraged clinical data from Caucasian and various Asian demographics to construct a physiologically based probabilistic model of sustained OATP1B1 substrates in blood plasma serum. They were able to construct substrate curves for these different ethnic groups, highlighting the differences for some sustained OATP1B1 substrates (five of these being statins). Ultimately, they stated that differing allele frequencies of single transporters are insufficient to explain all ethnic variability in transporter activity for some substrates and that multiple transporters' encoding alleles should be further studied.[35]

Corticosteroids are commonly used to treat inflammation as they mimic a naturally occurring chemical known as cortisol. Cortisol is a powerful stress hormone, known not only for its antiinflammatory and immunosuppressive properties but also for its regulation of blood pressure, increase in blood glucose, and enhancement of metabolic processes. Prednisolone is a corticosteroid commonly used to treat chronic autoimmune inflammatory illnesses, such as arthritis, colitis, and multiple sclerosis. Overuse of prednisolone has been associated with lowered bone density, eyesight issues, diabetes-related complications, and high blood pressure.[36] In a 2001 study, the pharmacokinetics and pharmacodynamics of prednisolone were studied among a cohort of different racial groups and sexes. This was done by measuring total and free plasma prednisolone concentrations, as well as examining immune cell trafficking. Magee and colleagues determined that immunosuppressive effects required higher dosages in White participants than in Black counterparts, regardless of sex. The authors noted some sex- and race-based differences in the pharmacodynamics and pharmacokinetics of prednisolone but called for more studies with repeated dosing to evaluate the clinical implications of their findings. They also postulated that the variance in drug response among different racial populations could obscure sex-based variability.[37]

Warfarin is a commonly prescribed anticoagulant drug. Due to the drug's narrow therapeutic index, it is a significant contributor to adverse drug effect hospitalizations. GWAS performed in individuals of European and Asian origin have identified three primary SNPs that can account for up to 30% of the variability in dose requirements.[38] However, GWAS-based prediction methods have historically been shown to have low portability outside the originally sampled population when it comes to making predictions on risk or dose.[39] In the case of warfarin, the SNPs identified in European and Asian populations do not explain the same level of variability in individuals of African ancestry. In a GWAS performed by Perea and colleagues, African American individuals were included for the first time in evaluating the contribution of genetic variants to the variation in safe warfarin dosing. This study identified a SNP not previously associated with warfarin dosage that explained 5% of the variability in African American populations and was common in African American populations, with a frequency of 25% in the studied cohorts. The authors concluded that safe and equitable pharmacogenetic dosing should be achieved through the identification and inclusion of population-specific variants.[38]

Equitable use of therapeutics is a complex, multifaceted issue with underlying social and genetic ancestry-based drivers. A lack of diversity in cohorts used to test therapeutics leads to incomplete information regarding their safety and efficacy across demographics. A similar trend can be observed in the development of diagnostics. A study conducted by Sarnowski et al. investigated the effects of rare and common genetic variants associated with hemoglobin A1c (HbA1c) in the misdiagnosis of diabetes and mismanagement of glycemia (glucose in the blood). HbA1c is a measurement of average blood glucose over the previous 2–3 months, which corresponds to the turnover time for red blood cells (RBCs). Blood exposure to glucose causes irreversible binding of sugars to proteins, which in turn leads to several vascular issues. However, variations in RBC turnover can impact the accuracy of this readout. Sarnowski and colleagues leveraged whole genome sequencing data for over 10,000 individuals across four ancestries (European, African American, Hispanic, and Asian) from the National Heart, Lung, and Blood Institute's Trans-Omics for Precision Medicine (TOPMed) program. The authors determined that a variant associated with decreased HbA1c readout could lead to the underdiagnosis of diabetes in African American and Hispanic individuals. The researchers concluded that the consideration of these variants is necessary to make precision medicine-based diagnoses.[40]

Conclusion

Therapeutics, diagnostics, clinical work, and laboratory methodologies stem from preclinical research. Animal models have been conventionally used to translate biomedical research from the laboratory benchtop to clinical settings. However, historically, this approach has not been efficient and costs billions of US dollars annually.[41] There are substantial biological differences between animals and humans that contribute to the insufficiency of animal models in accurately studying human diseases, including the mechanisms by which social and genetic ancestry impact human health and disease.[19] In addition, the development of novel animal models can be time-consuming and unsuitable in time-sensitive contexts such as a pandemic.[30,42] The conventional use of animals thus leaves biomedical research at a deficit for efficiently capturing heterogeneous disease presentations among different ancestral human populations. With these deficiencies in mind, there is a critical need to enhance the translatability of preclinical work.

Preclinical nonanimal models have not only supplemented animal model research but have also provided disease mechanism information absent in animals.[19] The use of primary human samples in preclinical models may have greater physiological relevance. Primary cells isolated from patients better capture the genetic heterogeneity present in human disease, along with socioeconomic and demographic impacts, such as environmental exposures. Immortalized cell lines offer several benefits for modeling, including ease of maintenance, genetic characterization of the cell line by other research groups, and comparability among work that implements the same line.[13]

While there are myriad strengths in incorporating diverse human samples into preclinical work, it would be misguided to say these models are perfectly standardized or offer perfect translatability. Table 1 summarizes the limitations of ancestry-based studies and the use of human samples.

Limitations within scientific works are not unique to these models, but they should be identified and disclosed when conducting this work. There are also several methods for mitigating these limitations:

- Clearly state in the methods section the context in which the ancestral population descriptors are being leveraged. Care should be taken when extrapolating ancestry-related conclusions from one demographic group to another.
- Being inclusive in recruitment efforts.
- Approaching studies with a representative sample of patients for the given disease prevalence helps alleviate the limitations of portability.
- Consider incorporating multiple ancestry-determination techniques, such as genotyping and social surveys, to fully capture population predispositions to environmental, societal, and genetic influences.
- Consulting reference materials and guidelines established by outside agencies, as well as collaborating with experts in adjacent fields such as the social sciences.
- Collaborating with patient advocacy groups to ensure informed consent from donors and their communities.
- Taking into account donor interests when generating cell lines helps avoid repeating the historical exploitation of vulnerable populations.
- Reaching out to suppliers and vendors to determine if ancestry information is available.
- Collaborating with antiracism coalitions or advocacy groups to include antiracism statements in publications.

The responsible incorporation of diverse human populations in preclinical studies may strengthen benchtop-to-bedside translation and should include two main responsibilities at a minimum: (1) the incorporation of a representative human population for the given

TABLE 1 Limitations of ancestry-based in vitro research.

Research context	Limitations
Cell culture work	Cell lines accumulate mutations over time with their expansion, which can interfere with comparability.[13]
	Phenotypic changes that allow cell lines to be maintained long-term in vitro may cause cell lines to function differently from the cell population they are meant to model. For example, genetic changes that promote cell proliferation needed for expansion could interfere with investigations into how ancestry shapes proliferative capacity.[13,43]
	Ancestry of cell lines can be mislabeled. Studies, particularly those related to cancer, have found mislabeling of ancestral backgrounds for several cell lines.[44] In addition, it is well known that mammalian cell lines are subject to cross-contamination among one another, and investigators rarely test for this when conducting experiments.[43]
	The diversity of primary human samples can be limited based on recruitment efforts. Attitudes surrounding scientific mistrust, stemming from the historical unethical experimentation on vulnerable populations, can create an additional barrier to recruitment.
Incorporation of genetic ancestry in studies	Genetic ancestral population descriptors do not always match what is clinically observed. For example, an intervention study on cardiovascular (CV) risk found that the genetics of West African individuals demonstrated no predisposition to CV events, despite the prevalence of CV pathologies within this population.[45]
	Statistical analysis and software used to assess genetic ancestry are not standardized or uniformly ubiquitous.
	Admixture mapping is not standardized, and variability in the number of population origins used as references can increase cross-validation error.
	Polygenic risk scores, a measure of disease risk based on genes, have low portability outside of the ancestral group from which they are sampled. This impacts prediction accuracy and restricts the context in which these scores can be used.[39]
	Misuse of ancestral population descriptors can have extreme consequences and lead to damaging, unfounded claims about groups of people.
Incorporation of social ancestry in studies	Sociocultural settings and timeframes limit social ancestral population descriptors (race, ethnicity, nationality). Therefore these descriptors may be difficult to apply outside of the aforementioned contexts.
	Social ancestral population descriptors have not been proven to have intrinsic biological validity but have been leveraged as proxies for the environment.
	Misuse of ancestral population descriptors can have extreme consequences and lead to damaging, unfounded claims about groups of people.[46]

study and (2) transparency in the use of ancestral population descriptors. Lack of representation within preclinical modeling can have severe consequences and propagate health inequities. Clinical technologies developed with an unrepresentative demographic are unable to capture the breadth of disease presentations—indeed, a major limitation of animal-based preclinical models. This misrepresentation then permeates the public distribution of health and wellness. However, when leveraged properly, the use of ancestral population descriptors can be a powerful tool for improved translatability. This effort is multifaceted and requires the involvement of a multitude of disciplines to propel the field of biomedical research further.

References

[1] National Institutes of Health. National Human Genome Research Institute. Use of population descriptors in genomics; 2024. Available at: https://www.genome.gov/about-genomics/policy-issues/population-descriptors-in-genomics. Accessed 31 March 2025.

[2] Mulligan CJ. Systemic racism can get under our skin and into our genes. *Am J Phys Anthropol.* 2021;175(2):399–405. doi:10.1002/ajpa.24290.

[3] Patel RA, Musharoff SA, Spence JP, et al. Genetic interactions drive heterogeneity in causal variant effect sizes for gene expression and complex traits. bioRxiv. United States: bioRxiv. 2021. doi:10.1101/2021.12.06.471235.

[4] Wesselink A, Kira K, Hatch EE, et al. Residential proximity to major roads and fecundability in a preconception cohort. *Environ Epidemiol.* 2020;4(6). doi:10.1097/EE9.0000000000000112.

[5] American Heart Association. Newsroom. Food insecurity linked to increased heart risks for people with peripheral artery disease. American Heart Association, Dallas TX; 2024. Available at: https://newsroom.heart.org/news/food-insecurity-linked-to-increased-heart-risks-for-people-with-peripheral-artery-disease Accessed 31 March 2025.

[6] White AJ, Fisher JA, Sweeney MR, et al. Ambient fine particulate matter and breast cancer incidence in a large prospective US cohort. *J Natl Cancer Inst.* 2024;116:53–60.

[7] Cooper RS. Race in biological and biomedical research. *Cold Spring Harb Perspect Med.* 2013;3(11). doi:10.1101/cshperspect.a008573.

[8] Mathieson I, Scally A. What is ancestry? *PLoS Genet.* 2020;16(3). doi:10.1371/journal.pgen.1008624.

[9] Korunes KL, Goldberg A. Human genetic admixture. *PLoS Genet.* 2021;17(3). doi:10.1371/journal.pgen.1009374.

[10] Zakharia F, Basu A, Absher D, et al. Characterizing the admixed African ancestry of African Americans. *Genome Biol.* 2009;10(12). doi:10.1186/gb-2009-10-12-r141.

[11] DeGroot A, McMurry J, Moise L. Prediction of immunogenicity: in silico paradigms, ex vivo and in vivo correlates. *Curr Opin Pharmacol.* 2008;8(5):620–626. doi:10.1016/j.coph.2008.08.002.

[12] Richtr M, Piwocka O, Musielak M, et al. From donor to the lab: a fascinating journey of primary cell lines. *Front Cell Dev Biol.* 2021;9. doi:10.3389/fcell.2021.711381.

[13] Kaur G, Dufour JM. Cell lines: valuable tools or useless artifacts. *Spermatogenesis.* 2012;2(1):1–5. doi:10.4161/spmg.19885.

[14] Randolph HE, Fiege JK, Thielen BK, et al. Genetic ancestry effects on the response to viral infection are pervasive but cell type specific. *Science.* 2021;374(6571):1127–1133. doi:10.1126/science.abg0928.

[15] Thompson EE, Dang Q, Mitchell-Handley B, et al. Cytokine-induced molecular responses in airway smooth muscle cells inform genome-wide association studies of asthma. *Genome Med.* 2020;12(1). doi:10.1186/s13073-020-00759-w.

[16] Loo AEK, Ho R, Halliwell B. Mechanism of hydrogen peroxide-induced keratinocyte migration in a scratch-wound model. *Free Radic Biol Med.* 2011;51(4):884–892. doi:10.1016/j.freeradbiomed.2011.06.001.

[17] Jha A, Moore E. Collagen-derived peptide, DGEA, inhibits pro-inflammatory macrophages in biofunctional hydrogels. *J Mater Res.* 2022;37(1):77–87. doi:10.1557/s43578-021-00423-y.

[18] Runft S, Färber I, Krüger J, et al. Alternatives to animal models and their application in the discovery of species susceptibility to SARS-CoV-2 and other respiratory infectious pathogens: a review. *Vet Pathol.* 2022;59(4):565–577. doi:10.1177/03009858211073678.

[19] Jang KJ, Otieno MA, Ronxhi J, et al. Reproducing human and cross-species drug toxicities using a liver-chip. *Sci Transl Med.* 2019;11(517). doi:10.1126/scitranslmed.aax5516.

[20] Karamanos NK, Theocharis AD, Piperigkou Z, et al. A guide to the composition and functions of the extracellular matrix. *FEBS J.* 2021;288(24):6850–6912. doi:10.1111/febs.15776.

[21] Macarak EJ, Wermuth PJ, Rosenbloom J, Uitto J. Keloid disorder: fibroblast differentiation and gene expression profile in fibrotic skin diseases. *Exp Dermatol.* 2021;30(1):132–145. doi:10.1111/exd.14243.

[22] Karri V, Kumar V, Ramos D, Oliveira E, Schuhmacher M. Comparative in vitro toxicity evaluation of heavy metals (lead, cadmium, arsenic, and methylmercury) on HT-22 hippocampal cell line. *Biol Trace Elem Res.* 2017;184(1):226–239. doi:10.1007/s12011-017-1177-x.

[23] Lin W. Comparison of urine heavy metals in exclusive menthol and non-menthol cigarette users by race/ethnicity: the 2015–2016 National Health and Nutrition Examination Survey Special Sample. *Tob Prev Cessat.* 2023;9(June):1–5. doi:10.18332/tpc/167389.

[24] Tapia JL, McDonough JC, Cauble EL, et al. Parabens promote protumorigenic effects in luminal breast cancer cell lines with diverse genetic ancestry. *J Endocr Soc.* 2023;7(8). doi:10.1210/jendso/bvad080.

[25] Wang T, Chen P, Zheng M, et al. In vitro loading models for tendon mechanobiology. *J Orthop Res.* 2018;36(2):566–575. doi:10.1002/jor.23752.

[26] Schroeder ME, Batan D, Rodriguez AG, et al. Osteopontin activity modulates sex-specific calcification in engineered valve tissue mimics. *Bioeng Transl Med.* 2023;8(1). doi:10.1002/btm2.10358.

[27] Bisogno LS, Yag G, Bennett BD, et al. Ancestry-dependent gene expression correlates with reprogramming to pluripotency and multiple dynamic biological processes. *Sci Adv.* 2020;6(47):abc3851.

[28] Fleury M, Annabi MS, Voisine M, et al. Impact of sex and sex hormones on pathophysiology and progression of aortic stenosis in a murine model. *Physiol Rep.* 2022;10(16). doi:10.14814/phy2.15433.

[29] Holmes EC, Goldstein SA, Rsmussen AL, et al. The origins of SARS-CoV-2: a critical review. *Cell.* 2021;184(19):4848–4856. doi:10.1016/j.cell.2021.08.017.

[30] Takayama K. In Vitro and Animal Models for SARS-CoV-2 research. *Trends Pharmacol Sci.* 2020;41(8):513–517. doi:10.1016/j.tips.2020.05.005.

[31] Harris AR, Panigrahi G, Liu H, et al. Chromatin accessibility landscape of human triple-negative breast cancer cell lines reveals variation by patient donor ancestry. *Cancer Res Commun.* 2023;3(10):2014–2029. doi:10.1158/2767-9764.crc-23-0236.

[32] Ober C, Sperling A, von Mutius E, Vercelli D. Immune development and environment: lessons from Amish and Hutterite children. *Curr Opin Immunol.* 2017;48:51–60. doi:10.1016/j.coi.2017.08.003.

[33] Ortega VE, Meyers DA. Pharmacogenetics: implications of race and ethnicity on defining genetic profiles for personalized medicine. *J Allergy Clin Immunol.* 2014;133(1):16–26. doi:10.1016/j.jaci.2013.10.040.

[34] Lee HH, Ho RH. Interindividual and interethnic variability in drug disposition: polymorphisms in organic anion transporting polypeptide 1B1 (OATP1B1; SLCO1B1). *Br J Clin Pharmacol.* 2017;83(6):1176–1184. doi:10.1111/bcp.13207.

[35] Li R, Barton HA. Explaining ethnic variability of transporter substrate pharmacokinetics in healthy Asian and Caucasian subjects with allele frequencies of OATP1B1 and BCRP: a mechanistic modeling analysis. *Clin Pharmacokinet.* 2018;57(4):491–503. doi:10.1007/s40262-017-0568-7.

[36] Yasir M, Goyal A, Sonthalia S. Corticosteroid adverse effects. In: *StatPearls.* Treasure Is, FL: StatPearls Publishing; 2024.

[37] Magee MH, Blum RA, Lates CD, Jusko WJ . Prednisolone pharmacokinetics and pharmacodynamics in relation to sex and race. *J Clin Pharmacol.* 2001;41(11):1180–1194. doi:10.1177/00912700122012733.

[38] Perera MA, Cavallari LH, Limdi NA, et al. Genetic variants associated with warfarin dose in African-American individuals: a genome-wide association study. *Lancet.* 2013;382(9894):790–796. doi:10.1016/S0140-6736(13)60681-9.

[39] Ding Y, Hou K, Xu Z, et al. Polygenic scoring accuracy varies across the genetic ancestry continuum. *Nature.* 2023;618:774–781. Available at: https://www.nature.com/articles/s41586-023-06079-4.

[40] Sarmwski C, Leong A, Raffield LM, et al. Impact of rare and common genetic variants on diabetes diagnosis by hemoglobin A1c in multi-ancestry cohorts: the trans-omics for precision medicine program. *Am J Hum Genet.* 2019;105(4):706–718. doi:10.1016/j.ajhg.2019.08.010.

[41] Keen J. Wasted money in United States biomedical and agricultural animal research. *Brill.* 2019;244–272. doi:10.1163/9789004391192_011.

[42] Fan C, Wu Y, Rui X, et al. Animal models for COVID-19: advances, gaps and perspectives. *Signal Transduct Target Ther.* 2022;7(1). doi:10.1038/s41392-022-01087-8.

[43] Kessler MD, Bateman NW, Conrads T, et al. Ancestral characterization of 1018 cancer cell lines highlights disparities and reveals gene expression and mutational differences. *Cancer.* 2019;125(12):1945–2133. doi:10.1002/cncr.32020.

[44] Hooker SE, Woods-Burnham L, Bathina M, et al. Genetic ancestry analysis reveals misclassification of commonly used cancer cell lines. *Cancer Epidemiol Biomarkers Prev.* 2019;28(6):1003–1009. doi:10.1158/1055-9965.EPI-18-1132.

[45] Rao S, Segar MW, Bress AP, et al. Association of genetic West African ancestry, blood pressure response to therapy, and cardiovascular risk among self-reported black individuals in the systolic blood pressure reduction intervention trial (SPRINT). *JAMA Cardiol.* 2021;6(4):388–398. doi:10.1001/jamacardio.2020.6566.

[46] Carlson J, Henn BM, Al-Hindi DR, Ramachandran S. Counter the weaponization of genetics research by extremists: geneticists must rethink how they conduct their research and how they communicate results. *Nature* 2022;610:444–447. Available at: https://www.nature.com/articles/d41586-022-03252-z.

PART III

Achieving a practical shift toward nonanimal-based research and testing

CHAPTER

12

Overcoming inertia against nonanimal research and testing in the United States

Catharine E. Krebs, Emily Anderson, Janine McCarthy, Eryn Slankster-Schmierer, Jarrod Bailey, and Elizabeth Baker

Physicians Committee for Responsible Medicine, Washington, DC, United States

Abstract

The emerging developments in the United States signal a growing shift toward human-focused, nonanimal research and testing. This is true in Congress, in academic research centers, at agencies like the National Institutes of Health, the Environmental Protection Agency, and the Food and Drug Administration, as well as in the biotechnology, pharmaceutical, cosmetic, and chemical industries. However, systemic barriers remain, characterized by institutional inertia that inhibits change and upholds the status quo of animal use. In this chapter, we characterize this inertia as it relates to the science, policy, and training contexts. We describe various stakeholders and where and how they can apply pressure toward change, providing a comprehensive set of recommendations that will improve translatability and foster a more effective health research enterprise.

Keywords: Biomedical research; Drug development; Regulatory testing; Research policy; National Institutes of Health; Food and Drug Administration; Alternative methods; Nonanimal methods; Translatability; Institutional inertia

Introduction

Biomedical researchers and regulatory scientists are increasingly turning to nonanimal methods that use human cells, tissue, and data to model human biology and disease, develop therapeutics, and test chemicals and other compounds. The rationale for the need to shift toward nonanimal methods has been described throughout this book. The emerging developments in the United States signal a major change in support of this shift: in Congress, in biotechnology, pharmaceutical, cosmetic, and chemical industries, and at agencies like the Food and Drug Administration (FDA) and National Institutes of Health (NIH). On April 10, 2025, the FDA announced a plan to phase out animal testing requirements for

Moving Toward Nonanimal Approaches in Medical Research and Testing
https://doi.org/10.1016/B978-0-443-26514-3.00013-8

pharmaceuticals, beginning with monoclonal antibodies.[1] On April 29, 2025, the NIH announced a new initiative to prioritize human-based research technologies while reducing animal use.[2]

Despite this exciting progress, biomedical research and testing still rely heavily on other animal species, with as many as 100 million dogs, cats, monkeys, mice, rats, and other animals being used in US laboratories each year.[3,4] Systemic barriers inhibit progress toward reducing these numbers and replacing animals with more effective human-centered approaches. In this context, *scientific inertia* has been defined as "a certain degree of conservatism in scientific practice regarding the development and use of non-animal approaches to replace animal experimentation."[5] Government agencies, legislators, academic scientists, companies and industry groups, nongovernmental organizations (NGOs), the public, and the media play a part in upholding this inertia and must, therefore, also contribute to overcoming it.

This chapter discusses who these stakeholders are in the US biomedical research and regulatory testing context and how they can enact change. Then, scientific, policy, and training barriers to the shift toward nonanimal methods are characterized, along with ongoing progress and remaining challenges in overcoming them. The shift to nonanimal research and testing in the United States is not a matter of *if* but *when*. This chapter serves as a roadmap toward that goal, which will improve clinical translation and patient outcomes, promote safer products, and reduce animal use.

Stakeholders, pressure points, and mechanisms of change

Stakeholders, people with the power to make change and/or those who are affected by change, have important roles in driving progress in nonanimal methods and overcoming inertia. Different stakeholders have various ways they can influence research policies and practices to support a transition away from animal use (Table 1). When working to overcome inertia, it is helpful to understand who the key stakeholders are, which pressure points they can influence, and how they can use their power to create change.

Government agencies

Agencies set requirements and expectations for research through regulations (or rules) and other policies (guidelines, guidance, statements, waivers), which provide specifics to implement laws enacted by Congress. Regulations and policies that encourage the use of nonanimal methods can change practices for entire industries or research areas. Conversely, regulations and policies that require animal methods over nonanimal methods can prevent progress in the broader use of nonanimal methods. Regulatory agencies like the FDA and EPA can help overcome inertia against nonanimal research by ensuring their regulations and policies adequately allow for or encourage nonanimal approaches and do not specify requirements for animal tests. In addition, research agencies like the NIH can prioritize the development and use of nonanimal methods within strategic plans, which provide coordination and direction to research projects, infrastructure, and training initiatives.

TABLE 1 Stakeholders who can impact change toward nonanimal methods.

Stakeholders	Mechanisms of change
Government agencies	Providing funding for nonanimal methods projects, infrastructure, and training
	Setting research policies that allow for or encourage nonanimal method use
	Developing, and using nonanimal methods and providing training in them
Legislators	Creating and changing laws to support nonanimal methods
	Funding nonanimal methods programs at government agencies
	Directing agency activities and policies
Members of the public	Contacting legislators and agencies
	Responding to public input opportunities
	Buying cruelty-free products
Academic researchers	Developing, and using nonanimal methods and training other researchers in them
	Publishing and presenting research and perspectives about nonanimal methods and their advantages
	Serving as grant and journal reviewers
	Lobbying legislators and agencies
Companies and industry groups	Developing and using nonanimal methods
	Publishing and presenting research and perspectives about nonanimal methods and their advantages
	Lobbying legislators and agencies
Nongovernmental organizations	Leading grassroots advocacy efforts
	Lobbying legislators
	Providing resources to support nonanimal methods, projects, infrastructure, and training
	Publishing and presenting research and perspectives about nonanimal methods and their advantages
Scholarly and mass media	Providing a venue for scholarship, perspectives, and reporting that supports nonanimal methods

Legislators

Congress can dictate how federal funds are spent via annual funding (or appropriations) bills. Because agencies rely on funding from Congress to operate, Congress has the power to require, encourage, or prohibit certain activities. For instance, legislators can include language in funding bills that blocks funds for animal research, provides funding for programs to advance nonanimal methods, requires transparency measures and reporting, or includes other directives to spur change in support of nonanimal research. Reauthorization bills are also opportunities for Congress to revisit legislation and make improvements to programs and policy mandates to meet changing needs. Certain laws have expiration dates and are intended to be revisited every few years to keep programs running. Like funding bills, these reauthorizations allow legislators to direct agency operations and continue, launch, or end certain programs, which can be used to promote greater use of nonanimal methods. Advocacy from the public can help educate legislators, engage them in issues, and lead to policy change. For example, lobbying by advocacy groups on the need for the FDA to better integrate nonanimal methods in drug safety and efficacy testing led members of Congress to write to the agency in 2019, urging the FDA to modernize regulatory requirements and establish a program to qualify nonanimal drug development methods.[6] Ongoing public influence then helped ensure similar directives to the FDA were included in funding bills, and in 2020, the FDA launched an important qualification program to review and accept nonanimal methods for regulatory use.[7] This progress and increased attention to the use of animals in drug testing helped lead Congress to, for the first time, provide the FDA with dedicated funding for reducing animal testing through nonanimal methods in 2023.[8]

Aside from passing legislation, there are nonlegislative strategies that can be used to elevate attention to nonanimal methods. Sign-on letters are a common way for legislators to communicate with agencies about pressing topics outside of the legislative process. By collecting signatures from their colleagues, sign-on letters can demonstrate broad congressional support for requests. Other opportunities include hearings, which are meetings held in Congress that usually involve the questioning of witnesses to gain information about a given issue, bill, or government activity. Hearings on topics related to research and testing could serve as platforms for legislators to question witnesses about advancing nonanimal science.

Members of the public

The public can influence federal agencies by submitting input on various agency activities. The federal government communicates with the public about government policies, requirements, and meetings via the Federal Register. Public input opportunities are baked into the federal rulemaking process, meaning agencies are required to alert the public of their intent to change or add to existing regulations and allow the public to participate in the decision-making process. Acting on these opportunities for public input can impact agency activities. The public can also have a direct influence on legislators, especially those who represent them. Legislators are elected by the people they represent and have a duty to take their concerns seriously. Electing policymakers who are committed to reducing animal experimentation is a key part of moving the needle. Beyond voting, any member of the public can effect change by engaging with their members of Congress. Meeting with congressional staff is an effective way of discussing issues and building relationships that can lead to legislative support for nonanimal methods. Sending personalized emails and calling congressional offices are also effective options. Regardless of the avenue, it is important for constituents to let legislators know they want to advance animal-friendly initiatives. As consumers, members of the public can also influence change by purchasing cruelty-free (not tested on animals) products whenever possible.

Academic researchers

Academic researchers can help create change because they bring important scientific expertise and unique research perspectives. Researchers working with nonanimal methods can provide first-hand accounts of how nonanimal approaches have aided their research and how their work would be less effective if they used animals, as well as demonstrate the practical impact of policies on scientific practices. Researchers can express this point of view effectively during informal interactions with their institutional peers, collaborators, and other researchers within their field, but also during opportunities to influence decision-making within their workplaces or in broader scientific communities. For instance, researchers using nonanimal methods should apply to be considered for advisory groups, which agencies form to help advise on policies and programs, or to become grant or journal reviewers. It is essential that experts in nonanimal methods are represented in these groups to counter any bias against or lack of knowledge about nonanimal methods, and there are frequent calls for new members.[9] Publishing papers, presenting at scientific conferences, and training the next generation of researchers are also great strategies to spread information about the advantages of nonanimal methods. Researchers can also bring their knowledge to meetings with legislators or provide testimonies at legislative hearings or briefings, offering an authoritative perspective about the path forward for effective science policy reforms.

Companies and industry groups

Many health-related industries, including the biotechnology, pharmaceutical, cosmetic, and chemical industries, are regulated by the federal government. Like academic researchers, industry stakeholders can speak to the real-world impact of nonanimal methods in the development of medicines and other products. Companies can develop effective nonanimal approaches and employ them for regulatory testing (when legally allowed). They can provide perspective on the economic benefits of implementing nonanimal approaches in their work, demonstrating how these methods improve efficiency and save money by being more predictive for humans. Companies are also influential constituents, able to lobby legislators for policies that benefit their industry and the local and national economy, either independently or as part of a trade group's broader effort. When necessary, companies have pursued legal action when nonanimal methods are not appropriately accepted by government regulatory agencies such as the FDA, as in the case of *Vanda Pharmaceuticals, Inc. v. FDA*, No. 23-5200 (D.C. Cir. 2024)[10] (for an industry perspective from Vanda Pharmaceuticals, see Chapter 14). Because economic factors are important for many legislators, companies can also influence legislative stances and are essential stakeholders in encouraging a shift to nonanimal methods in federal laws and policies.

Nongovernmental organizations

NGOs operate independently to advance their missions, but they also rely on collaborations with a range of stakeholders. NGOs can help drive strategies to overcome inertia by playing a coordinating role, including building relationships across stakeholder groups to work toward shared goals, organizing meetings, trainings, and other events focused on advancing nonanimal methods, developing policy solutions, and working to implement these solutions. NGOs can also provide funds for nonanimal methods projects, infrastructure, and training. Many NGOs working to reduce the use of animals in experiments engage the public to take action in support of their advocacy. For instance, NGO lobbying efforts can benefit from having constituents and scientific experts contact legislators or join congressional meetings. NGOs can also respond to public input opportunities and publish research and perspectives about the advantages of nonanimal methods. NGOs often have connections within agencies and relevant industries and a broad membership of supporters. These relationships can be leveraged in various strategies and contribute to a multifaceted understanding of how to best affect change to support nonanimal methods.

Scholarly and mass media

Scholarly and mass media, encompassing scientific journals, science and traditional news outlets, as well as other digital, print, and broadcast media outlets, can play an important role in overcoming inertia. They can provide a venue for scholarship, perspectives, and reporting that support nonanimal methods, highlighting their advantages and the scientific, economic, and ethical shortcomings of animal experiments. In doing so, they can shape public opinion, which is crucial for influencing other important stakeholders, such as agencies and Congress.

Building on progress in biomedical research, drug development, and regulatory testing

Overcoming scientific barriers

Key to advancing the broader use of nonanimal research approaches for disease modeling and product testing will be overcoming scientific and operational barriers. While broad considerations are discussed here, specific methodological barriers are discussed in Part II of this book.

Many nonanimal methods are relatively novel, such as those that can be categorized as new approach methodologies (NAMs): in vitro, in chemico, and in silico approaches that have the potential to improve understanding of human health and disease mechanisms and reduce reliance on animals.[11] While many researchers are eager to investigate new methods and incorporate them into their projects, others are more cautious. Therefore, further investment is needed to establish wider confidence in their utility to model human biology and clinical characteristics. Historically, benchmarking nonanimal methods has involved comparisons to animal test results, but increasingly, stakeholders are recognizing this as problematic due to the lack of physiological relevance and validity testing of animal-based methods. Formal validation practices for NAMs in regulatory toxicology are moving away from comparisons to nonhuman animal test results and instead are incorporating human data, resulting in the acceptance of human-specific approaches that outperform animal tests.[12–15] Patient-derived data should also be used as benchmarks to corroborate findings from human disease models and establish model validity. Other considerations that are crucial for validation efforts include fitness for purpose, human biological relevance, technical characterization, data integrity and transparency, and independent review.[12] Validation of NAMs should continue to be refined and tailored according to each use case to help build confidence among researchers, industry, and regulators. Sequencing, imaging, and other data-driven approaches can accelerate model validation by enabling deep resolution and high-throughput comparisons between experimental platform readouts and gold-standard clinical characteristics.[16] In addition, agencies and companies could work together to anonymize existing human data to be made available for use in model evaluation.

Reproducibility across experiments and research labs is crucial for building trust in NAMs and maximizing the translatability of findings. Animal-based methods have long suffered from poor reproducibility, presenting an opportunity for NAM-centered research to learn from these errors. To improve reproducibility, the NIH and other research funders should establish standard in vitro reporting guidelines, require all grantees to follow these standards, and encourage publishers to require these standards. Several examples of in vitro reporting guidelines exist that the NIH could build from or point toward.[17–20] Prospective study design and registration can also help make progress toward these goals and ensure rigor and reproducibility. Funders should aim to raise awareness and understanding of prospective registration and develop and implement programs to evaluate the impacts of prospective registration on research quality, especially in the context of NAMs and other novel methodological technologies.[21] More on building confidence in NAMs is described in Chapter 17.

Facilities designed to breed, house, and experiment on animals, the specialized animal laboratory equipment and supply industry, scientific conferences, professional societies, and academic programs can all act as infrastructure dedicated (and sometimes financially incentivized) to maintaining the animal research status quo. As much as animal research infrastructure helps perpetuate animal use, infrastructure dedicated to nonanimal approaches must be established and expanded to support the shift toward more effective human-centered research and testing. Evidence indicates that biomedical researchers are eager to use 3D human in vitro models, but the full potential of these technologies is held back by high entry costs and a lack of ready-to-use systems.[22] Therefore, to help make NAMs and other translational models more accessible and operational for researchers, coordinated infrastructure is crucial. Such infrastructure should include mechanisms to disseminate NAMs resources, technology, and expertise (e.g., through consortia), support for small- to mid-scale physical laboratory infrastructure, as well as larger-scale infrastructure like core facilities that can develop and run NAMs assays and provide training.[23] In addition, because in vitro models rely on primary tissue and cell lines, key to this infrastructure will be accessible and diverse biobanks. Large-scale patient cohorts and biobanks must be accessible to researchers and model developers and should be representative of diverse cell and tissue types, phenotypes, genotypes, and demographics. More on the use of human tissues and biobanks is detailed in Chapter 10. Funders and research institutions must play a major role in bolstering translational infrastructure and streamlining research operations.

Overcoming policy barriers

Biomedical research

The NIH is the largest funder of biomedical research in the world, with an annual budget of over $47 billion in public funds, 83% of which goes to funding extramural research at universities, medical schools, and other research institutions across the United States.[24] The agency plays a major role not just in funding research projects, but also in training the next generation of biomedical researchers and supporting research infrastructure like core laboratories, biobanks, and databases. NIH priorities influence the success of drug development, diagnostics, and public health guidance in the United States and around the world. With such a large responsibility, it is crucial that the NIH invests in the most promising and effective science. The agency has even acknowledged that its investment in animal experiments creates "a substantial opportunity cost, if some models poorly represent the corresponding human disease."[25]

In 2022, a working group of the Advisory Committee to the Director (ACD) of the NIH was chartered to explore high-priority areas for investment in NAMs.[26] In 2023, the group released its final report, which included recommendations for catalyzing the development and use of NAMs, such as supporting NAMs infrastructure, providing training, and tracking investments to identify gaps and support new initiatives.[23] The group described NAMs as "offering tremendous promise for helping us better understand fundamental biology to advance human health" while simultaneously providing the potential to reduce or replace the use of animals and to "enable research to be done more quickly, by more researchers, and at a more affordable cost."[23] In 2024, the agency accepted the group's recommendations and began implementing some of them through a new program called Complement Animal Research in Experimentation (Complement-ARIE).[27,28] Finally, in April 2025, the NIH announced a new initiative to prioritize human-based research technologies while reducing animal use, detailing several measures reflected in the group's recommendations, including expanded funding, training, and infrastructure for nonanimal approaches, as well as a new office to coordinate NIH-wide and interagency efforts.[2] Altogether, these steps represent a watershed moment in the shift toward more human-relevant medical research that does not use animals.

These efforts build on programs like the decade-old Tissue Chips program led by the National Center for Advancing Translational Sciences[29] (NCATS) and the interagency center that develops and evaluates animal alternatives in toxicology, the National Toxicology Program Interagency Center for the Evaluation of Alternative Toxicological Methods (NICEATM).[30] NCATS is the NIH hub for innovation in highly effective animal-free medical research. Because it focuses on translational research—research that bridges experiments with clinical interventions—NCATS recognizes the disadvantages of animal models, especially their inability to replicate human biology and predict clinical responses, and has thus been a pioneer in the development of nonanimal approaches. NICEATM is also instrumental in advancing NAMs in biomedical and toxicological research. NICEATM evaluates NAMs for regulatory applications, promotes their scientific validation and regulatory acceptance, fosters national and international harmonization in their use, and provides training and demonstrations for scientists. By building on these programs and progress to refocus medical research on humans, the NIH can accelerate the capabilities of science and healthcare as we know it.

The NIH has made major progress toward the goals of improving translation and reducing animal use, but further commitments, policy changes, high-level strategies, and structural shifts are needed. One of the first steps that will be crucial for achieving these goals is for the NIH to publicly report the number of animals used in the research it funds. There is no established system for precisely tracking this number; instead, the NIH requires that facilities provide crude estimates every four years rather than exact annual numbers,[31] and those numbers are only accessible to the public through Freedom of Information Act requests that take months to fulfill. Also unclear is the number of projects and dollars spent on animal-based and nonanimal-based research. In 2012, an NIH official stated that 47% of NIH-funded research had an animal-based component,[32] and a 2016 NIH blog revealed that approximately 40% of R01 grants (research project grants) used mice.[33] During a December 2022 ACD meeting, a figure showing the percentage of competitive awards using NAMs (excluding clinical research) each year since 1981 was presented, showing a steady increase in NIH investment in NAMs, with the 2021 percentage resting just below 8%.[34] These data are crucial not only for justifying large-scale coordinated efforts like Complement-ARIE but also for identifying gaps in research, promising areas for continued investment, and monitoring the progress of these efforts. A measure in the new initiative to prioritize human-based research—to "publicly report on research spending annually to measure progress toward reduction of funding for animal studies and an increase in funding for human-based approaches"—will begin to provide some transparency.[2] Understanding how many animals are used in research and testing will provide basic public accountability on animal use, federal research spending, and progress toward animal reduction and replacement goals.

Another important step will be incorporating progress and recommendations made by ongoing efforts at NCATS, NICEATM, Complement-ARIE, and the ACD NAMs Working Group into agency-wide priorities and strategies. The NIH-Wide Strategic Plan is a key coordinating and goal-setting document that is mandated by Congress to be developed every six years to provide direction to biomedical research, facilitate agency-wide collaboration, leverage scientific opportunities, and advance biomedicine.[35] Prioritizing and coordinating the broader development and use of NAMs and other nonanimal approaches through this strategic plan and subsequent investments will enable better understanding and improvement of human health and reduce the use of animals in biomedical research. It is also crucial that the agency implement comprehensive stakeholder engagement as it develops and implements strategic plans. For example, in preparation for the NCATS Strategic Plan for 2024–2029, the Center demonstrated open and comprehensive strategic planning, with stakeholder input spanning the earliest stages.[36] Similarly, the Common Fund hosted a series of listening sessions to gather broad stakeholder input on the goals and structure of the Complement-ARIE Program.[37] These exercises helped NCATS and Complement-ARIE understand diverse research perspectives and integrate practical strategies for overcoming barriers and maximizing research impact and should be implemented by other NIH institutes, centers, offices, programs, and the NIH-Wide Strategic Plan team.

NIH strategic plans set goals and priorities that translate into new program concepts, funding opportunities, and highlighted topics. To evaluate the scientific merit and potential impact of incoming grant applications, the agency forms scientific review groups coordinated by the Center for Scientific Review (CSR). If reviewers prefer animal-based methods or lack the expertise to properly evaluate nonanimal methods, it can result in unfair assessments of animal-free research—a phenomenon called animal methods bias, which is discussed at length in Chapter 15. In recognition of this issue, the final report of the ACD NAMs Working Group suggested that reviewers need to understand the unique value of NAMs when evaluating proposals and accordingly recommended that the NIH "promote training for grant reviewers to better understand how to evaluate the use of NAMs in fundamental and applied research grants."[23] Although CSR cannot train reviewers on how to evaluate NAMs, NCATS and programs like Complement-ARIE can provide optional education and outreach to the scientific community so the value of NAMs and other translational approaches is fully appreciated. In addition, funding opportunity announcements can specify review criteria to ensure that the unique value of NAMs is properly evaluated by scientific review groups in a constructive and equitable manner. Thus it is important that NCATS, Complement-ARIE, and other institutes, centers, offices, and programs investing in NAMs research explicitly state and implement such criteria. Other measures that can help ensure NAMs are fairly evaluated include broadening the pool of NAMs expertise available for scientific review groups and training reviewers to identify, address, and report incidences of animal methods bias. Many of these measures have been proposed in the new NIH initiative to prioritize human-based research while reducing animal use,[2] although the timeline and details of their implementation remain to be determined at the time of writing.

Regulatory testing

The importance of agency leadership in overcoming the entrenched acceptance of animal studies cannot be overstated. Through its NAMs work plan, the EPA maps out and implements efforts to move away from animal use.[38] The FDA does not have an equivalent plan but has released a strategy to phase out animal testing of drugs, along with an accompanying roadmap.[1,39,40] Regulated industries look to agencies to understand which tests must be conducted and submitted to gain approval for selling certain products. Companies have decades of experience using animal-based approaches to support applications that are accepted by regulators. Despite an influx of scientifically valid and commercialized nonanimal approaches, animal tests continue to be submitted to varying degrees, depending on the agency and sector. Without clear communication from agencies about which nonanimal tests will be accepted under specific circumstances, companies have little incentive to change. Conversely, companies may even be incentivized *not* to change, as delays in product approvals can cost them significant amounts of money. Any ambiguity in method acceptance or deviation from favored animal tests may expose companies to potential financial risk. For the pharmaceutical industry, a delay of just one day on the market can cost a company up to $500,000.[41]

Change requires clarity on the acceptance of nonanimal approaches. Where agency regulations require animal use, regulations should be updated because regulations carry more weight than other nonbinding policy communications. However, as the process of changing regulations can be slow and burdensome, agencies sometimes communicate updates on the acceptance of nonanimal approaches through other written policies, such as guidance, guidelines, orders, waivers, and statements. These may be acceptable where agencies are clear that nonanimal approaches will be accepted or animal studies are not needed, and may include measures such as: equalizing the playing field by ensuring that nonanimal approaches are accepted alongside animal tests; prioritizing nonanimal approaches over animal-based approaches; waiving prescribed animal tests; or restricting animal experiments. The following examples demonstrate how approaches other than regulatory change can be used.

- Neutralizing FDA drug development and testing regulations is one way that stakeholders have sought to equalize the playing field.[42,43] By replacing references to "animal" data with "nonclinical," the agency would clearly communicate to drug sponsors that it will accept in vitro, computational, or in vivo data.
- Through recommendations in nonbinding agency guidance, the FDA has prioritized in vitro data for eye irritation in narrow applications.[44] The FDA recommends that the eye irritation potential of reformulated topical drug products be evaluated using appropriate in vitro or ex vivo methods. In addition, through guidance, the FDA has communicated clearly that Phase 0 studies in humans are acceptable prior to certain animal experiments.[45]
- After an internal analysis indicating that EPA decision-making was not influenced by animal tests submitted for acute dermal toxicity testing, the EPA announced it would provide waivers for the test.
- In 2018, California became the first state to restrict the sale of cosmetics that had been tested on animals. Since then, multiple states have enacted similar laws, and now cosmetic animal testing is being phased out in favor of nonanimal approaches.

Some programs have been established in an effort to provide clarity around the acceptance of methods, such as the FDA's Innovative Science and Technology for Advancing New Drugs.[46] The idea behind qualification programs is to provide a path for companies to work with the agency to achieve acceptance of a method for a certain set of circumstances, such that when a method is qualified, the agency will accept the method without the need for additional validation studies. However, while many companies have sought to utilize this pathway, nearly four years after its inception, no methods have been qualified through the program. And because companies are global entities seeking to sell in multiple markets around the world, progress in the acceptance of nonanimal approaches will need to be made around the world. Fortunately, regulators in the United States are already working at the international level—through the Organisation for Economic Co-operation and Development for chemicals and pesticides and through the International Council on Harmonization of Technical Requirements for Pharmaceuticals for Human Use for drug development and testing. While regulators are not required to accept the output of these international collaborations in lieu of traditional animal data, nonanimal approaches are being integrated into guidelines that have the potential to reduce animal use around the globe.

Overcoming training barriers

Biomedical research

Training and education play a key role in overcoming barriers to the broader use of nonanimal research approaches and constitute one of the highest priority needs for advancing NAMs.[23,47] Education begins early: exposing interested elementary and high school students to nonanimal methods can help inspire broader thinking for those who enter scientific research, weaken the mistaken belief that animal use is necessary, and broaden interest in nonanimal methods.[48] Mid-stage and senior researchers who may not have received NAM training would benefit from such opportunities, helping to build awareness of these new approaches and their advantages. Early-career researchers, however, "may be more open to new solutions than senior scientists, who have adopted conventional approaches and succeeded in the existing system,"[49] and thus may be more receptive to new nonanimal methods than senior scientists. In addition, as the largest and most diverse cohort of scientists, early-career researchers play an important part in improving the scientific ecosystem and are vital stakeholders leading efforts to catalyze change in research culture and practice. On the other hand, barriers to using NAMs and other nonanimal methods may be especially difficult to overcome for early-career researchers, partially because of hierarchical academic structures that lead to increased precarity and reduced empowerment among young scientists. This may also be the case because early-career researchers rely on previous generations of scientists for training, who are less familiar with NAMs simply because animal-based approaches were the preferred complex systems for studying disease mechanisms when they were trained.

To support the broader development and use of nonanimal methods, funders, research institutions, and publishers should incentivize and reward the rigorous use of and training in these approaches. Institutional training mechanisms should be established for graduate students and postdoctoral researchers, and reviewers of training and career development grants should be able to assess the value of proposals involving innovative NAMs training. In addition, funders and research institutions should offer and fund more concentrated NAMs training opportunities, such as summer schools (one- to two-week intensive summer programs for advanced undergraduates, graduate students, and postdoctoral researchers, offering specialized, experiential, collaborative education) and determine other ways of centrally disseminating and supporting skill-building. Other efforts to reward and encourage research using nonanimal methods, support global research integrity and bioethics initiatives, and aggregate resources, funding, and training opportunities will also help overcome this inertia. Numerous resources are already available for students and researchers interested in human-centered science, including university courses, scientific literature, hands-on training, networking groups, and webinars. These resources and events are essential to continue educating the next generation of scientists on the need to shift away from animal use in the life sciences and develop the skills to contribute to this shift. Table 2 provides a nonexhaustive list of programs and resources for early-career scientists that support the development and use of nonanimal methods.

Regulatory testing

Toxicology is often segregated from biomedical science in conferences, textbooks, and other educational contexts, despite many overlapping mechanisms. Cross-discipline training and educational opportunities between toxicology, biomedical, and regulatory science could facilitate awareness, uptake, and ultimately acceptance of NAMs and other nonanimal methods. For example, adverse outcome pathways (AOPs) are a conceptual framework for organizing, synthesizing, and presenting specialized scientific knowledge regarding the link between the perturbation of a biological pathway by a stressor and a consequent adverse outcome frameworks.[51] However, the concept is primarily reserved for chemical risk assessment and included in very few early career educational programs. As a result, many early

TABLE 2 Training resources for early-career scientists focused on nonanimal methods.

Resource	Training characteristics			Modalities						Target audience				
	Competencies[a]	Format[b]	Fee[c]	Course	Conference	Funding	Professional development	Seminar series	Workshop	Undergraduate students	Graduate students	Postdoctoral fellows	Early-career researchers	Others
The 3Rs Collaborative	B	O						✓				✓	✓	✓
3R-Center Tübingen	Bo	Bo	✓		✓		✓	✓	✓		✓	✓	✓	
Alternatives Research and Development Foundation	Bo	O				✓						✓	✓	✓
Columbia University Tissue Talks	Bo	O						✓		✓	✓	✓	✓	✓
Complement-ARIE[d]	Bo	O				✓							✓	✓
Emulate	Bo	O						✓			✓	✓	✓	✓
Human Advanced In Vitro Model Initiative	Bo	O				✓						✓	✓	✓
International Foundation for Ethical Research	Bo	O				✓					✓			
Johns Hopkins Center for Alternatives to Animal Testing	Bo	O		✓		✓		✓			✓	✓	✓	✓
Joint Research Center Summer School on Non-Animal Approaches in Science	Bo	IP	✓		✓	✓	✓		✓		✓	✓	✓	
Microphysiological Systems World Summit	Bo	IP	✓		✓	✓	✓		✓	✓	✓	✓	✓	✓
Replacing Animal Research	Bo	O				✓				✓	✓	✓	✓	✓
Society of Toxicology	Bo	IP	✓		✓					✓	✓	✓	✓	✓

Continued

TABLE 2 Training resources for early-career scientists focused on nonanimal methods—cont'd

Resource	Training characteristics			Modalities						Target audience				
	Competencies[a]	Format[b]	Fee[c]	Course	Conference	Funding	Professional development	Seminar series	Workshop	Undergraduate students	Graduate students	Postdoctoral fellows	Early-career researchers	Others
Summer Immersion on Innovative Approaches in Science	Bo	IP	✓		✓	✓	✓		✓	✓	✓	✓	✓	
Animal-Free Workshop Series for Early-Career Researchers	Bo	O					✓	✓	✓	✓	✓	✓	✓	
World Congress on Alternatives and Animal Use in the Life Sciences	Bo	IP	✓		✓	✓	✓		✓	✓	✓	✓	✓	✓
American Society for Cellular and Computational Toxicology	R	Bo	✓		✓		✓	✓			✓	✓	✓	✓
Animal-Free Safety Assessment	R	Bo		✓				✓	✓			✓	✓	✓
Institute for In Vitro Sciences	R	Bo	✓					✓	✓				✓	✓
National Toxicology Program Interagency Center for the Evaluation of Alternative Toxicological Methods	R	Bo						✓	✓				✓	✓
New Approach Methodology Used for Regulatory Application	R	O		✓				✓					✓	✓

[a]*B, biomedical; R, regulatory; Bo, both.*
[b]*IP, in-person; O, online; Bo, both.*
[c]*Checkmark indicates that the resource requires a registration fee, travel, and/or lodging.*
[d]*Beginning in Fiscal Year 2027, Complement-ARIE will include a training component.*[50]

career scientists are not aware of AOPs, even though they offer a guided framework for developing research and testing projects (including graduate theses) that build off existing information and avoid duplication of research efforts.[52] This cross-disciplinary approach could provide an effective, sustainable, and dependable tool for NAM development, chemical safety assessment, and other health and environmental applications.[53]

Poor communication between scientists and regulators, along with a lack of understanding of each other's needs, can be a significant barrier to the development and use of NAMs. Improved relationships could be fostered through training and educational venues, including conferences, roundtables, webinars, and listening sessions. Scientists must be aware of the path to regulatory use and the ways it will maximize the impact of a developed NAM. Regulators should establish the needs of their agency and sector, and scientists should develop or modify their tools with these needs in mind, addressing specific regulatory questions and concerns not typically encountered in other laboratory settings. For nonanimal methods to reach their potential in replacing animal tests, regulators should also understand the intricacies of new methods to be confident in their results.

Availability of accessible and affordable supplemental training is crucial for the progression of science at any career level.[54] As thorough as an educational degree program may be, there will always be new technology and information emerging that toxicologists must understand. Certification programs, such as those provided by the American Board of Toxicology,[55] can be used to ensure toxicologists stay up to date on the latest toxicological information. Conferences are a great place for toxicologists and regulators to learn about the latest NAMs. In addition to highlighting the latest in NAM technologies, many conferences offer curated continuing education courses on NAMs. While conferences provide high-quality, cutting-edge research and technologies, they can be expensive. Low-cost or free and flexible options provide training for those with financial, travel, and scheduling limitations and may be able to fill training gaps (Table 2).

Conclusion

While promising progress has been made in shifting US medical research and testing from primarily animal-based approaches to nonanimal ones, formidable barriers remain. This systemic inertia prevents improved translation of research and testing to human outcomes and requires multistakeholder strategies. Government agencies, legislators, members of the public, academic researchers, companies and industry groups, NGOs, and scholarly and mass media should all play a role in overcoming this inertia. Potential recommendations discussed in this chapter are summarized in Table 3. (Chapter 13 discusses progress and remaining inertia around the world.)

Although these solutions distinguish between biomedical research and regulatory testing, there are many commonalities and overlapping opportunities that could mutually benefit from cross-sector, transdisciplinary collaboration. This further highlights the systemic nature of the inertia against nonanimal research and testing in the United States—if any of these facets improve, it can affect the whole enterprise. Stakeholder relationships and power structures should be considered when pursuing these strategies. For example, agency-funded training programs are likely to be better resourced than NGO-funded ones; but if there is a

TABLE 3 Recommendations for overcoming scientific, policy, and training barriers to the shift toward nonanimal research and testing methods.

Barriers		Recommendations
Scientific		Advancing the validation, reproducibility, and accessibility of nonanimal methods
Policy	Biomedical research	Federal investment in and prioritization of the development and use of nonanimal methods
		Tracking and publicly reporting the number of animals used, the number of projects, and the dollars spent on animal- and nonanimal-based research
		Improving awareness and understanding of nonanimal methods and their translational value in the scientific community, including among reviewers
	Regulatory testing	Updating regulations that require animal use to allow for or encourage nonanimal method use
		Implementing nonbinding policies to communicate updates on the acceptance of nonanimal methods.
		Working with companies to achieve acceptance of a qualified nonanimal method
Training	Biomedical research	Incentivizing and rewarding training in nonanimal methods
		Funding institutional training programs, concentrated training opportunities, and other skill-building education for early-career researchers
	Regulatory testing	Implementing cross-discipline continuing education opportunities between toxicology, biomedical research, and regulatory science
		Improving communication between scientists and regulators
		Establishing accessible and affordable training opportunities for toxicologists and regulatory scientists

lack of will to implement such programs at the agency level, NGO efforts can fill that gap. Nonagency actors can also achieve agency change by applying pressure in other venues, such as through Congress or the media. Through coordinated, multistakeholder action, the shift to nonanimal research and testing will occur in a timely and effective manner, improving clinical translation and patient outcomes, promoting safer products, and reducing animal use.

References

[1] U.S. Food and Drug Administration. FDA announces plan to phase out animal testing requirement for monoclonal antibodies and other drugs; 2025. https://www.fda.gov/news-events/press-announcements/fda-announces-plan-phase-out-animal-testing-requirement-monoclonal-antibodies-and-other-drugs. Accessed 9 May 2025.

[2] U.S. National Institutes of Health. NIH to prioritize human-based research technologies; 2025. https://www.nih.gov/news-events/news-releases/nih-prioritize-human-based-research-technologies. Accessed 9 May 2025.

[3] Animal and Plant Health Inspection Service. Research facility annual usage summary report. aphis.usda.gov; 2024. https://direct.aphis.usda.gov/sites/default/files/fy2024-research-animal-use-summary.pdf. Accessed Aug 14, 2026.

[4] Carbone L. Estimating mouse and rat use in American laboratories by extrapolation from Animal Welfare Act-regulated species. *Sci Rep*. 2021;11(1):493. doi:10.1038/s41598-020-79961-0.
[5] Lohse S. Scientific inertia in animal-based research in biomedicine. *Stud Hist Philos Sci*. 2021;89:41–51. doi:10.1016/j.shpsa.2021.06.016.
[6] Letter to Norman Sharpless, Acting Commissioner, Food and Drug Administration. 2019. Available at: https://boyle.house.gov/media-center/press-releases/boyle-leads-letter-fda-save-animals-unnecessary-laboratory-tests. Accessed 26 June 2025.
[7] U.S. Food and Drug Administration. Innovative science and technology approaches for new drugs (ISTAND) pilot program; 2025. Available at: https://www.fda.gov/drugs/drug-development-tool-ddt-qualification-programs/innovative-science-and-technology-approaches-new-drugs-istand-pilot-program. Accessed 26 June 2025.
[8] Physicians Committee for Responsible Medicine. Congress takes steps toward reducing animal experimentation in 2023 omnibus spending package; 2023. Available at: https://www.pcrm.org/news/good-science-digest/congress-takes-steps-toward-reducing-animal-experimentation-2023-omnibus. Accessed 26 June 2025.
[9] Krebs CE, Geissler S, Herrmann K, et al. Exploring animal methods bias in biomedical research funding: workshop proceedings and action steps. *NAM J*. 2025;1:100004. doi:10.1016/j.namjnl.2024.100004.
[10] Justia US Law. Vanda Pharmaceuticals, Inc v. FDA No 23-5200(D.C.Cir.2024); 2024. Available at: https://law.justia.com/cases/federal/appellate-courts/cadc/23-5200/23-5200-2024-12-17.html. Accessed 12 May 2025.
[11] NIH Complement Animal Research In Experimentation (Complement-ARIE). Landscape analysis report 2024: new approach methodologies in biomedical research; 2024. https://commonfund.nih.gov/sites/default/files/Complement-ARIE-Landscape-Analysis-29-Feb-2024-508.pdf. Accessed 9 August 2024.
[12] van der Zalm AJ, Barroso J, Browne P, et al. A framework for establishing scientific confidence in new approach methodologies. *Arch Toxicol*. 2022;96:2865–2879. doi:10.1007/s00204-022-03365-4.
[13] Parish ST, Aschner M, Casey W, et al. An evaluation framework for new approach methodologies (NAMs) for human health safety assessment. *Regul Toxicol Pharmacol*. 2020;112:104592. doi:10.1016/j.yrtph.2020.104592.
[14] Kleinstreuer NC, Hoffmann S, Alépée N, et al. Non-animal methods to predict skin sensitization (II): an assessment of defined approaches. *Crit Rev Toxicol*. 2018;48(5):359–374. doi:10.1080/10408444.2018.1429386.
[15] Clippinger AJ, Raabe HA, Allen DG, et al. Human-relevant approaches to assess eye corrosion/irritation potential of agrochemical formulations. *Cutan Ocul Toxicol*. 2021;40(2):145–167. doi:10.1080/15569527.2021.191 0291.
[16] Loewa A, Feng JJ, Hedtrich S. Human disease models in drug development. *Nat Rev Bioeng*. 2013;1–15. doi:10.1038/s44222-023-00063-3.
[17] Hartung T, De Vries R, Hoffmann S, et al. Toward good in vitro reporting standards. *ALTEX*. 2019;36(1):3–17. doi:10.14573/altex.1812191.
[18] Organisation for Economic Co-operation and Development. Guidance document on good in vitro method practices (GIVIMP); 2018. https://www.oecd.org/en/publications/guidance-document-on-good-in-vitro-method-practices-givimp_9789264304796-en.html. Accessed 24 February 2023.
[19] The RIVER working group. *Reporting in Vitro Experiments Responsibly—The RIVER Recommendations*. Center for Open Science;2023. doi:10.31222/osf.io/x6aut.
[20] Pamies D, Leist M, Coecke S, et al. Guidance document on good cell and tissue culture practice 2.0 (GCCP 2.0). *Altern Lab Anim*. 2022;39(1):30–70. doi:10.14573/altex.2111011.
[21] ACD Working Group on Enhancing Rigor, Transparency, and Translatability in Animal Research. Final Report; 2021. https://www.acd.od.nih.gov/documents/presentations/06112021_ACD_WorkingGroup_FinalReport.pdf.
[22] Busek M, Aizenshtadt A, Amirola-Martinez M, Delon L, Krauss S. Academic user view: organ-on-a-chip technology. *Biosensors*. 2022;12(2):126. doi:10.3390/bios12020126.
[23] Chang HY, Jorgenson L, Baines A, et al. Catalyzing the development and use of novel alternative methods. National Institutes of Health; 2023:42. https://www.acd.od.nih.gov/documents/presentations/Working_Group_Report.pdf. Accessed 25 July 2024.
[24] National Institutes of Health. Budget. National Institutes of Health (NIH); 2024. https://www.nih.gov/about-nih/what-we-do/budget. Accessed 27 August 2024.
[25] NIH Advisory Committee to the Director. ACD Working Group on Enhancing Rigor, Transparency, and Translatability in Animal Research; 2020. https://www.acd.od.nih.gov/working-groups/eprar.html. Accessed 1 April 2025.

[26] NIH Advisory Committee to the Director. ACD Working Group on Catalyzing the Development and Use of Novel Alternative Methods to Advance Biomedical Research. NIH Advisory Committee to the Director; 2022. https://www.acd.od.nih.gov/working-groups/ai.html. Accessed 4 September 2024.

[27] Bertagnolli MM Statement on catalyzing the development of novel alternative methods. NIH.gov; 2024. https://www.nih.gov/about-nih/who-we-are/nih-director/statements/statement-catalyzing-development-novel-alternatives-methods. Accessed 21 March 2024.

[28] NIH Common Fund. Complement animal research in experimentation (complement-ARIE) program; 2024. https://commonfund.nih.gov/complementarie. Accessed 1 June 2024.

[29] NIH National Center for Advancing Translational Sciences. Tissue chip projects & initiatives; n.d. https://ncats.nih.gov/research/research-activities/tissue-chip/projects. Accessed 4 September 2024.

[30] US National Toxicology Program. NICEATM: alternative methods. National Toxicology Program; n.d. https://ntp.niehs.nih.gov/whatwestudy/niceatm. Accessed 4 September 2024.

[31] Office of Laboratory Animal Welfare. *Public Health Service Policy on Humane Care and Use of Laboratory Animals*. Office of Laboratory Animal Welfare; 2015. https://olaw.nih.gov/policies-laws/phs-policy.htm. Accessed 7 February 2022.

[32] U.S. Institute of Medicine, U.S. National Research Council. Emerging legal trends impacting animal research. In: *International Animal Research Regulations: Impact on Neuroscience Research: Workshop Summary*. US: National Academies Press; 2012:23. https://www.ncbi.nlm.nih.gov/books/NBK100123/. Accessed 3 September 2024.

[33] Lauer M A look at NIH support for model organisms, part two. Open mike; 2016. https://nexus.od.nih.gov/all/2016/08/03/model-organisms-part-two/. Accessed 26 September 2023.

[34] Chang H, Jorgenson L Proposed ACD novel alternative methods working group. In: *Presented at: 125th Meeting of the Advisory Committee to the Director (ACD) National Institutes of Health*. 2022. https://www.acd.od.nih.gov/documents/presentations/12082022_Proposed_ACD_Novel_Alternative.pdf. Accessed 14 August 2023.

[35] Rep. Bonamici S [D O 1. H.R.34]—114th Congress (2015-2016): 21st century cures act. 2016. https://www.congress.gov/bill/114th-congress/house-bill/34. Accessed 3 August 2023.

[36] National Center for Advancing Translational Sciences. NCATS strategic plan; n.d. https://ncats.nih.gov/about/ncats-overview/strategic-plan. Accessed 23 July 2024.

[37] NIH Common Fund. Executive summary of the NIH listening sessions on the complement-ARIE program concept; 2023. https://commonfund.nih.gov/complementarie/strategicplanning/listeningsessions. Accessed 4 September 2024.

[38] U.S. Environmental Protection Agency. New approach methods work plan. EPA.gov; 2023. https://www.epa.gov/chemical-research/new-approach-methods-work-plan. Accessed 4 September 2023.

[39] U.S. Food and Drug Administration. Animal welfare, testing and research of FDA-regulated products; 2025. https://www.fda.gov/consumers/consumer-updates/animal-welfare-testing-and-research-fda-regulated-products. Accessed 1 April 2025.

[40] U.S. Food and Drug Administration. Roadmap to reducing animal testing in preclinical safety studies; 2025. https://www.fda.gov/media/186092/download?attachment. Accessed 9 May 2025.

[41] Gentz k. *How Much Does a Day of Delay in a Clinical Trial Really Cost?* Applied Clinical Trials Online; 2024. https://www.appliedclinicaltrialsonline.com/view/how-much-does-a-day-of-delay-in-a-clinical-trial-really-cost-. Accessed 17 April 2026.

[42] Baker EJ, Beck NA, Berg EL, et al. Advancing nonclinical innovation and safety in pharmaceutical testing. *Drug Discov Today*. 2019;24(2):624–628. doi:10.1016/j.drudis.2018.11.011.

[43] Physicians Committee for Responsible Medicine. Modernizing the models and tools used to develop and test new drugs; 2021. https://www.pcrm.org/news/good-science-digest/modernizing-models-and-tools-used-develop-and-test-new-drugs. Accessed 12 May 2025.

[44] U.S. Food and Drug Administration Center for Drug Evaluation and Research. Nonclinical safety evaluation of reformulated drug products and products intended for administration by an alternate route: guidance for industry and review staff good review practice; 2015. https://www.fda.gov/files/drugs/published/Nonclinical-Safety-Evaluation-of-Reformulated-Drug-Products-and-Products-Intended-for-Administration-by-an-Alternate-Route.pdf. Accessed 1 April 2025.

[45] U.S. Food and Drug Administration. Guidance document: exploratory IND studies; 2006. https://www.fda.gov/regulatory-information/search-fda-guidance-documents/exploratory-ind-studies. Accessed 1 April 2025.

[46] FDA Center for Drug Evaluation and Research. Innovative science and technology approaches for new drugs (ISTAND) pilot program; 2024. https://www.fda.gov/drugs/drug-development-tool-ddt-qualification-programs/innovative-science-and-technology-approaches-new-drugs-istand-pilot-program. Accessed 1 April 2025.

[47] Moruzzi A, Shroff T, Keller S, Loskill P, Cipriano M. Training the next generation of researchers in the organ-on-chip field. *Educ Sci.* 2023;13(2):144. doi:10.3390/educsci13020144.

[48] NAVS.org. Transforming science classrooms; n.d. Available at: https://navs.org/ending-classroom-dissection/. Accessed 26 June 2025.

[49] Kent BA, Holman C, Amoako E, et al. Recommendations for empowering early career researchers to improve research culture and practice. *PLOS Biol.* 2022;20(7):e3001680. doi:10.1371/journal.pbio.3001680.

[50] NIH Common Fund. Complement-ARIE program description; 2024. https://dpcpsi.nih.gov/sites/default/files/2024-01/1-1PM-OSC-Concept-Complement-ARIE-Rutter-Woychik-onepager-508.pdf. Accessed 15 April 2025.

[51] AOP-Wiki. n.d. https://aopwiki.org/info_pages/2/info_linked_pages/9. Accessed 14 August 2023.

[52] Willett C "Adverse outcome pathways:" A perfect framework on which to build a thesis. In: *Abstracts of the 12th World Congress on Alternatives and Animal Use in the Life Sciences, Niagara Falls, 2023*. Vol 11. ALTEX;2023. doi:10.58847/ap2302

[53] Carusi A, Davies MR, De Grandis G, et al. Harvesting the promise of AOPs: an assessment and recommendations. *Sci Total Environ.* 2018;628-629:1542–1556. doi:10.1016/j.scitotenv.2018.02.015.

[54] Slankster-Schmierer E. Modernizing toxicology: the importance of accessible NAM training. *Toxicol Lett.* 2025;406: 38–39. doi:10.1016/j.toxlet.2025.02.011.

[55] American Board of Toxicology. n.d. https://www.abtox.org/about-abt/. Accessed 26 August 2024.

CHAPTER

13

Overcoming inertia against nonanimal approaches around the world

Mohammad Abdulkader Akbarsha[a,b,c], *Lauren Dalat de Sousa Coelho*[d], *Abhay Deshpande*[c,e], *Yogendra Kumar Gupta*[c,f,g], *Yoko Hirabayashi*[d,h], *Hajime Kojima*[d,h,i], *Francesca Pistollato*[j], *Marcella Miranda Siqueira Furtuoso Rodrigues*[d], *Jordana Andrade Santos*[d], *and Marize Campos Valadares*[d]

[a]Jamal Mohamed College (Autonomous), Tiruchirappalli, India
[b]Mahatma Gandhi-Doerenkamp Center, Bharathidasan University, Tiruchirappalli, India
[c]Indian Society for Alternatives to Animal Experiments, Bharathidasan University, Tiruchirappalli, India
[d]Federal University of Goiás, Goiânia, Brazil
[e]Jai Research Foundation, Vapi, Gujarat, India
[f]National Poison Information Centre, All India Institute of Medical Sciences, New Delhi, India
[g]National Scientific Coordinator, Pharmaco-vigilance Programme of India (PVPI), Indian Pharmacopoiea Commission, Ghasiabad, India
[h]Center for Biological Safety and Research, National Institute of Health Sciences, Kawasaki, Japan
[i]Sanyo-Onoda City University, Sanyo-Onoda, Japan
[j]Research and Toxicology, Humane World for Animals, Brussels, Belgium

Abstract

Over the past few decades, significant progress has been made in replacing animal testing with scientifically robust alternatives, particularly for routine toxicity assessments. Reliable nonanimal approaches now exist for endpoints such as skin and eye irritation, phototoxicity, and genotoxicity, with growing potential for

Moving Toward Nonanimal Approaches in Medical Research and Testing
https://doi.org/10.1016/B978-0-443-26514-3.00006-0

broader applications. Despite these advances, challenges remain—especially in replicating complex biological processes and overcoming cultural and institutional resistance to change. The transition to nonanimal approaches requires not only scientific innovation but also societal and regulatory shifts. This chapter explores global efforts to advance innovative nonanimal methods in research and safety testing, focusing on developments across India, Brazil, the European Union, and Japan. The chapter reviews key policy and legislative initiatives, funding strategies, and educational efforts that support the integration of nonanimal methods, highlighting how these regions are working to overcome barriers and accelerate the move toward more ethical, human-relevant science.

Keywords: Research policy; Regulatory testing; New approach methodologies; India; Brazil; Japan; European Union

Introduction

Over the past three decades, remarkable strides have been made in developing alternative nonanimal methods, also called new approach methodologies (NAMs), particularly for routine toxicity and safety testing of chemicals and biological substances. These advancements bring us closer to a future where animal testing may no longer be necessary. Reliable alternative methods already exist for assessing corrosive substances, skin and eye irritants, skin phototoxicity, skin penetration, and genotoxicity.[1] These methods offer valuable insights without the need for animal testing. Looking ahead, it is conceivable that in the near to mid-term, the majority of safety assessments could be conducted through a strategic combination of nonanimal testing methods, sophisticated computer models, and integrative approaches. This progress also extends to basic research, where nonanimal models can effectively contribute to expanding our scientific understanding.

While significant progress has been made in developing and validating nonanimal methods, certain scientific obstacles continue to limit their complete replacement of animal testing. For example, in the area of skin sensitization, a combination of nonanimal methods can now effectively identify chemicals as likely skin sensitizers or nonsensitizers in most cases.[2] However, there are still complex areas in toxicology and biomedical research where nonanimal methods struggle to fully replace traditional approaches. One of the primary challenges lies in the complexity of biological systems. Replicating this level of complexity in a single in vitro test is unlikely. Instead, a combination of different tests and computational models would be necessary to approximate the intricacies of real-life human biology.[3]

Moreover, in transitioning from traditional animal-based safety assessments and research to nonanimal approaches, potential cultural or societal shifts are required. While innovative nonanimal approaches offer clear scientific benefits, moving away from established (animal) methods can be challenging due to nontechnical obstacles. These include resistance to change, comfort with existing practices, uncertainties about regulatory acceptance, and concerns about how to integrate new methods with existing data. In addition, there may be apprehension about losing continuity in data comparisons between new and previously generated results.[4]

In the following sections, we summarize both past and recent policy and legislative initiatives, as well as future prospects and educational efforts, across India, Brazil, the European Union (EU), and Japan, to show how these regions are working to promote the adoption of nonanimal methods in research and testing, overcoming the inertia against their widespread use and acceptance.

Overcoming inertia against nonanimal approaches in India

Mohammad Abdulkader Akbarsha, Abhay Deshpande, and Yogendra Kumar Gupta

India was declared independent from British rule on August 15, 1947 and became a sovereign democratic republic state on January 26, 1950. Animal welfare, as an aspect of national interest, could only be promulgated thereafter. The country enacted the *Prevention of Cruelty to Animals Act* in 1960, with the *Animal Welfare Board* under Section 4A, and the *Committee for the Purpose of Control and Supervision of Experiments on Animals (CPCSEA; now CCSEA)* under Section 15(A). The Animal Welfare Board ensures that animal welfare laws are followed, provides grants to animal welfare organizations, and advises the government on animal welfare issues. CCSEA is a statutory committee bound to take all necessary measures to ensure that animals are not subjected to unnecessary pain or suffering before, during, and after their use in experiments. Establishments engaged in biomedical research must register with CCSEA, constitute an institutional animal ethics committee (IAEC), have animal facilities inspected, and clear research projects with the IAEC before commencing experiments.[5] Thus, the care for and comfort of experimental animals are statutorily provided, but in view of the developing economy, implementation of animal welfare provisions is frequently compromised.[6]

In India, the regulatory authorities with respect to various products are as follows:

1. Drugs are regulated under the *Drugs and Cosmetics Act, 1940* (DC Act), including the *Drugs and Cosmetics Rules, 1945* (DC Rules) under this Act, as well as the *New Drugs and Clinical Trials Rules, 2019*, which were amended in 2023.
2. Cosmetics are regulated by the Central Drugs Standards Control Organization under the DC Act, the DC Rules, and the *Cosmetics Rules, 2020*.
3. Pesticides and chemicals are regulated by the *Central Insecticides Board and Registration Committee,* as well as the *Food Safety and Standards Authority of India.*

Inertia in adopting nonanimal approaches

In the past, academic institutions and not-for-profit nongovernmental organizations (NGOs) were slow to adopt the 3Rs. Until around the turn of the 21st century, there was a prevalent form of resistance to and criticism of the 3Rs, alternatives, and NAMs because of the long history of the deep-rooted classic animal experimentation approach to drug discovery and risk assessment in the Claude Bernardian faith. There is still some resistance from academic institutions and scientists, which can be partially explained by conservatism (a lack of inclination to change to modern tools) and a lack of confidence in new nonanimal approaches. This is reflected in the recommendations of a meeting convened by the Indian National Science Academy in 2011,[7] which was described by a news item in *Nature*: "The National Academy of Sciences of India has urged the prime minister to intervene to prevent the implementation of new rules on using animals in research."[8] Nevertheless, the inertia has diminished over the years, especially with the turn of the century, due to the perseverance of supporters of NAMs.

NAMs are now routinely used by the research and development (R&D) labs of Indian industry and contract research organizations (CROs) at the lead identification stage, where the most promising candidate compounds are identified during initial screening assays to be further developed.

Many of the top industrial R&D centers and CROs have been working on NAMs-based discovery and regulatory solutions for quite a long time. For example, in vitro absorption, distribution, metabolism, and excretion tests have been routinely applied by corporate and institutional R&D centers and CROs for discovery and regulatory safety evaluations for over two decades. The application of bioinformatics and in silico tools in lead identification has also been in place for several years. A few nonanimal tests are adopted from Organisation for Economic Co-operation and Development (OECD) test guidelines, such as in vitro testing of cosmetics. However, while regulatory prescriptions may encourage the use of NAMs, many still rely on animal testing dossiers, and animal testing is still widely used in regulatory testing of drugs, chemicals, and other products in India. Furthermore, the industry and CROs are not limited in service to only one country but rather extend across the globe and are subject to each country's regulations, which can slow the progress toward NAMs acceptance in India and globally.

Regulators open up their doors for NAMs-driven drug testing and hazard assessment

With the US Environmental Protection Agency formalizing the application of NAMs through 27 well-defined guidelines and six test methods,[9] at least three of the six tests have moved away from animal requirements in the EU. The EU, OECD, US National Toxicology Program, and several other centers for the validation of alternative methods around the world are committed to developing alternatives to animals, much to the advantage of the pharmaceutical industry in India. The EPA list was updated in 2026.

In 2017, the Central Insecticide Board and Registration Committee (CIB&RC) in India published its guidance document on chemical pesticides, allowing for the acceptance of alternate approaches.[10] The CIB&RC approved new guidelines for chemical pesticides and biopesticides, amended as per the 450th RC minutes,[11] which state: "The replacement alternatives not involving experiments on animals would be considered and accepted in those cases where data from valid 3Rs (Replacement, Reduction and Refinement) methods recognized by OECD and other renowned Regulatory Authorities is provided wherever applicable, case to case basis." The Government of India prohibited the testing of cosmetics on animals in G.S.R. 346(E) dated May 21, 2014 by inserting the following rule in the DC Rules: "148-C: Prohibition of testing of cosmetics on animals—No person shall use any animal for testing cosmetics."[12] Cosmetics tested in or on animals shall not be imported or exported. The OECD nonanimal guidelines for skin and eye irritation and corrosion, skin sensitization, and other endpoints are to be followed using weight-of-evidence and integrated approaches to testing and assessment in order to assess the toxicity potential of novel cosmetic ingredients or finished products.[13]

As Soumya Swaminathan, Director General of the Indian Council of Medical Research, and colleagues[14] stated in 2019:

> "If India is to emerge as the innovation center for [drug discovery and development] research, the current state of the art should be embraced by both public and private sectors. Government, academia, and industry collaborations need to be encouraged. In addition, regulatory bodies should be in a closed loop with both public and private sectors for any policy changes that may be required in the future in support of nonanimal technologies. Government funding agencies should fund private sectors with these kinds of innovations in alternatives to animal technology start-ups to give the necessary boost for many of the concepts to grow to a prototype stage."

As a sequel to this development, the Government of India, Ministry of Health and Family Welfare, released a Gazette of India Notification No. 136, dated March 09, 2023,[15] proclaiming acceptance of the NAMs data:

> "The general requirements of non-clinical studies have been specified in the Second Schedule (II). Non-clinical testing methods to assess the safety and efficacy of a new drug or investigational new drug include the following, namely:
>
> 1 Cell-based assays;
> 2 Organ chips and micro physiological systems.
> 3 Sophisticated computer modelling;
> 4 Other human biology-based test methods;
> 5 Animal Studies."

The regulatory conundrum

In India, there is still no legislative ban on animal testing for drugs, which means that the existing drug approval process may not change noticeably for some time to come. Furthermore, although many alternatives to animal testing are available, such as cell-based assays, in silico models, computational studies, 3D organoids, organs-on-chips, and human biology-based test methods, these solutions are still in their nascency and may have their own limitations, especially with regard to their ability to assess safety or toxicity. Industry, start-up companies, and scientists continue to develop novel and sophisticated alternatives,[16–18] but it is quite another challenge to convince end-users to switch to using them.

For several decades, global and Indian regulators have been comfortable granting registrations based on animal studies submitted in compliance with specific, globally harmonized guidelines that have defined end points. Unfortunately, these regulations are difficult to adapt to NAMs-driven experiments when evaluated by OECD GD 34.[19] Such a scenario tends to force regulators to be cautious in granting approvals for NAMs-driven studies.

Education and training

Education and training on NAMs in India are not directly connected with the concerned regulatory authorities. The agencies primarily provide grants for research, while the majority of contributions to education and training come from scientific societies, research institutes, and NGOs. The society providing the largest contribution is the Indian Society for Alternatives to Animal Experiments, operating from Bharathidasan University, Tiruchirappalli. Its seed was sown as the Doerenkamp-Zbinden Foundation-supported megaproject, Mahatma Gandhi-Doerenkamp Centre for Alternatives, established in 2009. In 2016, the University Grants Commission funded it as the National Centre for Alternatives to Animal Experiments (2017–2022). Since 2024, the Doerenkamp-Zbinden Foundation has supported the Centre and in the meantime, the Society was established. The Society and Centre are concerned with sensitization, propagation, education, and training in alternative methods and NAMs. Innumerable conferences and teaching and training programs have been conducted since 2009, with an emphasis on in vitro and in silico alternatives, alternative model organisms, and NAMs.[20–30] The Centre and Society are a part of a global 3Rs network associated with many international organizations, including the World Congress on Alternatives and Animal

Use in the Life Sciences, the European Society for Alternatives to Animal Testing, and the International Microphysiological Systems Society.

Other examples of NAMs education and training in India are as follows. An exclusive session on alternatives was conducted at the 99th Indian Science Congress in 2012,[31] with tremendous popularity and outcomes. People for Ethical Treatment of Animals India organized a webinar series on the "Use of New Approach Methodologies for the Risk Assessment of Pesticides" in 2023.[31] Humane Society International India, the Centre for Predictive Human Model Systems,[18] International Life Sciences Institute India, and Jai Research Foundation have also organized programs and activities independently and in collaboration with scientific societies like the Society for Alternatives to Animal Experiments-India. In addition, the Institute of Chemical Technology in Mumbai and OncoSeek Bio in Hyderabad and Vizag have been organizing international conferences since 2017. The national workshop on "Handling of Reconstructed 3-Dimensional Human Tissue Model,"[32] the Integrated Discrete Multiple Organ Culture hands-on workshop,[33] and a comprehensive hands-on training in alternative model organisms[34] were the first of their kind in Asia. The efforts of the Mahatma Gandhi-Doerenkamp Centre have helped propagate modern, nonanimal tools for risk assessment to the Indian community.

With the turn of the 21st century, India has made appreciable progress with the use of NAMs and the replacement of animals in biomedical research and regulatory testing. In light of the commendable roles of academic and service organizations, adequately supported by the regulatory community, India is poised to continue making steady progress in the implementation of NAMs in drug discovery and risk assessment.

Overcoming inertia against nonanimal approaches in Brazil

Jordana Andrade Santos, Lauren Dalat de Sousa Coelho,
Marcella Miranda Siqueira Furtuoso Rodrigues, and Marize Campos Valadares

In Brazil, substantial efforts have been made to establish laws regulating animal welfare and testing, driven by international regulatory influence and growing ethical demands from society. The historical context of animal research in Brazil traces back to the 1970s with the enactment of Law No. 6638/1979, which aimed to standardize animal vivisection for educational and scientific purposes but did not cover all experimental procedures and was never enforced.[35] Subsequently, Law No. 9605/1998, also known as the Law on Environmental Crimes, was enacted.[36]

In 2008, the Arouca Law (Law No. 11,794/2008) marked a significant step in regulating animal use in Brazil, leading to the establishment of the National Council for the Control of Animal Experimentation (CONCEA). CONCEA ensures compliance with laws on humane animal treatment, accredits animal-using institutions, and promotes nonanimal alternatives in education, testing, and research.[37–39] This legislation encouraged initiatives like the first hands-on training on OECD guidelines in 2011, hosted by the Laboratory of Education and Research in In Vitro Toxicology (Tox In), promoting collaboration among regulatory agencies, universities, industries, and CROs.[40]

In 2012, the Oswaldo Cruz Foundation and the Brazilian Health Regulatory Agency established the Brazilian Center for the Validation of Alternative Methods (BraCVAM), responsible for validating alternative methods in Brazil.[41] Concurrently, the National Network of Alternative Methods (RENAMA) was established in 2012 by the government to optimize resources and promote alternative methods across the country.[42] Collaborating with BraCVAM and CONCEA, RENAMA contributes to the validation and implementation of alternatives, overcoming technical barriers and promoting technological autonomy.[41,42]

Normative Resolution No. 17/2014 by CONCEA established the framework for the national expansion of nonanimal methodologies, promoting regulatory acceptance by ANVISA.[43] CONCEA Resolutions (No. 18/2014[44] No. 35/2015[45], No. 31/2016[46], and No. 45, 2019[47]) secured national recognition for OECD-adopted alternative methods, including various toxicological endpoints,[46,48] as eye irritation and corrosion; skin irritation, corrosion, sensitization, and permeation; phototoxicity; acute toxicity; genotoxicity; reproductive toxicity; and pyrogenicity.[49] In Brazil, the recommendation for alternative methods is based on endpoints rather than specific products. This approach allows recommended methods to be applied broadly to pharmaceutical products, reflecting an adaptable framework. However, specific regulations mandating toxicity assessment have not yet adopted nonanimal methods for certain products like antimicrobials and disinfectants.[49]

In 2015, an initiative of the Brazilian Government established the Regional Platform of Alternative Methods to the Use of Experimental Animals (PReMASUL). The primary mission of this platform is to offer short-term courses to disseminate knowledge about "Alternative Methods to Animal Experimentation" among the countries that make up the Southern Common Market (MERCOSUR). This initiative trained hundreds of professionals in Latin America, consolidating the regional commitment toward nonanimal approaches in scientific research and testing.[50]

Another Resolution, by CONCEA, No. 53/2021,[51] expressly prohibits the use of animals in educational activities. Alternative methodologies are needed to meet educational goals while adhering to ethical guidelines.[52] In Brazil, the Education Network, REDEC, offers alternatives to the use of animals in didactic activities, utilizing information technology for interactive courses, digital games, and simulators to facilitate learning and experimentation.[53–55]

CONCEA, No. 58/2023, imposed a partial ban on cosmetic testing on animals. The resolution prohibits the use of vertebrate animals in scientific research, development, and quality control of cosmetics.[56] In addition, it mandates the use of alternative methods for research and quality control, as well as testing for the toxicity and efficacy of cosmetics. However, this resolution also allows other regulatory entities jurisdiction in specific cases, opening up possibilities for animal testing. Progress has been made on Bill No. 3062 for banning animal use in cosmetics, approved by the Brazilian Senate in 2022 and awaiting approval by the Chamber of Deputies.[57] If enacted, Bill No. 3062 will complement Normative No. 58/2023, ensuring a complete ban on animal use for cosmetics testing.[42,56,57]

Over more than a decade, the CONCEA council and experts have collaborated to create the Brazilian Guide for the Production, Maintenance, or Use of Animals for Teaching or Scientific Research Activities. This guide offers clear and ethical directives for researchers regarding the appropriate use of animals in teaching and research when necessary.[58] The remarkable initiatives of RENAMA, BraCVAM, PReMASUL, CONCEA, and REDEC reflect the commitment of the Brazilian community to the transition and incorporation of alternative methods to animal experimentation in the educational landscape. These collective actions signify a robust allegiance to innovation, ethics, and sustainability, signaling a progressive shift in the country's educational and scientific approach.

Billions of dollars have been invested in developing and validating NAMs to substitute animal use in safety science. However, challenges remain in understanding assay outcomes, building confidence, increasing adoption, and ensuring regulatory acceptance. Since 2017, the Brazilian Congress of Toxicology (CBTox) has emphasized 21st-century toxicology strategies, offering hands-on training on OECD guidelines. In the last three editions of CBTox and the Latin American Congress of Clinical and Laboratorial Toxicology, there has been an intensified focus on NAMs for animal testing, signaling a shift toward the replacement of animal experimentation.[59–62] In this context, language barriers persist for South American researchers in deeper discussions about the theme. In 2022, Tox In initiated webinars on "Science In Vitro," delivered by experts who speak Portuguese or Spanish as their mother languages, in partnership with the Animal Free Safety Assessment Collaboration, the Institute for In Vitro Sciences, Unilever, BraCVAM, the Society of Environmental Toxicology and Chemistry Latin America, and Innvitro, engaging researchers and students from Mexico, Colombia, Uruguay, Paraguay, Argentina, Chile, Portugal, and Brazil.[63–65]

In 2023, the first Brazilian Congress on Alternative Methods to Animal Use in Research and Teaching was held, bringing together national and international experts to discuss alternative methods to promote ethical practices in research and education.[66–68] In recent years, a collective effort within the Brazilian scientific community has been directed toward nonanimal approaches in research and testing. This effort has led to the opportunity to host, for the first time in the Southern Hemisphere, the World Congress on Alternatives and Animal Use in the Life Sciences in Rio de Janeiro, Brazil, in 2024. This is a significant opportunity not only for Brazil but for the entire Latin America and the Caribbean to discuss the progress made and to establish new connections and partnerships to address the important task of advancing the replacement of animals in medical research and testing.[68]

Brazil has made significant strides in advancing nonanimal methodologies, solidifying its leadership in Latin America. However, several challenges persist, including the need for broader regulatory integration and acceptance of alternative methods across various sectors. Expanding training and capacity-building efforts is critical, with an emphasis on hands-on education and multilingual knowledge exchange to foster expertise in NAMs. Furthermore, enhancing public awareness and policy advocacy is essential to sustain the momentum of ethical and scientific progress. Ongoing investment in funding and infrastructure is vital to support technological development and the validation of alternative methods. Strengthening international collaboration will also be crucial in accelerating Brazil's transition to nonanimal science. Overcoming these obstacles is vital for positioning Brazil as a global leader in modern, ethical, and innovative research and safety assessment.

Overcoming inertia against nonanimal approaches in the EU

Francesca Pistollato

The journey toward reducing animal testing in the EU has been evolving for decades, driven by both scientific advancements and ethical considerations. The roots of this commitment can be traced back to the 1980s, when the EU first pledged to explore ways to replace animal use in experiments and to improve the welfare of animals still needed for research.

Policy and legislative progress in the EU

In 1986, the commitment to reducing animal testing materialized into a landmark piece of legislation: Directive 86/609/EEC.[69] It marked the EU's first comprehensive effort to regulate the treatment of animals used for scientific purposes. This directive introduced minimum standards for animal welfare and care, emphasizing the need to reduce suffering and ensure proper housing conditions. In addition, it set out strict requirements for individuals and institutions involved in animal research, requiring authorization before conducting such activities. Crucially, the directive encouraged the use of alternative, nonanimal research methods wherever they were available and feasible. It also supported the development of new techniques aimed at reducing the reliance on animals. When the use of animals was deemed unavoidable, the directive mandated that the smallest number of animals be used and that efforts be made to minimize their pain and distress. However, it's worth noting that this early legislation did not explicitly reference the widely recognized 3Rs principles (replacement, reduction, and refinement) that are now central to animal research ethics.

By the late 2000s, it became evident that the original directive from 1986 needed a thorough revision to keep pace with new scientific and ethical standards. In response, the European Commission proposed a revised version, leading to the adoption of Directive 2010/63/EU in 2010.[70] This updated legislation was far more detailed and explicitly integrated the principles of the 3Rs, making them a cornerstone of the EU's approach to animal testing.

Directive 2010/63/EU went beyond simply clarifying previous regulations. It also laid out legal obligations for both the EU and its Member States to actively support the development and validation of alternative methods. As part of this effort, the directive established a framework for the European Union Reference Laboratory for Alternatives to Animal Testing, known as EURL ECVAM.[71] This institution became central to driving innovation in nonanimal approaches across Europe, providing scientific validation and promoting the acceptance of NAMs within regulatory frameworks.

In addition to Directive 2010/63/EU, several other key pieces of legislation have reinforced the EU's stance on reducing animal testing. Notably, the REACH regulation[72] 69 emphasizes that animal testing should only be a last resort and mandates the minimization of duplicate tests. REACH also strengthens the legal basis for pursuing alternative methods to assess the safety of chemicals and other substances. The Cosmetic Products Regulation 1223/2009[73] enforces a ban on animal testing for cosmetics. In addition, there are various laws governing plant protection products, biocides, and laboratory practices. Collectively, these regulations form a robust legal landscape that continually pushes for the reduction and eventual elimination of animal testing in the EU.

EURL ECVAM: a key player in advancing alternatives

The European Centre for the Validation of Alternative Methods (ECVAM) was first established in 1991 under the European Commission's Joint Research Centre (JRC).[74] However, with the introduction of Directive 2010/63/EU, ECVAM's role was significantly expanded and formally enshrined in EU law, transforming it into EURL ECVAM, the EU's reference laboratory for alternative testing methods.

EURL ECVAM's mission is multifaceted. It leads the promotion of alternative testing methods in regulatory contexts and across biomedical research. This includes coordinating the validation of new methods throughout Europe, a process that involves both its own laboratories and a network of validation labs known as EU-NETVAL.[75] In addition, EURL ECVAM is a central hub for information dissemination, hosting databases like the Tracking System for Alternative Methods towards Regulatory acceptance[76] and the Dataset on Alternative Methods to Animal Experimentation,[77] which track and share information on alternative methods.

The institution also plays a pivotal role in fostering dialog among diverse stakeholders. Through the ECVAM Stakeholder Forum,[78] EURL ECVAM collaborates with industry associations, animal welfare organizations, academic researchers, and EU regulatory bodies such as the European Chemicals Agency,[79] the European Food Safety Authority,[80] and the European Medicines Agency.[81] These collaborations ensure that the development, validation, and regulatory acceptance of alternative methods proceed smoothly and efficiently.

EURL ECVAM's efforts go well beyond validating alternative testing methods. In collaboration with the JRC, EURL ECVAM is actively involved in advancing nonanimal approaches across various scientific areas. Their contributions include representing the Directorate-General for Environment in the OECD's Test Guidelines Programme, conducting research on in vitro and computational methods, and operating both a Good Laboratory Practice-compliant in vitro testing facility and a high-throughput screening lab. They also focus on promoting alternative methods, offering specialized training, and providing direct support to other European Commission services and agencies. In addition, EURL ECVAM organizes expert workshops, participates in initiatives like the European Partnership for Alternative Approaches to Animal Testing,[82] and engages in international collaborations through the International Cooperation on Alternative Test Methods.[83] Their work also involves shaping international standards, legislation, and guidance while contributing to various EU research projects.[84]

Promoting education and training on 3Rs and nonanimal approaches

In an effort to map available education and training courses on the 3Rs, in 2019, EURL-ECVAM compiled a catalog, including 569 courses and resources available at high school, university, and professional levels worldwide, covering both explicit and implicit aspects of the 3Rs.[85] This collection provided a foundation for the JRC to explore how to further support and integrate the 3Rs into educational and training contexts.

It was acknowledged that while there is a wealth of 3Rs courses and resources available, there is still a need to enhance and expand education and training opportunities, particularly at the secondary school level and in higher education. Consequently, efforts have been made to explore ways to integrate the 3Rs principles and related materials into school and university curricula. EURL ECVAM has developed a collection of learning scenarios and teaching resources designed to inspire and assist educators in incorporating the 3Rs into their teaching. In addition, they have created several massive open online courses focused on the 3Rs.[86,87]

The JRC Summer School offers a unique chance to learn from leading experts about innovative nonanimal approaches, exchange knowledge with peers, and explore career development paths.[88] Along the same lines, the global education hub for animal-free education, launched by TPI Utrecht and PETA UK, serves as a collaborative platform for sharing and developing educational resources. This initiative aims to promote animal-free education and research methods, highlighting the importance of innovative nonanimal models and the collective effort needed to shape the future of education in this field.[89,90]

Public attitudes toward the use of animals in science: two European citizens' initiatives

It is worth noting that there is a growing trend toward the 'democratization of science,' the process of making scientific knowledge, tools, and decision-making processes more accessible, inclusive, and participatory for a broader range of interest-holders beyond traditional academic and institutional boundaries. With this, it is becoming increasingly important to understand public attitudes toward existing scientific practices and actively engage society in these discussions.[91]

The European Citizens' Initiative (ECI) "Stop Vivisection"[92] called for the repeal of Directive 2010/63/EU and demanded the complete replacement of animal testing with alternative methods.[93] This initiative should be seen by regulators and the scientific community as an important opportunity to explore new approaches to public engagement on the ethical and scientific issues surrounding animal testing.

The relatively new "Save Cruelty Free Cosmetics" ECI[94] calls on the European Commission to take decisive action in three key areas to eliminate animal testing while upholding safety and environmental standards. First, the initiative seeks to strengthen the existing ban on animal testing for cosmetics by ensuring that no animal tests are conducted on cosmetics ingredients at any stage, for any purpose. This would involve legislative changes aimed at guaranteeing consumer, worker, and environmental safety without resorting to animal testing. Second, the initiative advocates for a transformation in EU chemicals regulation. It calls for a system that manages chemical safety effectively while protecting human health and the environment without adding new requirements for animal testing. The focus is on modernizing regulation to reduce reliance on animal tests. Finally, the ECI pushes for a broader shift in the EU's scientific approach by urging the Commission to commit to a plan that phases out all animal testing within the existing legislative term. This would involve mapping out a strategy to embrace innovative, nonanimal methods, ultimately steering the EU toward a future where science and ethics are fully aligned. In essence, the initiative aims to guide the EU toward more humane, innovative, and sustainable practices in both regulatory frameworks and scientific research, with the goal of completely eliminating animal testing.

In response to the 'Save Cruelty-Free Cosmetics' campaign, the European Commission outlined key actions aimed at reducing animal testing. In particular, regarding EU chemicals regulation, the Commission is working on a roadmap to guide efforts toward chemical safety assessments free from animal testing.[95] This roadmap will involve collaboration with various stakeholders[96] and is expected to be delivered in 2026.

To modernize science, the Commission plans to continue funding the development of alternative testing methods and initiate actions to accelerate the reduction of animal testing in research, education, and training. The Commission has also integrated nonanimal testing approaches into the 2025–2027 strategic plan of Horizon Europe,[97] and a dedicated policy action under the European Research Area is being developed to support nonanimal methods in biomedical research and pharmaceutical testing.[98]

Overcoming barriers to nonanimal approaches in drug safety evaluation in Japan

Hajime Kojima and Yoko Hirabayashi

Validation and adoption of nonanimal test methods in Japan

The Japanese Centre for the Validation of Alternative Methods (JaCVAM) was established within the National Institute of Health Sciences (NIHS) in 2005.[99] To avoid redundant studies and accelerate research, the International Cooperation on Alternative Test Methods (ICATM) was established in March 2009 by the Interagency Coordinating Committee on the Validation of Alternative Methods (ICCVAM), ECVAM, JaCVAM, and Health Canada.[100] The goal of ICATM is to foster international cooperation in validation studies, independent peer review, and harmonized recommendations, ensuring global acceptance of alternative methods and strategies.

History of official nonanimal test method formulation in JaCVAM

JaCVAM was established to promote alternative animal testing methods in regulatory studies. JaCVAM's activities support the application and approval of the manufacture and sale of pharmaceuticals and other products, as well as the revision of cosmetic product standards. To achieve this, JaCVAM evaluates the utility, limitations, and suitability of test methods for regulatory studies and contributes to validation studies when needed. In addition, JaCVAM cooperates and collaborates with similar organizations in related fields in both Japan and internationally.

As part of its mission of validation via international cooperation, JaCVAM evaluates test methods developed in Japan through method-validation studies and independent peer review. This work has produced over 20 validation and peer-reviewed reports, including those from other countries and the OECD, supported by the ICATM. NIHS experts and test developers have formalized 16 OECD TGs, including animal test methods and four major revisions of TGs related to human health.[101] JaCVAM has contributed to developing new test methods for regulatory pharmaceutical use, including the in chemico reactive oxygen species (ROS) assay, cardiotoxicity assays with human-induced pluripotent stem cell-derived cardiomyocytes (hiPSC-CMs), alternative assays for the detection of reproductive and developmental toxicity, in vitro skin irritation test methods, and nonanimal pyrogenicity assays.

ROS assay in ICH S10

The International Council for Harmonisation of Technical Requirements for Pharmaceuticals for Human Use (ICH) guidelines for evaluating the photosafety of pharmaceuticals (ICH S10) were adopted in 2015[102] It applies to new active pharmaceutical ingredients, excipient clinical formulations for dermal applications, and photodynamic therapy products. This guideline includes nonanimal test methods such as the in vitro 3T3 neutral red uptake assay[101] and the in chemico ROS assay, which was developed by Professor Satomi Onoue of the University of Shizuoka[103] and validated by JaCVAM and members of the Japanese Pharmaceutical Manufacturers Association. The ROS assay detects ROS production when a chemical substance is irradiated with simulated sunlight. ROS can be detected using various methods such as spectrophotometry, fluorescence, chemiluminescence, and electron spin resonance. The ROS assay has now been approved under the OECD TG495.[99,101]

Assays with hiPSC-CMs for ICH E14/S7B questions and answers (Q&A)

Pharmaceuticals are tested for potential proarrhythmic effects at various stages of drug development according to ICH S7B, a guideline providing information on nonclinical assays and integrated risk assessment principles for evaluating the potential for delayed ventricular repolarization (QT interval prolongation).[104]

In drug development under ICH S7B, human ether-a-go-go related gene (hERG) testing in cultured cells is used for early screening, and telemetry testing in large animals (e.g., dogs and monkeys) is used to later identify compounds before clinical trials. The hERG test assesses the inhibitory effect on the IKr channel associated with QT interval prolongation on electrocardiograms, providing unambiguous results by measuring only current inhibition, with automated devices widely available. However, several ion channels other than IKr channels are involved in ventricular muscle action potentials, and discrepancies between hERG test results and clinical outcomes have been reported. Therefore, relying solely on hERG test results may lead to the exclusion of valuable pharmaceutical candidates in the early stages of development and may not accurately assess the risk of QT interval prolongation in humans. Consequently, in vitro test systems with high clinical predictability and the ability to integrate risk assessment at an early stage are needed.

In December 2013, the Human iPS Cell Application Safety Evaluation Consortium (CSAHi) was established under the auspices of the Japan Pharmaceutical Manufacturers Association. Since its establishment, CSAHi has engaged in developing various safety evaluation techniques using hiPSC-CMs, focusing on safety issues related to the heart, liver, and nerves, and confirming their applicability for practical use. Particularly, a cardiotoxicity evaluation system using hiPSC-CMs has been promoted, which utilizes a multielectrode array (MEA) extracellular potential waveform to predict risk with high accuracy. The MEA scoring system captures the electrical activity of different ion channels in hiPSC-CMs in a composite waveform, assessing the QT interval and detecting early afterdepolarization, which causes lethal arrhythmias that cannot be detected in animal experiments. Notably, in validation studies conducted in Japan, the MEA scoring system demonstrated a clinical predictive accuracy of 83%, surpassing the 72% accuracy of the hERG test.[105,106] However, ion channel expression varies significantly depending on the hiPSC-CM type; therefore, the selection of the hiPSC-CM type and lot number is crucial for improving the predictive accuracy of cardiotoxicity assessment using MEA.[107] Cardiotoxicity assessment using hiPSC-CMs has been studied via CSAHi activities in collaboration with Japan iPS Cardiac Safety Assessment based on NIHS and the US Food and Drug Administration for international validation tests.[108] Consequently, the Q&A of ICH E14/S7B, including the use of hiPSC-CMs, was published in 2022.[109]

Alternative assays for the detection of reproductive and developmental toxicity for human pharmaceuticals in ICH S5(R3)

The ICH guidelines for the Detection of Reproductive and Developmental Toxicity for Human Pharmaceuticals, ICH S5(R3), approved in 2021, include Section 4.2.2: Alternative Approaches for Addressing Embryo-Fetal Development (EFD) Risk; and Annex 4.2.2.1: Use of Alternative Assays. Many alternative in vitro, ex vivo, and nonmammalian in vivo assays have been developed to detect potential hazards to EFD.[110] These assays have been used as pharmaceutical discovery screens for EFD adverse effects and to understand the mechanism of toxicity, aiding in translating nonclinical data to human risk (especially for human-specific targets). Continued use of alternative assays for these purposes is encouraged, and if properly qualified, they can potentially defer or (in certain circumstances) replace conventional in vivo assays, potentially reducing animal use. Based on the direction of scientific development at the time of this writing, several alternative assays will be used for regulatory purposes within a tiered or battery approach. These testing strategies will be qualified within a specific context of use, defined by the chemical applicability domain of the assay and the characterization of the biological mechanisms covered by the assay.

In Japan, researchers have already developed the Hand1-Luc embryo stem cell test (EST) for Developmental and Reproductive Toxicity studies[111] and validated its use via JaCVAM[99] based on the validated EST by ECVAM.[112] This test method was proposed for inclusion in an OECD TG. Unfortunately, the proposal was withdrawn as test developers later abandoned the OECD process.

In vitro skin irritation test methods for ISO 10993-23

The International Organization for Standardization (ISO) 10993–23:2021 is part of the ISO 10993 series, which outlines the biological evaluation standards for medical devices.[113] ISO 10993–23 specifies the requirements for in vitro testing of skin irritation potential in medical devices using validated in vitro test methods, such as reconstructed human epidermis models that mimic human skin structure and function. These models provide a reliable platform for irritation assessment without animal testing, aligning with global regulatory trends favoring in vitro approaches. The significance of in vitro skin irritation testing using EpiDerm and EpiSkin is highlighted in ISO 10993–23, impacting product development.[114]

Based on previous models, the Division of Medical Devices in NIHS and JaCVAM conducted validation studies on skin irritation test methods using the LabCyte EPI-MODEL 24 SIT (J-TEC Corporation) for medical devices. The assay using this model may be acceptable under ISO 10,993–23.

Nonanimal pyrogenecity assays

Pyrogens are fever-inducing substances that, when introduced into the bloodstream, can cause fever or more severe symptoms, such as septic shock and death, in humans and other mammals.[115] Various pyrogen testing methods have been introduced over the years, including the rabbit pyrogen test (RPT) (an in vivo test) in the 1940s, which measures fever response as an endpoint, and the limulus amebocyte lysate test (an in vivo test) in the 1970s, which uses horseshoe crab hemolymph to specifically detect endotoxins. Nonanimal pyrogen testing has been gaining favor, such as the monocyte activation test (MAT), which was included as a compendial method in the Ph. Eur. in 2010, an alternative method in USP in 2012, and a pyrogen test in the ISO/TR 21,582:2021 guidelines for pyrogen detection in medical devices[116,117]; and the recombinant factor C test in 2018[116] a nonanimal-based in vitro pyrogen test that completely replaces the rabbit test.

Although the lack of harmonization between pharmacopoeias is a major hurdle, the European Pharmacopoeia Commission's pyrogenicity strategy aims to eliminate the RPT from the Eur. Ph. by July 2026 and publish a general chapter on pyrogenicity.[117] (For a deeper discussion about the phasing out of the RPT in Europe, see Chapter 19.) Nonanimal test methods for pyrogenicity assays have been evaluated in the Japanese Pharmacopoeia: PyroMAT (Sigma-Aldrich) was reviewed, and MycoMAT (MiCAN Technologies, Inc.) and Lumi MAT (FUJIFILM Wako Pure Chemical Corporation) have been validated by JaCVAM. The Japanese Pharmacopoeia plans to accept the MAT in 2026.

Summary

In response to requests from international organizations such as ICH, ISO Medical Devices, and Pharmacopoeia, Japanese researchers have played a significant role in the development of several alternatives to animal testing. Not all efforts have been successful, and the withdrawal of the Hand1 Luc EST for developmental and reproductive toxicity studies from the OECD process was disappointing. However, Japan will contribute to developing and assessing NAMs worldwide.

Conclusions

The journey toward widespread adoption of nonanimal testing methods faces both cultural and scientific challenges. Historically, regulatory authorities have shown a tendency to be cautious when considering alternative methods for safety assessments. This cautious approach stems from the longstanding reliance on traditional animal-based methods, which have been seen as tried-and-true standards. However, over time, the scientific community has increasingly recognized the potential and reliability of alternative testing methods. Today, there is broad consensus that nonanimal approaches are valuable tools, especially when they undergo rigorous validation and peer review by reputable institutions like EURL ECVAM, BraCVAM, and JaCVAM.

The barriers to broader adoption of nonanimal methods are both cultural and scientific. Culturally, the regulatory sector needs to continue embracing the growing body of evidence that supports the effectiveness of alternative methods, moving beyond the conservative reliance on traditional testing. Scientifically, the challenge is to advance toxicological and biomedical research to the point where nonanimal models can accurately and comprehensively predict the full spectrum of potential adverse effects.

To overcome these obstacles, a multifaceted approach is needed. This includes ongoing innovation in alternative testing methods, greater investment in research that tackles complex biological processes, and stronger collaboration between regulators, researchers, and industry. By addressing both cultural resistance and scientific limitations, the vision of replacing animal testing with reliable and ethical alternatives can move closer to reality, paving the way for a more humane and scientifically sound future worldwide.

Acknowledgments

We thank supporters of JaCVAM and the JaCVAM Steering Committee. In addition, we thank Editage (www.editage.jp) for English language editing.

Funding

This review was funded by AMED, grant no. 25mk0121281j0002.

References

[1] Pistollato F, Madia F, Corvi R, et al. Current EU regulatory requirements for the assessment of chemicals and cosmetic products: challenges and opportunities for introducing new approach methodologies. *Arch Toxicol*. 2021;95(6):1867–1897.

[2] Skin Sensitisation - European commission [internet]. [cited 2024 Aug 21]. Available from: https://joint-research-centre.ec.europa.eu/projects-and-activities/reference-and-measurement/european-union-reference-laboratories/eu-reference-laboratory-alternatives-animal-testing-eurl-ecvam/alternative-methods-toxicity-testing/validated-test-methods-health-effects/skin-sensitisation_en.

[3] US. Department of Health and Human Services. National toxicology program. Integrated approaches to testing and assessment. Available from: https://ntp.niehs.nih.gov/whatwestudy/niceatm/comptox/ct-its/its. Accessed Aug 21, 2024.

[4] Sewell F, Alexander-White C, Brescia S, et al. New approach methodologies (NAMs): identifying and overcoming hurdles to accelerated adoption. *Toxicol Res*. 2024;13(2):tfae044.

[5] The prevention of cruelty to animals act, 1960. Available at: https://lawbhoomi.com/prevention-of-cruelty-to-animals-act-1960/. Accessed April 15, 2025

[6] Sharma B, Sharma P. Rights of animals at practice in India. *J Contemp Issues Law*. 2017;3(7):1–23.

[7] Tandon PN, Muralidhar K, Gupta YK. Recommendations. In: *Use of Animals in Scientific Research and Education*. Indian National Science Academy: New Delhi; 2012:217–218.

[8] Jayaraman KS. Row in India over rules on animal experiments. *Nature*. 1998;395:108.

[9] U.S. Environmental Protection Agency. List of alternative test methods and strategies (or new approach to methodologies). Updated July 23, 2026. Available at: https://www.epa.gov/assessing-and-managing-chemicals-under-tsca/list-alternative-test-methods-and-strategies-or-new#list. Accessed Aug 15, 2026.

[10] Indian Pesticide Guidelines: Available at: https://prsindia.org/files/bills_acts/bills_parliament/1970/Draft%20Pesticide%20Managemnt%20Bill%202017.pdf. Accessed April 15, 2025.

[11] Minutes of the 450th meeting of the registration committee, agenda item no. 10.32. Available at: https://www.scribd.com/document/821888106/450-rc-minutes.

[12] Akbarsha MA, Mascarenhas B (2019) Cosmetic regulation and alternatives to animal experimentation in India. In: Kojima H, Seidle T, Speilmann H (Eds), Alternatives to Animal Testing. Proceedings of Asian Congress 2016. Springer Open. pp 57–62.

[13] CDSCO. The Cosmetics Rules 2020. Available at: https://cdsco.gov.in/opencms/opencms/en/Acts-and-rules/Cosmetics-Rules/ Accessed April 15, 2025.

[14] Swaminathan S, Kumar V, Kaul R. Need for alternatives to animals in experimentation: an Indian perspective. *Indian J Med Res*. 2019;149:584–592.

[15] Gazette of India, Ministry of Health and Family Welfare, No 136, dated March 09, 2023. Available at: https://www.stationeryprinting.tn.gov.in/extra_ordinary_lists.php?id=MjAyMw==.

[16] Mohanan PV, ed. *Microfluidics and Multi Organs on Chip*. Singapore: Springer; 2022.

[17] Rahman SZ, Anam A, Pandey A, Bisht R, Akbarsha MA. Meeting report: Indian society for alternatives to animal experiments: sixth annual meeting and international conference. *ALTEX*. 2024;41(2):324–328.

[18] Centre for Predictive Human Modelling Systems. Microphysiological Systems Researchers' Database [database of Indian scientists working in the fields of organoids, organ-on-chips and computational modeling] Available at: https://aic.ccmb.res.in/mps-researchers-database/ Accessed April 15, 2025.

[19] OECD Guidance Document No 34: guidance document on the validation and international acceptance of new or updated test methods for Hazard assessment. Available at: https://ntp.niehs.nih.gov/sites/default/files/iccvam/suppdocs/feddocs/oecd/oecd-gd34.pdf. Accessed April 15, 2025.

[20] Zeeshan MD, Kunnathodi F, Akbarsha MA. 3R approach and the science of animal alternatives. *Scientia*. 2010; 6(1):17–26.

[21] Akbarsha MA, Zeeshan MD, Pereira S. CPCSEA National Conference, India. *ALTEX*. 2011;28(2):153–155.

[22] Akbarsha MA, Zeeshan MD, Meenakumari KJ. Tenth in vitro toxicology workshop at MGDC. *ALTEX*. 2012;29(3):353.

[23] Shanmugam KB, Zeeshan M, Bhattacharya S, et al. Scientists in India sensitized about Systems Biology approach to toxicity testing: hands-on workshop on Computational Systems Biology and Dose-Response Modelling. *ALTEX Proc*. 2013;2(1/13):83–92.

[24] Akbarsha MA, Zeeshan MD, Meenakumari KJ (2013) Alternatives to Animals in Education, Research, and Risk Assessment: an Overview with Special Reference to Indian Context. *ALTEX* Proceedings 2, 1/13, Proceedings of Animal Alternatives in Teaching, Toxicity Testing and Medicine. Available at: https://proceedings.altex.org/data/2013-01/rISC_001_Akbarsha2.pdf. Accessed April 15, 2025.

[25] Koduri C, Akbarsha MA. End to animal testing of cosmetics and house-hold products. *ALTEX*. 2014; 31(2):236–237.

[26] Gayathri L, Dhanasekaran D, Akbarsha MA. Scientific concepts and applications of integrated discrete multiple organ co-culture technology. *J Pharmacol Pharmacother*. 2015;6:63–70. doi:10.4103/0976-500X.155481.

[27] Akbarsha MA, Mascarenhas B, Pellevoisin C. First National Congress on Alternatives to Animal Testing and post-congress workshops in India. *ALTEX*. 2018;35(2):258–260.

[28] Akbarsha MA, Raisuddin S, Kapoor DM. National conference on alternatives to animal experiments (NCAAE-2018) in commemoration of launching of the Society for Alternatives to Animal Experiments (SAAE) in India. *ALTEX*. 2019;36(3):472–476.

[29] Palahe P, Pendse S, Pardhi V, et al. Setting the trend for the alternatives movement in India: the second national conference of the Society for Alternatives to Animal Experiments-India. *ALTEX*. 2020;37(3):503–506.

[30] Akbarsha MA, Zeeshan MD, Pereira S. Alternatives discussed at Indian Science Congress. *ALTEX*. n.d.;29 (2/12):216–218.

[31] Webinar Series on the Use of New Approach Methodologies for the Risk Assessment of Pesticides. Available at: https://www.thepsci.eu/india-pesticide-webinars/ Accessed April 15, 2025

[32] Zeeshan MD, Akbarsha MA. First IdMOC workshop conducted in India. *ALTEX*. 2011;28(2):167–168.

[33] Zeeshan MD, Riyasdeen A, Pellevoisin C, Akbarsha MA. News: IND: first workshop on reconstructed 3-dimensional human tissue. *ALTEX*. 2013;30(4):559.

[34] Siddique YH, Beg T, Varshney H, et al. National Workshop on Alternatives to Higher Animals in Toxicology and Biomedical Science. *ALTEX*. 2024;41(3):488–490. doi:10.14573/altex.2403151.

[35] Brasil. Lei n° 6.638, de 8 de maio de 1979. Estabelece normas para a prática didático-científica da vivissecção de animais e determina outras providências [Internet]. Brasília, DF: Presidência da República; 1979 [cited 2024 Feb 28]. Available from: https://www.planalto.gov.br/ccivil_03/leis/1970-1979/l6638.htm.

[36] Brasil. Lei n° 9.605, de 12 de fevereiro de 1998. Dispõe sobre as sanções penais e administrativas derivadas de condutas e atividades lesivas ao meio ambiente [Internet]. Brasília, DF: Presidência da República; 1998 [cited 2024 Feb 28]. Available from: https://www.planalto.gov.br/ccivil_03/leis/l9605.htm.

[37] Brasil. Lei n° 11.794, de 8 de outubro de 2008. Regula o inciso VII do §1° do art. 225 da Constituição Federal e estabelece procedimentos para o uso científico de animais [Internet]. Brasília, DF: Presidência da República; 2008 [cited 2024 Feb 28]. Available from: https://www.planalto.gov.br/ccivil_03/_ato2007-2010/2008/lei/l11794.htm.

[38] Dalben D, Emmel JL. A Lei Arouca e os direitos dos animais utilizados em experimentos científicos. *Rev Eletr Inic Cient*. 2013;4(4):280–291.

[39] Rivera E, Hernandez R, Carissimi AS, Pekow C. Laboratory animal legislation in Latin America. *ILAR*. 2016;57(3):293–300. doi:10.1093/ilar/ilw017I.

[40] Universidade Federal de Goiás. Workshop Internacional Prático sobre Métodos Alternativos: curso Aplicado pelo Institute for In Vitro Sciences, IIVS [Internet]. [cited 2024 Mar 1]. Available from: https://farmacia.ufg.br/n/23138-workshop-internacional-pratico-sobre-metodos-alternativos-curso-aplicado-pelo-insti-tute-for-sciences-iivs.

[41] Presgrave O, Moura W, Caldeira E, et al. Brazilian Center for the Validation of Alternative Methods (BraCVAM) and the process of validation in Brazil. *Altern Lab Anim*. 2016;44(1):85–90.

[42] de Ávila I, Fentem J, Villela I, et al. Brazilian National Network of Alternative Methods (RENAMA) 10th Anniversary: meeting of the Associated Laboratories, May 2022. *Altern Lab Anim*. 2024;52(1):60–68.

[43] CONCEA. Resolução Normativa n° 17, de 3 de julho de 2014 [cited 2025 Apr 15]. Available at: https://www.gov.br/mcti/pt-br/acompanhe-o-mcti/concea/arquivos/pdf/legislacao/resolucao-normativa-no-17-de-3-de-julho-de-2014.pdf/view?TSPD_101_R0=086f3e6665ab2000ffba9f40547975e651b6139836e6217612d95baaa5beb4b422c0867436d8b8da08093bdc5f14300043c12fef91e30ed-267766de429701a9fc24d3dd70f06784186b8c2a00b378f753370fd01b31f6f8c13003b21646f0da3.

[44] CONCEA. Resolução Normativa n° 31, de 18 de agosto de 2016: reconhece métodos alternativos ao uso de animais em atividades de pesquisa [Internet]. Brasília, DF: Conselho Nacional de Controle de Experimentação

Animal; 2016 [cited 2025 Apr 15]. Available from: https://www.gov.br/mcti/pt-br/acompanhe-o-mcti/concea/arquivos/pdf/legislacao/resolucao-normativa-no-31-de-18-de-agosto-de-2016.pdf/view.

[45] Brasil. Resolução da Diretoria Colegiada - RDC nº 35, de 7 de agosto de 2015: dispõe sobre a aceitação dos métodos alternativos de experimentação animal reconhecidos pelo Conselho Nacional de Controle de Experimentação Animal - CONCEA [Internet]. Brasília, DF: Agência Nacional de Vigilância Sanitária; 2015 [cited 2025 Apr 15]. Available from: https://www.gov.br/mcti/pt-br/acompanhe-o-mcti/concea/arquivos/pdf/legislacao/resolucao-rdc-no-35-de-7-de-agosto-de-2015-anvisa.pdf/view.

[46] CONCEA. Resolução Normativa nº 18, de 24 de setembro de 2014: reconhece métodos alternativos ao uso de animais em atividades de pesquisa no Brasil, nos termos da Resolução Normativa nº 17, de 03 de julho de 2014, e dá outras providências [Internet]. Brasília, DF: Conselho Nacional de Controle de Experimentação Animal; 2014 [cited 2025 Apr 15]. Available from: https://www.gov.br/mcti/pt-br/acompanhe-o-mcti/concea/arquivos/pdf/legislacao/resolucao-normativa-no-18-de-24-de-setembro-de-2014.pdf/view.

[47] CONCEA. Resolução Normativa nº 45, de 22 de outubro de 2019: reconhece método alternativo ao uso de animais em atividades de pesquisa no Brasil [Internet]. Brasília, DF: Conselho Nacional de Controle de Experimentação Animal; 2019 [cited 2025 Apr 15]. Available from: https://www.gov.br/mcti/pt-br/acompanhe-o-mcti/concea/arquivos/pdf/legislacao/resolucao-normativa-no-45-de-22-de-outubro-de-2019.pdf/view.

[48] Villela IV, Machado MS. Brazil's regulatory context for using new approach methodologies (NAMs) on the registration of products. *Front Toxicol*. 2022;4:903027.

[49] Strickland J, Haugabrooks E, Allen DG, et al. International regulatory uses of acute systemic toxicity data and integration of new approach methodologies. *Crit Rev Toxicol*. 2023;53(7):385–411.

[50] Ministério da Ciência, Tecnologia, Inovações e Comunicações (MCTI). PReMASUL - Regional Platform of Alternative Methods to the Use of Experimental Animals [Internet]. Available from: https://www.gov.br/mcti/pt-br/acompanhe-o-mcti/premasul. Cited 2023 Apr 03.

[51] CONCEA. Resolução Normativa nº 53, de 19 de maio de 2021 [cited 2025 April 15] Available at: https://antigo.mctic.gov.br/mctic/opencms/legislacao/outros_atos/resolucoes/Resolucao_CONCEA_n_53_de_19052021.html.

[52] CONCEA. Resolução Normativa nº 54, de 10 de janeiro de 2022 [cited 2025 April 15]. Available at: https://antigo.mctic.gov.br/mctic/opencms/legislacao/outros_atos/resolucoes/Resolucao_Normativa_CONCEA_n_54_de_10012022.html.

[53] Garcia RC, Jukes N, Bones V, et al. Brazil starts to ban animal use in higher education: a positive and progressive development. *Altern Lab Anim*. 2018;46(4):235–239. doi:10.1177/026119291804600402. PMID: 30365336.

[54] Bachinski R, Alves G, Freitas de Almeida Souza M, et al. On the journey toward humane education in Brazil: first request for a total ban of harmful animal use in professional and higher education. *Altern Lab Anim*. 2017;45(5):287–293. doi:10.1177/026119291704500501. PMID: 29112455.

[55] Bachinski R, Tréz T, Alves GG, et al. Humane education in Brazil: organisation, challenges and opportunities. *Altern Lab Anim*. 2015;43(5):337–344. doi:10.1177/026119291504300508. PMID: 26551290.

[56] CONCEA. Resolução Normativa nº 58, de 23 de fevereiro de 2023 [cited 2026 April 15]. Available at: https://antigo.mctic.gov.br/mctic/opencms/legislacao/outros_atos/resolucoes/Resolucao_Normativa_Concea_n_58_de_23022023.html.

[57] Brasil. Câmara dos Deputados. Projeto de Lei nº 3062, de 22 de dezembro de 2022: altera a redação dos artigos 14, 17 e 18 da Lei nº 11.794, de 8 de outubro de 2008, que institui a Lei Arouca, para vedar a utilização de animais em ensino, pesquisas e testes laboratoriais para desenvolvimento de produtos cosméticos em humanos, e para aumentar os valores de multa em caso de infração [Internet]. Brasília: Câmara dos Deputados; 2022 [cited 2024 Apr 03]. Available from: https://www.camara.leg.br/proposicoesWeb/fichadetramitacao?idProposicao=597587.

[58] Brasil. Ministério da Ciência, Tecnologia e Inovação. Guia Brasileiro de Produção, Manutenção ou Utilização de Animais em Atividades de Ensino ou Pesquisa Científica. Conselho Nacional de Controle de Experimentação Animal – 1ª ed. Brasília; 2023 [cited 2025 Apr 15]. Available from: https://www.gov.br/mcti/pt-br/composicao/colegiados/concea/paginas/publicacoes-legislacao-e-guia/guia-brasileiro-de-producao-manutencao-ou-utilizacao-de-animais-para-atividades-de-ensino-ou-pesquisa-cientifica?TSPD_101_R0=086f3e6665ab20006120d54131cf8757688ba7b7b76173f11b0e0dcc630a1de1e92815d9eabe073c089afc3f8e144800540138f7280f70e465304f7625d2da4f11bd5de2244336204a7db8a5f2bf74a2db25724f593fb5dccf22cbb399803550d238295500069837541b57e219bf46ba6dee042231f5447d.

[59] Science K. XX Congresso Brasileiro de Toxicologia 2017: as estratégias da Toxicologia do século XXI para um mundo mais seguro e sustentável [Internet]. Available from: https://www.doisamaiscosmetica.com.br/noticias/grupo-kosmoscience-participa-do-xx-congresso-brasileiro-de-toxicologia-em-goiania/. Accessed 2024 Apr 02.

[60] Sociedade Brasileira de Toxicologia (SBTox). XX Congresso Brasileiro de Toxicologia 2017: as estratégias da Toxicologia do século XXI para um mundo mais seguro e sustentável [Internet]. Available from: https://sbtox.org/xx-congresso-brasileiro-de-toxicologia-xx-cbtox/. Accessed 2024 Mar 02.

[61] Sociedade Brasileira de Toxicologia (SBTox). XXII Congresso Brasileiro de Toxicologia 2022: inovação em Ciências Toxicológicas [Internet]. Available from: https://www.sbtox.org/post/nova-data-congresso-brasileiro-de-toxicologia-cbtox-2022. Accessed 2023 Dec 12.

[62] Sociedade Brasileira de Toxicologia (SBTox). XXII Congresso Brasileiro de Toxicologia 2022: inovação em Ciências Toxicológicas. Hands-on Training – adotando Novas Abordagens em Metodológicas (NAMs) para avaliação da toxicidade inalatória: teoria e prática. Available from: https://www.cbtox2021.com.br/. Accessed 2023 Dec 12.

[63] Série de webinars Atualizações em novas abordagens metodológicas (NAMs) aplicadas à segurança humana e ambiental. Available from: https://ufg.br/e/32675-serie-de-webinars-atualizacoes-em-novas-abordagens-metodologicas-nams-aplicadas-a-seguranca-humana-de-ambiental. Accessed 2023 Dec 12.

[64] SEAC scientists collaborate to launch LATAM In Vitro Science Webinars. Available from: https://sers.unilever.com/news/2022/seac-scientists-collaborate-to-launch-latam-in-vitro-science-webinars/. Accessed 2023 Dec 12.

[65] Cosmeticsdesign.com. Brazil paves the way to end animal testing throughout LATAM. Available from: https://www.cosmeticsdesign.com/Article/2023/09/12/brazil-paves-the-way-to-end-animal-testing-throughoutlatam. Accessed 2023 Dec 12.
From the editors: since the writing of this chapter, the website Cosmeticsdesign.com has closed. The Editors offer an alternative reference: The Macao News: Brazil poised to end animal testing after landmark vote. July 16, 2025. Available at: https://macaonews.org/news/lusofonia/brazil-animal-testing/. Accessed August 15, 2026.

[66] I Congresso Brasileiro de Métodos Alternativos ao Uso de Animais na Pesquisa e no Ensino (I CBMAlt). Available from: https://prpi.ufg.br/n/ufg-evento-metodos-alternativos-uso-de-animais-pesquisa-cbmalt. Accessed 2025 Apr 12.

[67] UFG participa de evento sobre métodos alternativos ao uso de animais em pesquisa. Available from: https://prpi.ufg.br/n/ufg-evento-metodos-alternativos-uso-de-animais-pesquisa-cbmalt. Accessed 2024 Jan 21.

[68] Experiência da UFG em uso alternativo de animais em pesquisa é compartilhada em evento. Available from: https://jornal.ufg.br/n/176030-experiencia-da-ufg-em-uso-alternativo-de-animais-em-pesquisa-e-compartilhada-em-evento. Accessed 2024 Jan 21.

[69] Council Directive 86/609/EEC of 24 November 1986 on the approximation of laws, regulations and administrative provisions of the Member States regarding the protection of animals used for experimental and other scientific purposes [Internet]. Oj L Nov 24, 1986. Available from: http://data.europa.eu/eli/dir/1986/609/oj/eng.

[70] Directive - 2010/63 - EN - EUR-Lex [Internet]. [cited 2024 Aug 21]. Available from: https://eur-lex.europa.eu/eli/dir/2010/63/oj.

[71] EU Reference Laboratory for alternatives to animal testing (EURL ECVAM) - European Commission [Internet]. [cited 2024 Aug 21]. Available from: https://joint-research-centre.ec.europa.eu/projects-and-activities/reference-and-measurement/european-union-reference-laboratories/eu-reference-laboratory-alternatives-animal-testing-eurl-ecvam_en.

[72] European Chemical Agency. Role of animal testing in ensuring the safe use of chemical substances. Available at: https://echa.europa.eu/documents/10162/13630/reach_factsheet_animal_testing_en.pdf/fb82d34b-7e58-4446-a3fb-f87b242d6829. Accessed Aug 15, 2026.

[73] Regulation - 1223/2009 - EN - Cosmetic Products Regulation - EUR-Lex [Internet]. [cited 2024 Aug 21]. Available from: https://eur-lex.europa.eu/legal-content/EN/ALL/?uri=celex%3A32009R1223.

[74] European Commission. Joint research centre. Available at: https://commission.europa.eu/about/departments-and-executive-agencies/joint-research-centre_en. Accessed Aug 15, 2026.

[75] European union network of laboratories for the validation of alternative methods (EU-NETVAL) - European Commission. Available from: https://joint-research-centre.ec.europa.eu/projects-and-activities/reference-and-measurement/european-union-reference-laboratories/eu-reference-laboratory-alternatives-animal-testing-eurl-ecvam/alternative-methods-toxicity-testing/european-union-network-laboratories-validation-alternative-methods_en. Accessed Aug 21, 2024.

[76] Home | EURL ECVAM – TSAR. Available from: https://data.jrc.ec.europa.eu/dataset/b7597ada-148d-4560-9079-ab0a5539cad3. Accessed Aug 21, 2024.

[77] EURL ECVAM dataset on alternative methods to animal experimentation (DB-ALM). 2019 Jan 1 Available from: https://data.jrc.ec.europa.eu/dataset/b7597ada-148d-4560-9079-ab0a5539cad3 Accessed Aug 15, 2026.

[78] ESTAF - EURL ECVAM Stakeholder Forum - European Commission. Available from: https://joint-research-centre.ec.europa.eu/reference-measurement/european-union-reference-laboratories/eu-reference-laboratory-alternatives-animal-testing-eurl-ecvam/knowledge-sharing-three-rs/knowledge-networks/estaf-eurl-ecvam-stakeholder-forum_en Accessed Aug 21, 2024.
[79] ECHA. Available from: https://echa.europa.eu/it/ Accessed Aug 21, 2024.
[80] EFSA. Science, safe food, sustainability. 2021. Available from: https://www.efsa.europa.eu/en Accessed Aug 21, 2024.
[81] European Medicines Agency (EMA). 2024. Available from: https://www.ema.europa.eu/en/homepage Accessed Aug 21, 2024.
[82] European partnership for alternative approaches to animal testing - European Commission. Available from: https://single-market-economy.ec.europa.eu/sectors/chemicals/european-partnership-alternative-approaches-animal-testing_en Accessed Aug 21, 2024.
[83] International cooperation on alternative test methods (ICATM) - European Commission. Available from: https://joint-research-centre.ec.europa.eu/projects-and-activities/reference-and-measurement/european-union-reference-laboratories/eu-reference-laboratory-alternatives-animal-testing-eurl-ecvam/alternative-methods-toxicity-testing/advisory-and-consultation-bodies/international-cooperation_en. Accessed Aug 21, 2024.
[84] Alternatives to animal testing at the JRC: the 2023 status report is out - European Commission. 2024 Available from: https://joint-research-centre.ec.europa.eu/jrc-news-and-updates/alternatives-animal-testing-jrc-2023-status-report-out-2024-03-12_en Accessed Aug 21, 2024.
[85] Holloway M, Florean O, Busquet F, Farcal L, Tenie A. Three Rs education and training courses and resources. 2019 Mar 15 Available from: http://data.europa.eu/89h/d0569abb-b5ab-4b9e-9e16-d70cea7a89f9 Accessed Aug 21, 2024.
[86] Education and training - european commission. Available from: https://joint-research-centre.ec.europa.eu/projects-and-activities/reference-and-measurement/european-union-reference-laboratories/eu-reference-laboratory-alternatives-animal-testing-eurl-ecvam/three-rs-education-and-training_en. Accessed Aug 21.2024.
[87] The Three Rs and animal use in science rerun. Available from: https://www.europeanschoolnetacademy.eu/courses/course-v1:ThreeRs+AnimalsInScience+2021/about. Accessed Aug 21, 2024.
[88] JRC Summer school on nonanimal approaches in science - European Commission. Available from: https://joint-research-centre.ec.europa.eu/events/jrc-summer-school-nonanimal-approaches-science-2023-05-23_en Accessed Aug 21, 2024.
[89] Global Education Hub - TPI Utrecht - Utrecht University. Available from: https://www.uu.nl/en/research/life-sciences/tpi-utrecht/education/global-education-hub Accessed Aug 21, 2024.
[90] Janssens MRE, Salvatori D, Hogervorst J, et al. Developing a global education hub for animal-free innovation. *ALTEX*. 2025;42:354–357.
[91] Ormandy EH, Schuppli CA. Public attitudes toward animal research: a review. *Anim Open Access J MDPI*. 2014;4(3):391–408.
[92] Stop vivisection. Available from: https://citizens-initiative.europa.eu/initiatives/details/2012/000007/stop-vivisection_en Accessed Aug 21, 2024.
[93] Menache A. The European Citizens' Stop Vivisection Initiative and the revision of Directive 2020/63/EU. *Altern Lab Anim ATLA*. 2016;44(4):383–390.
[94] Save cruelty free cosmetics - commit to a europe without animal testing. Available from: https://citizens-initiative.europa.eu/save-cruelty-free-cosmetics-commit-europe-without-animal-testing_en Accessed Aug 21, 2024.
[95] Directorate-General for Internal Market, Industry, Entrepreneurship and SMEs (European Commission), Cronin M. Report of the European commission workshop on "the roadmap towards phasing out animal testing for chemical safety assessments": Brussels, 11-12 December 2023. Publications Office of the European Union; 2024. Available from: https://data.europa.eu/doi/10.2873/34576 Accessed Aug 21, 2024.
[96] Walder L, Pallocca G, Bastos LF, et al. EU roadmap for phasing out animal testing for chemical safety assessments: recommendations from a multi-stakeholder roundtable. *ALTEX*. 2025. doi:10.14573/altex.2503241.
[97] Directorate-General for Research and Innovation (European Commission). Horizon Europe strategic plan 2025-2027. Publications Office of the European Union; 2024. Available from: https://data.europa.eu/doi/10.2777/092911 Accessed Aug 21, 2024.
[98] European Research Area action "Towards medical sciences and regulatory testing without the use of animals."; 2023. Available at: https://www.eurogroupforanimals.org/files/eurogroupforanimals/2023-08/European%20

Research%20Area%20action%20"Towards%20medical%20sciences%20and%20regulatory%20testing%20 without%20the%20use%20of%20animals".pdf Accessed May 2, 2025.
[99] JaCVAM(2025) Available at: http://jacvam.jp/ Accessed April 15, 2025.
[100] ICATM (2025) Available at: https://ntp.niehs.nih.gov/whatwestudy/niceatm/iccvam/international-partnerships/icatm Accessed April 15, 2025.
[101] OECD Guidelines for the Testing of Chemicals (2025) Available at: https://www.oecd.org/en/topics/sub-issues/testing-of-chemicals/test-guidelines.html. Accessed April 25, 2025.
[102] UNECE, GHS Rev.10. 2025. Available at: https://unece.org/transport/documents/2023/07/standards/ghs-rev10.
[103] Onoue S, Igarashi N, Yamada S, Tsuda Y. High-throughput reactive oxygen species (ROS) assay: an enabling technology for screening the phototoxic potential of pharmaceutical substances. *J Pharm Biomed Anal.* 2008;46(1):187–193.
[104] ICH S7B: the Non-Clinical Evaluation of the Potential For Delayed Ventricular Repolarization (Qt Interval Prolongation) by Human Pharmaceuticals S7b (2005) Available at: https://www.ema.europa.eu/en/ich-s7b-non-clinical-evaluation-potential-delayed-ventricular-repolarization-qt-interval-prolongation-human-pharmaceuticals-scientific-guideline.
[105] Nozaki Y, Honda Y, Watanabe H, et al. CSAHi study-2: validation of multi-electrode array systems (MEA60/2100) for prediction of drug-induced proarrhythmia using human iPS cell-derived cardiomyocytes: assessment of reference compounds and comparison with non-clinical studies and clinical information. *Regul Toxicol Pharmacol.* 2017;88:238–251. PMID: 28634147(2017).
[106] Ando H, Yoshinaga T, Yamamoto W, et al. A new paradigm for drug-induced torsadogenic risk assessment using human iPS cell-derived cardiomyocytes. *J Pharmacol Toxicol Methods.* 2017;84:111–127. doi:10.1016/j.vascn.2016.12.003.
[107] Kodama M, Furutani K, Kimura R, et al. Systematic expression analysis of genes related to generation of action potentials in human iPS cell-derived cardiomyocytes. *J Pharmacol Sci.* 2019;140(4):325–330. doi:10.1016/j.jphs.2019.06.006.
[108] Kanda Y, Yamazaki D, Osada T, Yoshinaga T, Sawada K. Development of torsadogenic risk assessment using human induced pluripotent stem cell-derived cardiomyocytes: japan iPS Cardiac Safety Assessment (JiCSA) update. *J Pharmacol Sci.* 2018;138(4):233–239. doi:10.1016/j.jphs.2018.10.010.
[109] ICHE14/S7B Q&As: clinical and Nonclinical Evaluation of QT/QTc Interval Prolongation and Proarrhythmic Potential Questions and Answers (2022) Available at: https://database.ich.org/sites/default/files/E14-S7B_QAs_Step4_2022_0221.pdf.
[110] ICH S5(R3) detection of toxicity to reproductions for human pharmaceuticals (2020) Available at: https://www.ich.org/page/safety-guidelines#5-2.
[111] Le Coz F, Suzuki N, Nagahori H, Omori T, Saito K. Hand1-Luc embryonic stem cell test (Hand1-Luc EST): a novel rapid and highly reproducible *in vitro* test for embryotoxicity by measuring cytotoxicity and differentiation toxicity using engineered mouse ES cells. *J Toxicol Sci.* 2015;40(2):251–261. doi:10.2131/jts.40.251.
[112] Genschow E, Spielmann H, Scholz G, et al. Validation of the embryonic stem cell test in the international ECVAM validation study on three *in vitro* embryotoxicity tests. *Altern Lab Anim.* 2004;32(3):209–244. doi:10.1177/026119290403200305.
[113] ISO 10993-23:2021 Biological evaluation of medical devices. Part 23: tests for irritation (2025) Available at: https://www.iso.org/standard/74151.html.
[114] Kandárová H, Bendova H, Letasiova S, et al. Evaluation of the medical devices benchmark materials in the controlled human patch testing and in the RhE in vitro skin irritation protocol. *Toxicol. In Vitro* 2018;50:433–438. doi:10.1016/j.tiv.2018.02.009.
[115] Fennrich S, Hennig U, Toliashivili L, et al. More than 70 years of pyrogen detection: current state and future perspectives. *Altern Lab Anim.* 2016;44:239–253.
[116] ISO/TR 21582:2021 Pyrogenicity — principles and methods for pyrogen testing of medical devices, (2025) USP, Available at: https://www.iso.org/standard/71150.html.
[117] Cirefice G, Schütte K, Spreitzer I, et al. The future of pyrogenicity testing: phasing out the rabbit pyrogen test. A meeting report. *Biologicals.* 2023;84:101702.

CHAPTER

14

Perspectives from pharmaceutical industry regarding nonanimal-based research and testing

Sandra P. Smieszek, Christina Tyner, and Mihael H. Polymeropoulos

Vanda Pharmaceuticals, Inc., Washington, DC, United States

Abstract

Vanda Pharmaceuticals Inc. is a leading global biopharmaceutical company focused on the development and commercialization of innovative therapies to address high unmet medical needs and improve the lives of patients. Vanda's vision is to avoid in vivo animal studies and replace them with the use of alternatives to support scientific discovery, protecting and enhancing the welfare of animals used in scientific research while advancing research and development through the use of optimal human-specific models.

The commercial perspective has shifted toward nonanimal methods, driven by scientific rigor, efficiency, ethical implications, and cost-effectiveness. Similarly, the academic perspective has focused on the scientific innovation and mechanistic understanding of the use of alternative methods. We are also witnessing a societal shift toward more humane research practices that minimize animal suffering. However, broader industry adoption relies on these methods meeting regulatory approval. One potential route would be to reject the approach of global harmonization of animal methods, which relies on unsupported concepts often elevated to impose false gold standards.

Keywords: Nonanimal methods; FDA; Microphysiological systems

Rationale for using nonanimal methods

It is critical to confirm the safety of drug products before they come to market. As part of the drug development process, the US Food and Drug Administration (FDA) asks companies to perform multiple animal studies of different lengths and in different species to identify potential toxicities that could pose risks to humans. For example, one type of study the FDA has routinely mandated is a nonrodent toxicity study of nine months' duration, and it is typically conducted in young beagle dogs, although it can also be conducted in nonhuman

Moving Toward Nonanimal Approaches in Medical Research and Testing
https://doi.org/10.1016/B978-0-443-26514-3.00009-6

primates (e.g., monkeys) or minipigs. For this type of study, all of the animals must be euthanized at the conclusion of their study participation so that their tissues can be analyzed. In many, if not most, cases, these longer-term, nonrodent toxicity studies are unnecessary. Killing animals without a scientifically justified purpose is unethical and inhumane. Such studies should not be conducted routinely but instead should be the exception rather than the rule.

It is striking that, over the past two decades, advances in technology have revolutionized drug development; however, the FDA has not yet significantly revised its approach to animal toxicity studies. The toxicity studies required by the FDA are the same in 2025 as they were in 1997, and the FDA's stated basis for its policy is an analysis published in 1999, which did not reach any conclusion as to the potential human significance of any toxicological findings identified in longer-term studies. With the rapid advancements in microphysiological system (MPS) technology, human-based relevant models are now available that were not within reach when the FDA first approached animal toxicity study requirements. Now that they are available, regulatory bodies need to conduct a comprehensive and up-to-date analysis of the value of animal studies and how we can utilize alternative methods, in the name of ethical animal research and good science.

The emerging developments at the FDA have made progress toward these ends, including the FDA Modernization Act 2.0,[1] which allows validated nonanimal studies in drug approvals. While the opportunity for change is now there, almost no practical change in market approval processes has thus far occurred, leading legislators to introduce the FDA Modernization Act 3.0,[2] setting out a goal of one year from passage for updating practices. At the time of this writing, the bill is still in committee. In addition, in April 2025, the FDA announced a plan to phase out animal studies, beginning with monoclonal antibodies, and the NIH announced a new initiative to prioritize human-based research and reduce animal use. How quickly these plans will be realized, however, remains to be seen.

There are several applications for MPS in research.[3] Multiorgan MPS models can mimic human physiology in ways that traditional animal models cannot. MPS systems integrate human cells and tissues, enabling researchers to study efficacy, pharmacokinetics, pharmacodynamics, and toxicology in a highly controlled environment. The use of human cells inherently provides more relevant data regarding human response to therapeutic compounds, thus reducing the risk of misleading data that can arise from interspecies differences seen in animal models.

For example, MPS systems can simultaneously model drug absorption, distribution, metabolism, and excretion (ADME) processes across different organ systems, including the liver, kidney, gut, and heart. The integration of these organs into a single device connected by a continuous fluid channel allows for capturing the effects of organ-organ interactions on each of the organs. In comparison, animal models often fail to accurately predict human ADME characteristics due to species-specific differences in enzyme activity, metabolic rates, and immune responses.[4–7]

In traditional toxicology, animals are exposed to a drug, and researchers monitor organ-specific toxicity, but they often lack the ability to directly observe how organs interact in response to drug exposure. This is particularly problematic for understanding systemic toxicities, such as off-target effects that result from organ-organ communication or metabolic byproducts affecting other tissues. A four-organ chip provides a platform to study these interactions in a controlled, interconnected manner. By linking organs such as the liver, kidney, gut, and heart, researchers can study how a drug is metabolized in the liver, how its metabolites affect the kidneys, and how the gut microbiome may influence drug absorption or systemic toxicity.[3,8,9] One of the rationales for longer-duration animal studies is the hypothesis of accumulating damage. MPS sensitivity of detection allows for much earlier detection of organ injury, long before the physiology is perturbed and detected either clinically or through pathology. This allows for the development of toxicity to be observed at a much faster rate than it would be observed in an animal model. In addition, MPS models provide researchers with the ability to extract a wide range of organ-specific mechanistic data, which can be used to further investigate mechanisms of toxicity. The detailed evaluation that MPS models provide is not possible in animal models.

One of the key advantages of using MPS models is their ability to generate unbiased and reproducible data.[10,11] Animal studies are highly unpredictable as an investigative tool because they are biased and lack the resolution to be comprehensive. Variability in genetic backgrounds, environmental conditions, and behavioral factors can all introduce significant variability in outcomes. This inconsistency in animal models can require additional testing, which delays drug development timelines and increases costs.

In contrast, MPS models provide controlled, reproducible environments. Since these systems are engineered with specific human cells and tissues, they can be standardized across multiple labs and experiments. This standardization reduces the experimental variability seen in animal studies, allowing researchers to consistently reproduce results in different settings. For example, the four-organ chip can be fabricated with precise configurations to replicate human organ interactions, creating a more consistent experimental platform. Data generated from such systems exhibits a higher degree of reliability compared to the unpredictability often seen in animal studies.

Moreover, MPS models are free from some of the subjective biases that can occur in animal research. Animal testing frequently involves subjective assessments of symptoms and outcomes, which can lead to inconsistencies between different studies or among individual researchers. MPS models rely on objective, quantifiable metrics such as organ function, cellular responses, and molecular markers. This enhances the overall accuracy and objectivity of the data.

Species differences are a well-known problem in preclinical toxicology, and establishing the human relevancy of toxicity findings in animals remains a challenge.[7,12,13] The interspecies variability in immune responses, metabolic pathways, and organ function frequently leads to discrepancies in how a drug behaves in animals versus humans. For instance, metabolic enzymes that process drugs in animals are often vastly different from those in humans, which can lead to false-positive or false-negative results when predicting drug toxicity or efficacy.[7] Drug-induced liver injury is a serious and real threat to patients. The existing system of detection, utilizing repeated-dose animal toxicology testing, cannot adequately anticipate human liver toxicity.[14] Several studies have shown that results from animal studies often fail to predict accurate outcomes in humans.[15] For example, the clinical development of Hu5c8, a monoclonal antibody intended for the treatment of autoimmune disorders, was terminated due to unexpected thrombotic complications—life-threatening side effects that were not discovered during preclinical testing due to the lack of predictive models.[16] However, organ-on-a-chip models were able to recapitulate this thrombosis event, providing evidence of the importance of utilizing MPS models to fully evaluate safety in humans.[16] We have seen similar events with fialuridine and troglitazone, where failures of the existing predictive toxicology system resulted in detrimental effects—events that would have been picked up by organ-on-a-chip testing.

Gleevec, a treatment for chronic myelogenous leukemia (CML), provides another important example of when animal testing failed us. In misleading animal tests, Gleevec showed serious adverse effects in at least five species tested, including severe liver damage in dogs.[17] However, in human cell assays, there was no evidence of liver toxicity. Clinical trials proceeded and further confirmed the absence of liver toxicity in humans.[17] If the sponsor had decided to drop this program due to these animal findings, CML patients would not now have this treatment. Yet the use of human cell assays provided sufficient evidence that this observed toxicity was indeed due to species differences and that Gleevec can be used in humans without these adverse toxic effects. The outcome of animal trials not predicting human toxicity has been seen throughout the drug development process for hundreds of drug programs.[7] A critical advantage of MPS systems is their ability to model human genetic diversity, which is not possible with the existing animal models.[9]

A 2006 review of 76 animal studies found that only 37% were ever replicated in humans.[18] A second review of 221 animal studies found only a 50% agreement.[19] Ongoing studies have continued to replicate these findings. In an analysis of 2366 drugs, Dr. Jarrod Bailey concluded that "results from tests on animals (specifically rat, mouse, and rabbit models) are highly inconsistent predictors of toxic responses in humans and are little better than what would result merely by chance—or tossing a coin—in providing a basis to decide whether a compound should proceed to testing in humans."[4] Similar results were found for nonhuman primates and dogs.[5] Utilizing an MPS, these findings of species-specific differences in drug metabolism were further exemplified when comparing a rat liver-chip, human liver-chip, and dog liver-chip composed of primary hepatocytes.[20] In that study, an experimental compound was found to induce fibrosis in a rat liver-chip but did not alter hepatocyte function in human chips, whereas a third compound demonstrated toxicity in the dog liver-chip.[20] These results pointedly show the inability to properly assess human toxicity in a different species.

Given the poor predictive validity of animal toxicity studies, we cannot and should not rely on the results found in these studies to accurately investigate human toxicity. Not only is a poorly predictive test both wasteful and misleading, but it may also result in the dismissal of truly beneficial drugs that present no human toxicity concerns. MPS models provide predictive validity and allow for the modernization of monitoring drug effects.

The FDA has argued that animal models are necessary due to their ability to perform extensive histopathology assessments. However, it is illogical and inhumane to conduct such experiments with poor predictivity and reliability "just because." Rather, these types of extensive assessments are warranted only if toxicity is first observed in human models. MPS models address this issue by utilizing human-derived cells that retain their organ-specific functions, such as hepatocytes in a liver-on-a-chip model or cardiomyocytes in a heart-on-a-chip. These systems more accurately simulate human metabolic and physiological responses to drugs. A four-organ chip can model the complex interplay between the liver, kidney, gut, and heart in drug metabolism and clearance, providing an integrated view of a drug's pharmacokinetics and pharmacodynamics that is far more relevant to humans than the information derived from rodent or nonhuman primate studies. The data generated from these models is translated into the whole human body response, increasing the translational value of the data.

Human genetics serves as a fundamental source of evidence in drug discovery. The lack of genetic diversity in animal models presents a challenge when trying to predict how a drug will affect a diverse human population, particularly considering differences in genetic backgrounds, age, gender, or comorbidities.[12] Genetics play a fundamental role in how efficacious a drug is and its potential toxicity, particularly in the context of metabolism. Humans exhibit high genetic variability, which is not captured by studies that utilize inbred animals. We conducted a genome-wide homozygosity analysis on human and dog genomes and reported the highest degree of inbreeding in beagles, even compared to highly isolated human populations like the Amish, where consanguinity is high.[21] This raises serious questions about the applicability of dog models in preclinical research, especially when the goal is to gauge the subtle effects on the organism's physiology produced by candidate therapeutic agents. Can the FDA really argue that the use of these genetically uniform dogs is predictive of drug toxicity or efficacy in a general, genetically diverse human population?

A critical advantage of MPS systems is their ability to model human genetic diversity. These systems can incorporate patient-derived cells, allowing for the study of pharmacological effects across different genetic backgrounds and disease states. This approach significantly improves the predictive power of toxicology studies compared to the homogeneous nature of animal models, which cannot easily mimic the complexities of human metabolism.

Aside from the robust scientific evidence supporting the superiority of MPS models in predicting potential drug efficacy and toxicity in humans, the ethical considerations of reducing animal use are critical. The authors' company, Vanda Pharmaceuticals Inc. (Vanda), values all forms of life, including those of experimental animals, and does not believe in the unnecessary killing of experimental animals. The unreasonable persistence in uninformative and cruel experiments is a weakness of a bureaucratic system that is far removed from the evolution of science or societal beliefs.

Commercial advantages and remaining challenges in using NAMs

In 2018, the estimated mean cost of developing a new drug was $172.7 million; when accounting for the cost of failures, it becomes $515.8 million.[22] The reasons for 90% of clinical failures in drug development include lack of clinical efficacy (40%–50%), unmanageable toxicity (30%), poor drug-like properties (10%–15%), and lack of commercial needs and poor strategic planning (10%).[23] With a striking 80% of drug development failures stemming from a lack of human efficacy or unmanageable toxicity, it begs the question: are preclinical animal studies accurately predicting how drugs will behave in people? And if they are not, are we *removing* potentially life-saving drugs for humans from the drug development pathway due to animal study findings? It is apparent that this problem has its roots in the poor predictive utility of animal models for human disease.

Utilizing nonanimal methods provides a huge commercial advantage, increasing the probability that a drug candidate will make it to market. This is achievable not only through rapid screening of potential targets and reducing developmental costs but also by ensuring that findings are applicable to humans at each stage. In addition, MPS systems offer economic advantages by reducing the high costs associated with animal studies, such as housing, feeding, and caring for animals over prolonged study periods, as well as eliminating the costs of repeating studies due to interspecies variability. MPS models are an advanced, scientifically robust, and ethical alternative to animal toxicology testing, providing comprehensive, unbiased, and human-relevant data that should be leveraged in the pharmaceutical industry.

From a commercial perspective, overcoming the experimental and scientific challenges faced in moving toward nonanimal methods will require ensuring reproducibility, standardization, and validation; the involvement of academic stakeholders in research and testing; and the cooperation and support of regulatory bodies. For example, MPS data need to adopt a set of good laboratory practices guidelines that are attuned to the specificity and rigor of MPS experiments. Improving the reliability and utility of alternative methods by benchmarking nonanimal models will facilitate the transition to animal-free research. For a more in-depth discussion on these matters, see Chapter 17.

There is a collaborative relationship between stakeholders in research and testing. While academic stakeholders may focus on the foundational science behind alternative methods, industry must work with regulatory agencies to refine and validate these approaches for commercial applications. This may involve pushing for policy changes and funding, alongside nongovernmental organization stakeholders, to help implement and standardize these alternative methods. It may also involve working to shift political will in Congress to help achieve these important changes.

Regulatory bodies fail to recognize the superior ability of MPS technology to predict human toxicity and efficacy over animal models. By mandating or even preferring long and extensive animal toxicity studies, the FDA is ignoring a large body of published scientific evidence, which concludes that animal studies are not representative of human physiology. The pharmaceutical industry and government regulatory bodies must abandon the unscientific, low-resolution animal testing and adopt modern, human-based scientific methods to advance human drug safety. Patients need safe and effective drugs that are developed using modern approaches to assess safety and efficacy, instead of relying on animal tests that are based on outdated science. It is inhumane and unethical to adhere to the existing guidelines that require animal suffering and slaughter in the absence of any expectation that it will lead to clinically relevant findings.

Our experience at Vanda Pharmaceuticals

The goal of predictive toxicology in drug development is to predict toxicities in humans before humans are exposed to the drug. A drug may proceed to testing in humans when there is an absence of severe toxicities in animals, or it may be terminated from development if severe toxicities are present in animals. However, because the animal testing paradigm lacks predictive validity, it often fails to detect eventual toxicities (a false negative) or predicts toxicities that are not reproduced in humans (a false positive).

In 2018, the FDA informed Vanda that the company would be required to conduct a nine-month toxicity study in dogs before proceeding with tradipitant studies in humans lasting longer than three months. Vanda refused to comply. The company explained that there was no scientific justification for such a study, particularly given that tradipitant had already been studied extensively in thousands of animals and humans. The FDA insisted that the nine-month toxicity study was required based on the International Council for Harmonisation guidance, and it placed long-term studies on clinical hold—not because of any specific safety concern, but due to the absence of the long-term dog study. Vanda sued in federal court, but the court ruled in favor of the FDA, concluding that the statutory and regulatory framework mandated animal testing. Congress subsequently amended the FDCA to remove the language that supposedly imposed this mandate through the FDA Modernization Act 2.0.[1]

The FDA's continued insistence on these brutal and nonpredictive studies in dogs only serves pharmaceutical companies that simply want to streamline the drug development process and check a box for the FDA, even at the expense of missed safety signals or the erroneous abandonment of useful drugs.

We should, as a community of scientists and drug developers, abandon these practices and refuse to conduct these unethical and scientifically unjustified studies.

The future of NAMs in a commercial context

The future use of nonanimal methods is poised for significant transformations, specifically in the context of standardization. As regulatory agencies continue to endorse human-relevant models, the integration of standardized MPS technology, such as organ-on-a-chip, will play a crucial role in derisking drug development by predicting major organ toxicity and reducing failures that lead to late-stage drug attrition. A commercial incentive lies here in reducing costly clinical failures and improving human safety by utilizing human-relevant models. Human-relevant models can more accurately predict safety and efficacy, ensuring that only viable drug candidates advance in developmental programs and that potentially effective human therapeutics are not left at the bench due to animal studies that inaccurately predict human toxicity. Regulatory networks are needed to formalize the validation frameworks for these systems to ensure industry-wide reproducibility and acceptance. This standardization will drive commercial adoption and enable companies to replace animal models in the drug development process.

The FDA has relied on industry complacency for too long. We all have an ethical responsibility to reduce, refine, and replace animal experimentation to the maximum extent possible. While the FDA purports to aspire to the same goals, it has been slow to adopt new methods, and its actions, along with its inflexible demands that companies continue to conduct animal toxicity studies that have been shown to serve little scientific purpose, suggest otherwise. The boards of directors, executives, and employees of drug companies, animal advocacy organizations, scientists, and the public at large should join in asking the FDA to conduct a comprehensive and up-to-date analysis of the value of animal studies and abolish a one-size-fits-all approach to animal research that results in the unnecessary sacrifice of too many animals.

One potential route would be to reject global harmonization of regulatory policies that rely on unsupported concepts regarding the accuracy of animal methods and often elevate them to impose false gold standards. We predict that a change in this field will be achieved when a few pioneers break away from the "harmonization" pact and assert new standards. Moving away from the herd mentality of the need for animal toxicity experiments requires a new generation of leaders who will advance the field into new norms. The moment for this change is now: both the necessary technology and expertise exist, and some promising steps in Congress and at the FDA could pave the path for more humane, scientifically sound studies. What is lacking is the will of the industry to challenge bureaucratic thinking and the refusal to conduct uninformative and brutal experiments.

References

[1] S.5002 - FDA Modernization Act 2.0. 117th Congress (2021–2022); 2022. Available at: https://www.congress.gov/bill/117th-congress/senate-bill/5002. Accessed 9 August 2025.

[2] S.355 - FDA Modernization Act 3.0. 119th Congress; 2025. Available at: https://www.congress.gov/bill/119th-congress/senate-bill/355. Accessed 22 April 2026.

[3] Leung CM, de Haan P, Ronaldson-Bouchard K, et al. A guide to the organ-on-a-chip. *Nat Rev Methods Primers*. 2022;2(1):33. doi:10.1038/s43586-022-00118-6.

[4] Bailey J, Thew M, Balls M. An analysis of the use of animal models in predicting human toxicology and drug safety. *Altern Lab Anim*. 2014;42(3):181–199. doi:10.1177/026119291404200306.

[5] Bailey J, Thew M, Balls M. Predicting human drug toxicity and safety via animal tests: can *any* one species predict drug toxicity in any other, and do monkeys help? *Altern Lab Anim*. 2015;43(6):393–403. doi:10.1177/026119291504300607.

[6] Ingber DE. Human organs-on-chips for disease modelling, drug development and personalized medicine. *Nat Rev Genet*. 2022;23(8):467–491. doi:10.1038/s41576-022-00466-9.

[7] Van Norman GA. Limitations of animal studies for predicting toxicity in clinical trials. *JACC Basic Transl Sci*. 2019;4(7):845–854. doi:10.1016/j.jacbts.2019.10.008.

[8] Bell CC, Lauschke VM, Vorrink SU, et al. Transcriptional, functional, and mechanistic comparisons of stem cell–derived hepatocytes, HepaRG cells, and three-dimensional human hepatocyte spheroids as predictive in vitro systems for drug-induced liver injury. *Drug Metab Dispos*. 2017;45(4):419–429. doi:10.1124/dmd.116.074369.

[9] Kim J, Koo BK, Knoblich JA. Human organoids: model systems for human biology and medicine. *Nat Rev Mol Cell Biol*. 2020;21(10):571–584. doi:10.1038/s41580-020-0259-3.

[10] Rubiano A, Indapurkar A, Yokosawa R, et al. Characterizing the reproducibility in using a liver microphysiological system for assaying drug toxicity, metabolism, and accumulation. *Clin Transl Sci*. 2021;14(3):1049–1061. doi:10.1111/cts.12969.

[11] Truskey GA. Human microphysiological systems and organoids as in vitro models for toxicological studies. *Front Public Health*. 2018;6. doi:10.3389/fpubh.2018.00185.

[12] Avila AM, Bebenek I, Bonzo JA, et al. An FDA/CDER perspective on nonclinical testing strategies: classical toxicology approaches and new approach methodologies (NAMs). *Regul Toxicol Pharmacol*. 2020;114:104662. doi:10.1016/j.yrtph.2020.104662.

[13] Pound P, Ritskes-Hoitinga M. Is it possible to overcome issues of external validity in preclinical animal research? Why most animal models are bound to fail. *J Transl Med*. 2018;16(1):304. doi:10.1186/s12967-018-1678-1.

[14] Navarro VJ, Senior JR. Drug-related hepatotoxicity. *N Engl J Med*. 2006;354(7):731–739. doi:10.1056/NEJMra052270.

[15] Fabre K, Berridge B, Proctor WR, et al. Introduction to a manuscript series on the characterization and use of microphysiological systems (MPS) in pharmaceutical safety and ADME applications. *Lab Chip*. 2020;20(6):1049–1057. doi:10.1039/C9LC01168D.

[16] Barrile R, van der Meer AD, Park H, et al. Organ-on-chip recapitulates thrombosis induced by an anti-CD154 monoclonal antibody: translational potential of advanced microengineered systems. *Clin Pharmacol Ther.* 2018;104(6):1240–1248. doi:10.1002/cpt.1054.

[17] Akhtar A. The flaws and human harms of animal experimentation. *Camb Q Healthc Ethics.* 2015;24(4):407–419. doi:10.1017/S0963180115000079.

[18] Hackam DG, Redelmeier DA. Translation of research evidence from animals to humans. *JAMA.* 2006;296:1731.

[19] Perel P, Roberts I, Sena E, et al. Comparison of treatment effects between animal experiments and clinical trials: systematic review. *BMJ.* 2006;334:197.

[20] Jang KJ, Otieno MA, Ronxhi J, et al. Reproducing human and cross-species drug toxicities using a liver-chip. *Sci Transl Med.* 2019;11(517). doi:10.1126/scitranslmed.aax5516.

[21] Smeiszek S, Polhymeropoulos MH. Study of regions of homozygosity (ROH) patterns to evaluate the use of dogs' genome in human drug development. *J Human Earth Future.* 2022;(3). Available at: https://hefjournal.org/index.php/HEF/article/view/103. doi:10.28991/HEF-SP2022-01-02.

[22] Sertkaya A, Beleche T, Jessup A, Sommers BD. Costs of drug development and research and development intensity in the US, 2000–2018. *JAMA Netw Open.* 2024;7(6):e2415445. doi:10.1001/jamanetworkopen.2024.15445.

[23] Sun D, Gao W, Hu H, Zhou S. Why 90% of clinical drug development fails and how to improve it? *Acta Pharm Sin B.* 2022;12(7):3049–3062. doi:10.1016/j.apsb.2022.02.002.

CHAPTER

15

Overcoming animal methods bias and its impacts on nonanimal research assessments

Emily R. Trunnell[a] *and Catharine E. Krebs*[b]

[a]People for the Ethical Treatment of Animals, Washington, D.C., United States

[b]Physicians Committee for Responsible Medicine, Washington, D.C., United States

Abstract

As awareness, acceptance, and use of nonanimal research methods grow, critical barriers to their full adoption remain. One of these barriers is animal methods bias—a preference for animal-based methods or lack of expertise to adequately evaluate nonanimal methods, resulting in unfair assessments of nonanimal research. In this chapter, we explore the evidence for animal methods bias in research assessment and the historical and systemic factors contributing to it, including the structural incentives that drive the research ecosystem. Mitigation recommendations for researchers, funders, and publishers are provided, adapted from work by the Coalition to Illuminate and Address Animal Methods Bias, and research gaps in understanding and correcting this bias are identified. Additional research and advocacy on animal methods bias are needed to understand its full impact on the uptake and acceptance of nonanimal methods and the persistence of animal use in biomedical research.

Keywords: Bias; Peer review; Research assessment; Research grants; Scientific publishing

Introduction

As seen throughout this book, nonanimal biomedical research methods are increasingly effective and available to model human diseases and test drugs and other products; however, a variety of barriers stand in the way of their broader use. One such barrier can be characterized by the scientific community's reliance on and preference for animal-based methods. Centuries of animal use have led to well-established animal research infrastructure, economy, and regulatory frameworks, which incentivize and perpetuate continued animal use. Researchers may have scientific or ethical reasons for aiming to use nonanimal approaches, but animals are often more widely accepted or accessible, adding undue pressure to decisions regarding model choice. These systemic conditions require large-scale solutions, such as broad public and

Moving Toward Nonanimal Approaches in Medical Research and Testing
https://doi.org/10.1016/B978-0-443-26514-3.00012-6

private investments in nonanimal research and infrastructure and an overhaul of outdated and ineffective regulatory guidance, as detailed in the other chapters in Part III of this book.

Animal methods bias is a peer-review bias characterized by a preference for animal-based methods or lack of expertise to adequately evaluate nonanimal methods, resulting in unfair assessments of nonanimal research.[1,2] Animal methods bias can cause delays in publication or rejection of manuscripts or grant applications, effectively disincentivizing the use of these powerful tools. Initially described in the context of manuscript peer review, animal methods bias may affect any scenario where animal-free biomedical research is evaluated, including grant application peer review, hiring and tenure decisions, dissertation defenses, and strategic planning or prioritization of funds. To broaden the development and use of highly translatable human-based, animal-free methods, biomedical stakeholders—including individual researchers, peer reviewers, publishers, research faculty, research institutions, policymakers, and funders—must work to overcome this formidable barrier.

This chapter discusses what we know about animal methods bias, including empirical and anecdotal evidence gathered by the Coalition to Illuminate and Address Animal Methods Bias (COLAAB) (www.animalmethodsbias.org), an international collaboration of researchers and advocates aimed at characterizing and mitigating animal methods bias. Animal methods bias is then contextualized among other biases, including conservatism bias, other peer-review biases, and publishing bias. We describe how animal methods bias is not merely an issue of individual peer reviewers but a systemic problem requiring systemic solutions, touching on broader issues such as the publish-or-perish paradigm. We discuss ways for various biomedical stakeholders, including researchers, publishers, and funders, to address animal methods bias. Finally, as there is still much to learn about the issue, we address the need to prioritize the collection of additional empirical evidence.

Characterizing animal methods bias

The concept of animal methods bias predates its formal definition. In 2020, Donald Ingber published a perspective on the burgeoning field of organ chips framed around the idea that the technology has advanced enough for reviewers to request organ chip validation experiments rather than animal ones.[3] Organ chips and other nonanimal research technologies have advanced significantly over the last decade, but many researchers are still unfamiliar with them or their advantages. Thus the classic case of animal methods bias manifests as manuscript peer reviewers expecting or requesting in vitro mechanistic evidence to be validated in vivo—even for complex in vitro models that have been shown to better replicate human biology and pathophysiology and more accurately predict clinical outcomes compared to animal-based models.

Anecdotal accounts of such peer-reviewer requests regularly materialize on social media, in conference question-and-answer sessions, in personal communication with the authors of this chapter, and likely in many other places of which the authors of this chapter are unaware. A few examples to highlight here include X user @ChrisCroy posting on January 10, 2024: "I knew somebody who submitted a paper about an investigative compound with data from human volunteers. Reviewers all demanded mouse data, so they killed some mice for no reason and the reviewers were appeased." Along the same lines, X user @14Kountz posted on November 26, 2023: "This tweet might make some enemies, but: certain journals that de facto or explicitly require animal experiments to demonstrate that a microbe will do the conversion of X into Y in the gut when we already know it robustly does the same thing in culture are [morally] culpable." On July 10, 2023, X user @schumarcherbj posted: "The reviewer insists on showing it in mice," accompanied by an image of a mouse with someone holding up a miniature telephone to a mouse's ear. The first author of this chapter also had an experience in graduate school during which an advisor suggested she use animals in her projects instead of in vitro methods to increase publication odds:

> When I was a new neuroscience graduate student in 2012, I sat down with my advisor to brainstorm what studies could form the story of my dissertation. The laboratory I joined did both in vitro and in vivo experiments [the latter using rats]. When I began discussing some of my ideas for in vitro experiments, a skeptical and hesitant look came over my advisor's face. He told me that I'd be better off doing experiments on animals because journals want to see in vivo work, and this would make it easier for me to publish. The perception of animal methods bias during manuscript peer review was one of the first messages I received as a young scientist. I trusted this advice and, four years later, the entirety of my dissertation was based on experiments using mice and rats. It was only when I sat down to write it that I realized how little my studies related to human health and the human experience.

Thus, anecdotally, it seems that many researchers are personally familiar with this phenomenon and may even be hesitant to speak out about it due to fear of retribution or reputational consequences (as reflected in "This tweet might make some enemies").

A 2022 survey of biomedical researchers gathered the first empirical evidence of animal methods bias during the peer review of manuscripts.[1] Of 68 survey respondents, 21 indicated that they had carried out animal-based experiments solely to anticipate reviewer requests for them; in other words, outside the context of review, researchers felt the animal experiments were unnecessary. Of 68 survey respondents, 31 had been asked by peer reviewers to add animal experiments to their otherwise nonanimal study, and though only three of them thought the request was justified, half complied with the request. In open-ended answers, three respondents indicated that journals with higher impact factors are more likely to expect animal experiments or reject studies that do not include animal experiments. One respondent aptly noted that "there is no incentive to not do animal experiments," highlighting the pressure researchers face to publish (including in high-impact journals) and implicating animal methods bias as a formidable barrier to researchers who use nonanimal methods or want to avoid using animals. As later described in a 2023 workshop exploring this issue, the time cost of rejection and resubmission can have serious impacts on researchers' careers, especially early-career researchers.[2] Even when not necessarily impacting researchers' careers, animal methods bias can affect the quality or constructiveness of reviews, reduce a study's reach, lead to the conduct of animal experiments that would otherwise not have been performed, and perpetuate reliance on poorly translatable animal approaches.[2]

Anecdotal accounts indicate that animal methods bias can also occur during the peer review of grant applications. One respondent to the 2022 survey about animal methods bias in publishing stated: "In my experience, reviewers on grant panels are even more likely to ask for animal experiments than manuscript reviewers."[1] In 2024, the COLAAB hosted a workshop to explore this issue further, synthesizing personal anecdotes about experiences during grant application peer review from researchers who use nonanimal approaches.[4] Researchers described outright rejection of proposals employing nonanimal new approach methodologies (NAMs) and reviewers' insistence on employing animal-based methods for the validation of in vitro findings—even when no such model exists, as is the case for one researcher studying human placentation and reproductive signaling. They also described reviewers' preference for animal-derived data over more clinically relevant patient-based data.

One study[5] provided initial findings on animal methods bias in neuroscience grant funding at the US National Institutes of Health (NIH). The publication history of neuroscience review group members was analyzed to assess members' methods expertise and compared to the methods used in funded grants evaluated by those review groups. The authors of this study found that greater animal-based methods expertise in review groups strongly correlated with more funding for animal-based grants and less funding for studies using nonanimal methodologies, indicating that mismatches between review group expertise and grant proposal content may contribute to funding patterns that favor animal use and underscoring the need for diverse expertise in review groups.

Animal methods bias may occur during any scenario in which research is assessed—conference abstracts, dissertation, tenure, and hiring committees, regulatory drug applications, etc. (Fig. 1). Researcher experiences with animal methods bias highlight a poor understanding among some reviewers of the advantages of NAMs and other nonanimal approaches. It is feasible that this poor understanding stems from a lack of education and training in such approaches. Indeed, as NIH increasingly supports the development and use of NAMs, the agency has highlighted the importance of promoting the translational relevance and value of NAMs in the biomedical community.[6] Some reviewers, however, may hold personal preferences for animal-based methods or unfounded prejudices against alternatives. Either way, as a peer-review bias affecting the quality and fairness of the review of nonanimal approaches, it presents a barrier to the broader use of these approaches, standing in the way of improved translatability and reductions in animal use. It is thus important to understand its causes and the contexts in which it occurs to better determine how to address it. These contexts are further explored in the next section.

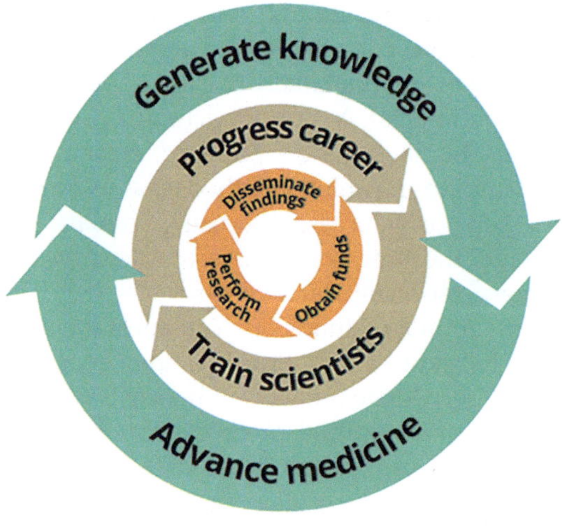

FIG. 1 The research lifecycle. Research is conducted cyclically, whereby previous findings inform novel projects, relying on the dissemination of findings (via publications, conference presentations, etc.) and on obtaining funding. Parallel to this, researchers are trained in new approaches, progress in their careers, and eventually train other researchers. Further parallel to this, scientific knowledge is generated, and, in the case of biomedical research, medicine is advanced. Each of these cycles requires that research be evaluated, often by scientific peers, sometimes by the public, the media, regulators, or other stakeholders. Anywhere research is assessed, there is a possibility for bias, including animal methods bias. The research lifecycle illustrates not only the systemic nature of the research enterprise—highlighting that any weaknesses or failures must be addressed systemically—but also that such weaknesses or failures can propagate through the system and feed into further ones in a snowball effect.

Contextualizing animal methods bias

Institutional inertia, conservatism, and peer-review bias

The persistence of animal research, despite the availability of alternatives, can be characterized by *technological lock-in*, where multiple constituencies and their vested interests perpetuate its dominance over other research methods and create bias against ideas that threaten its central place in research culture. Contributing factors include regulatory requirements for animal tests, the animal research industries and infrastructure, the wealth of existing animal data and the relative dearth of nonanimal data (which new findings are built upon and compared to), research culture heritage, and status quo perceptions among researchers that make it difficult to oppose or challenge the practice for fear of ostracization.[7,8]

A survey of Dutch animal researchers' reasons for using certain animal models found that an animal model's availability and the expertise of the research team in using that model were the leading factors driving model selection.[9] The authors concluded that "current selection of a specific animal model seems to be based on tradition rather than its potential predictive value for clinical outcome" and that "the use of animals to predict events in humans is 'locked-in' because animal experiments are embedded in a well-aligned set of institutions, such as regulations, norms, and values, that are taken for granted and are normatively endorsed," which ultimately slows down innovations.[9] Other scholarship has supported this theory,[10] and posited that "innovative methods do not sit well with the institutional logic of animal testing."[11]

Because experiments on animals are embedded into institutional culture, adherence to this paradigm and hesitance to abandon it become the conservative norm in science. As a result, researchers with conforming views are more likely to gain prominence, publications, funding, and other resources.[7,12] Animal researchers may thus hold disproportionate power over the conduct of science, more easily obtaining resources and influencing the further allocation of resources to others, including junior scientists. The placement of animal researchers in decision-making circles introduces the problem of *cognitive particularism*, wherein grant review committee members "make decisions based upon their membership in scientific schools of thought."[13] When upholding the more conservative status quo, cognitive particularism can suppress innovation because adventurous or frontier ideas are seen as too risky and threatening to institutional norms.[13]

Since peer reviewers determine what is published and funded, they become the arbiters of "good science." Those in established and inherently conservative fields define and retain the concept of quality and the distribution of resources.[14] This reduces the incentive for proposing and conducting nonconventional ideas and methodologies. Here, animal methods bias may take the form of self-censorship when researchers prefer nonanimal methods but use animals to meet the institutional norms of reviewers. They may inauthentically pay homage to animal research in their grant applications and manuscripts or preemptively include experiments on animals in their study designs to grease the wheels of the peer-review machine. Indeed, as mentioned previously, a 2023 survey found that 21 of 68 respondents carried out animal-based experiments not because they felt they were necessary but for the sole purpose of anticipating reviewer requests for them.[1] Because peer reviewers "evaluate research more favorably when findings support their prior beliefs," animal methods bias in this sense can "fuel a self-reinforcing dynamic in which censorship and self-censorship discourage empirical challenges to prevailing conclusions, encouraging a false consensus that further discourages dissent."[15]

Scholarship on grant peer review has demonstrated that cognitive particularisms or preferences not only uphold the status quo, but that a lack of familiarity with certain research fields or methods reduces trust in the peer-review system. A survey of Northern European researchers demonstrated low confidence in the ability of grant reviewers at national funding agencies to evaluate the quality of proposed research, partially due to a lack of matching expertise between the application and the reviewer.[16] Herein lies a challenge for researchers using nonanimal methods, which are often newer and less established than traditional animal models: proposals and manuscripts describing nonanimal methods are less likely to experience reviewer match, since reviewers are often chosen from representatives of conventional fields.

In summary, the persistence of animal use in science is complex and perpetuated by various factors, including tradition, inheritance, conservatism, technological lock-in, ubiquity, cognitive particularism in peer review, the need to obtain funding and publish, a lack of expertise match between scientists and those who are reviewing their work, and intermingling combinations of these elements. This has created an environment where experiments on animals are seen as being at the top of the biomedical research hierarchy. Until this "gold standard" is critically examined and challenged, the true potential of more humane and effective research methods may remain untapped.

Intersection with person-based and demographic biases

So far, this section primarily contextualizes animal methods bias within what Lee et al. defined as *content bias*[17]—bias resulting from personal perspectives on topics and methodologies. Other peer-review biases reflect certain person-based or demographic characteristics of investigators and their evaluators, such as prestige, seniority, affiliation, nationality, ethnicity, language, and gender.[18-24] The intersection of animal methods bias and person-based and demographic biases has yet to be directly studied, but previous analyses of research methodology and opinion provide clues as to how these concepts may interact.

An analysis comparing NIH funding rates for African American/Black researchers found that these investigators were less likely to propose work involving animals than their white counterparts, and the study identified topic choice as one of three key factors underlying lower funding rates for African American/Black scientists.[25] Surveys have also demonstrated that young people and those identifying as female are less likely to be supportive of experiments on animals.[8,26-29] Racism, sexism, and the hierarchical structure of academic research could mean that individuals in these demographics may experience biases in a compounded manner. When confronted with the pressure to use animals because of institutional inertia, pressure to obtain funding and publish, or direct requests from advisors, reviewers, or others, people with vulnerable or historically oppressed identities may be less empowered to push back on these directives or requests, making them more susceptible to their harmful effects. This hypothesis has not yet been empirically explored, though.

Different cultural views around the world, varying laws and regulations governing experiments on animals, and diverse levels of research infrastructure and resource availability, paired with the high entry costs of nonanimal technologies[30] may cause researchers in different global regions to have a variety of experiences with animal methods bias. Further scholarship in this area is needed to understand the intersection of animal methods bias with these and other person-based and demographic biases.

Publication bias

Publication bias was first defined as the failure to publish study results based on the direction or strength of the findings.[31,32] It has resulted in studies that reject the null hypothesis dominating the scientific literature, an absence of complete data in the scientific record, and, consequently, a threat to the rigor of systematic reviews. The Catalog of Bias later defined publication bias as "when the likelihood of a study being published is affected by the findings of the study."[33] Animal methods bias can be defined in parallel as when the likelihood of a study being published (or a proposal being funded) is affected by the *methods* of the study (or proposal).

Publication bias, whether it stems from the investigator's disinterest in publishing null studies or from negative perceptions by journal editors and reviewers, jeopardizes the accurate assessment of interventions and model validity. If experiments showing negative outcomes are never published, those null yet true results will never make their way into reports, narrative or systematic reviews, or metaanalyses, thus skewing the collective data on a particular topic and the combined knowledge distilled from it. Studies demonstrating positive—as opposed to negative/null—findings are also more likely to be cited in subsequent papers,[34] meaning those ideas or theories become the standard way of thinking in a field.[35] In this way, existing scientific literature may present an inaccurately positive portrayal of experiments on animals, where studies demonstrating the failure of those experiments to answer research questions or replicate human biology have been excluded from the record. Bias against null studies can therefore distort interpretations and opinions of experiments on animals, particularly regarding their precision and translatability, thus contributing to animal methods bias in the form of reviewers' expectations or requests for animal studies—including ones that have already been carried out but never published.

Publish or perish and research oversight

In addition to the compounding biases discussed in the previous sections, broader systemic issues in science and its oversight may also contribute to animal methods bias. In academia, especially, metrics of success are often grounded in the number of papers a scientist publishes, the ranks of the journals in which a researcher publishes, their position in the order of authorship on those papers, and how often their papers are cited.[36,37] This pressure to publish has been linked to a number of major issues in the field, such as questionable research practices and research misconduct, low reproducibility and translatability, greater conservatism, and a reduction in the quality of science.[38-43]

The entrenched "publish or perish" atmosphere can result in researchers attempting to publish at almost any cost to progress in their academic careers, shifting goals from advancing knowledge to seeking optimal bibliometrics. As previously mentioned, in response to a 2023 survey assessing animal methods bias in publishing, three researchers commented that journals with higher impact factors are more likely to expect animal experiments or reject studies that do not include animal experiments.[1] It may be the case, then, that the pressure to publish and the perception that animal use will result in publication in more desirable journals are factors contributing to the persistence of animal use. Authors who feel intense pressure to publish may more readily acquiesce to reviewers' requests to add experiments on animals to an otherwise nonanimal study or conduct these experiments before submission because of this perceived or valid expectation.

It has also been hypothesized that the nature of institutional research oversight and protocol approval favors animal use. Compared to the relatively faster and easier process of receiving protocol approval for experiments on animals by the Institutional Animal Care and Use Committee (IACUC), the lengthy and burdensome nature of Institutional Review Board (IRB) review for research with human subjects may lead more researchers to use animals because it entails fewer barriers.[44] When taken together with the "publish or perish" mindset, the relative ease of IACUC approval compared to IRB approval may result in researchers choosing to add animal experiments as opposed to human experiments to comply with a reviewer's request for validation studies and hasten their timeline to publication.

Mitigating animal methods bias

With a growing understanding of animal methods bias, it is not only possible but crucial to mitigate this formidable barrier to the broader use of effective nonanimal methods—to improve research translatability and avoid the ethical consequences of animal use. Since 2022, the COLAAB has been working to understand and address animal methods bias. At the time of writing, the COLAAB consists of an Evidence Working Group, tasked with characterizing animal methods bias and expanding the evidence base of this newly defined phenomenon, and a Mitigation Working Group, tasked with developing tools and engaging stakeholders to raise awareness and mitigate its harmful effects. In 2022, the COLAAB hosted a Workshop to Address Animal Methods Bias in Scientific Publishing,[2] and in 2024, a Workshop to Explore Animal Methods Bias in Biomedical Research Funding.[4] While future efforts will explore additional contexts, COLAAB findings have contributed to recommendations for researchers who use nonanimal approaches, as well as publishers and funders (Table 1). Efforts are ongoing to better understand the characteristics of animal methods bias and implement effective mitigation strategies. There are likely many more mitigation strategies of varying degrees of feasibility and effectiveness that have not yet been explored and that will benefit from further investigation.

To address animal methods bias during manuscript peer review specifically, the COLAAB has prepared a comprehensive resource for researchers using nonanimal methods: the Author Guide for Addressing Animal Methods Bias in Publishing (Author Guide),[47] available as a living resource (www.animalmethodsbias.org/author-guide). The Author Guide expands on many of the recommendations in Table 1, containing suggestions, tools, and databases that researchers can employ during study design, manuscript preparation and submission, and peer review. The Author Guide may also be useful for researchers facing animal methods bias in other contexts, such as preparing grant applications or doctoral dissertations.

TABLE 1 Recommendations for researchers, funders, and publishers to mitigate animal methods bias.

Domain	Researchers	Funders	Publishers
Awareness and training	Raise awareness of animal methods bias and the value of NAMs within your institution and scientific community.	Raise awareness of animal methods bias and the value of NAMs among reviewers and review staff.	Raise awareness of animal methods bias and the value of NAMs among editors and reviewers.
	Educate and train the scientific community in nonanimal research methods and their value in translational success, especially for early-career researchers.	Implement animal methods bias mitigation training for reviewers and review staff.	Implement animal methods bias mitigation training for editors and reviewers.
Appropriate scope	Seek funding from, and publish in, appropriate outlets.	Invest in NAMs-specific funding streams.	Remove any explicit requirements for animal in vivo experiments in journal author guidelines.
		Do not exclude NAMs from funding opportunities, nor seek animal-only proposals.	
Method suitability	In manuscripts and proposals, provide robust justification for the use and suitability of the chosen methods to address your specific research question and their impact on human health.	Implement evaluation criteria that assess methods based on their suitability for the research question, the context of use, translatability, and human relevance, ensuring impartiality toward both animal- and nonanimal-based approaches.	Incorporate reviewer guidelines that describe best practices for evaluating in vitro studies, such as PRIVAT[45] and RIVER.[46]
Review processes	In review groups, identify and address biased comments made by fellow reviewers.	Implement bias-reporting mechanisms.	Mandate that reviewers' requests for additional animal experiments be scrutinized by editors or other reviewers.
		Allow rebuttals to reviewers' decisions.	Encourage editors to look for animal methods biased reviews and to intervene when necessary, including by removing unhelpful reviews or instructing authors to address only certain parts of the reviews.

Continued

TABLE 1 Recommendations for researchers, funders, and publishers to mitigate animal methods bias—cont'd

Domain	Researchers	Funders	Publishers
Reviewer expertise	Volunteer for reviewer positions and make yourself available to funding agency review staff.	Ensure that review groups include members with experience using NAMs, as well as those who can adequately evaluate their suitability for the specific research question or context of use.	Ensure that invited reviewers have experience using NAMs and can adequately evaluate their suitability for the specific research question or context of use.
	Take the opportunity to provide manuscript reviewer suggestions.	Open review groups to early-career researchers, especially those who have experience in innovative, human-relevant technologies.	
Open science	Engage in open science practices, such as publishing in open-access journals, data sharing, and preregistration of studies.	Encourage and incentivize open science practices, such as publishing in open-access journals, data sharing, and preregistration of studies.	Allow open-access publishing, including open peer review.
			Offer registered report article types.
			Require authors to adhere to experimental reporting standards.
Policy and infrastructure	Advocate for policies and initiatives that support the development and use of NAMs, as well as the mitigation of animal methods bias.	Facilitate the investigation of animal methods bias evidence in review reports and funding rates.	Conduct internal investigations of peer-review reports for possible occurrences of animal methods bias across various research fields and demographics.
		Invest in research infrastructure, such as new centers for animal-free biomedical research, to facilitate the transition toward nonanimal approaches.	Explore the development and implementation of other policies to discourage animal methods bias during manuscript review.

Some of the COLAAB's mitigation strategies and recommendations are more systemic than others—broader solutions to overcoming barriers related to advancing NAMs and improving publishing and funding processes that also help address animal methods bias. For example, greater investment in NAMs does not just provide more opportunities for researchers who use them, setting them up so they do not always compete with animal-based proposals or become forced into review groups without proper expertise. It is also crucial for advancing NAMs' broader development and use, reducing animal use, and improving translatability. As another example, open science endeavors foster transparency, rigor, and reproducibility through practices like open access, preregistration, open peer-review reports, and reviewer identities.[48-50] In turn, open science practices may help establish confidence in NAMs and prevent poor-quality or biased reviews of nonanimal studies, although further investigation is necessary, especially to determine the effect of these practices on animal methods bias specifically.[47]

Addressing the metrics by which scientists are evaluated, and thus how they are incentivized to think and behave, may require a wholesale restructuring of the systems and institutions in which science operates. The topic has been frequently debated in peer-reviewed literature, commentaries, and the press. Test cases have been undertaken to assess how evaluative criteria might be re-envisioned. Developed in 2012, the San Francisco Declaration on Research Assessment (DORA) provides recommendations for funding agencies, institutions, publishers, and researchers to move away from the use of journal-based metrics, such as impact factors, in research evaluation.[51] It emphasizes the importance of understanding the scientific content of research outputs and qualitative factors, such as the research's influence on policy and practice.

Disentangling bibliometrics and funding rates from research assessment would significantly reduce publishing and funding anxieties for researchers and thus likely help to mitigate animal methods bias, as it would empower scientists to push back on requests for unnecessary experiments on animals by reducing the fear of manuscript or grant application rejection.

Another change might involve strengthening IACUC review of animal use proposals to reduce the discrepancies between IACUC and IRB processes, which make the former the path of least resistance.[44] More rigorous IACUC review, including thorough harm-benefit analyses and greater scrutiny to ensure that researchers have conducted adequate searches for alternatives to animal use, would reduce the likelihood of reviewer requests for additional experiments on animals.

There is also the possibility that, with the increased development, visibility, and representation of nonanimal methods, it will become more acceptable for faculty in departments with heavy animal use to voice their concerns about the practice and engage in more open dialog about the potential for animal replacement. Dissent from longstanding and conservative discipline norms may positively influence diversity, equity, and inclusion in academic environments; reduce animal research dogmatism and animal methods bias; and foster more interest in and uptake of nonanimal methodologies.

Future outlook

Much about the characteristics, prevalence, scope, and impacts of animal methods bias remains to be explored. Investigating reviewer reports may improve the objectivity of evidence compared to self-reported survey data. Although some journals have begun publishing review reports as part of open science initiatives,[52] they remain the minority. Funding agencies keep review reports private to protect researchers' intellectual property and review groups' deliberative decision-making, but there is precedent for agencies like NIH to provide access to independent meta-researchers.[4] Large language models and qualitative assessments may be suitable for investigations of review reports, whether conducted by independent meta-researchers or by publishers and funders themselves.

Support from medical research stakeholders, including scientists, the public, the press, and lawmakers, may be needed to generate buy-in from publishers and funders and enable such investigations. Surveying or interviewing manuscript and grant reviewers could be useful for understanding the experiences and rationales of reviewers who prefer animal-based approaches, providing a fuller picture of occurrences of animal methods bias. In funding contexts, further investigations are needed to understand how animal methods bias manifests in reviewer comments and reports, its prevalence, and its impacts on researchers and funding rates, including whether proposals using animal methods are more successful in obtaining funding than proposals that do not.[4] It will be crucial to understand factors contributing to poor review outcomes, including a proposal's framing or rationale for using nonanimal approaches, as well as reviewers' potential preference for animal methods or lack of expertise or awareness regarding NAMs.[4] In publishing contexts, the prevalence of reviewer requests for experiments on animals in otherwise nonanimal studies is not yet known, and more evidence is needed to show whether papers without experiments on animals are more likely to be rejected or subjected to a longer review process.[53]

It remains to be determined whether animal methods bias intersects with researcher demographic characteristics, such as career stage, race, gender, or nationality. Varying legal, regulatory, and biomedical infrastructure contexts worldwide may also impact researchers' experiences with animal methods bias. The 2022 COLAAB workshop exploring animal methods bias in publishing made it clear that different research fields—infectious disease, neuroscience, toxicology, etc.—have different needs for model systems, have made progress in the development and uptake of nonanimal models at different rates, and may differentially impact researchers' experiences with animal methods bias.[2] It will be important for future investigations to systematically explore group differences across research assessment scenarios to understand animal methods bias more thoroughly. A crucial knowledge gap remains regarding researchers' experiences with and reviewers' rationale for occurrences of animal methods bias in other contexts in which research is assessed. As mentioned above, animal methods bias may occur during any scenario in which methods are assessed across the research lifecycle (Fig. 1), but future investigations will need to fully uncover this scope.

As nonanimal research approaches continue to advance and are increasingly used in a variety of applications around the world, funders, publishers, and other research stakeholders must acknowledge and address animal methods bias as a formidable barrier. NIH has taken some promising steps in this direction. An NIH advisory group tasked with assessing NAMs and making recommendations to catalyze their development and use noted in its 2023 report that grant peer reviewers may be less familiar with NAMs and recommended that reviewers receive training to "better understand how to evaluate the use of NAMs in fundamental and applied research grants."[6] The agency later accepted the group's recommendations.[54] In 2024, a reviewer evaluation framework introduced by the NIH Center for Scientific Review recognized factors for ensuring that reviewers provide impartial, objective assessments of grant proposals—including a lack of methodological biases.[55] In 2025, NIH announced an initiative to prioritize human-based research technologies and reduce animal use, which described measures related to animal methods bias: training grant review staff to address any possible bias towards animal studies and integrating experts on human-based methods into review groups.[56] That same year, the UK government announced a similar initiative, a strategy to replace animals in science and support the development, validation and uptake of nonanimal methods.[57] These developments signal an important shift in how governments are thinking about unfounded biases against certain methods. The success of these initiatives remains to be seen, as well as how other major stakeholders will work to understand and address animal methods bias.[58]

Conclusions

Though animal methods bias has likely affected the use and proliferation of nonanimal methods for some time, the phenomenon has only recently been named and measured. The structural factors contributing to animal methods bias include the domination of conservative institutional norms, its interaction with other peer-review and publication biases, the pressure to publish to progress one's career, and the lower regulatory burden of animal-based research compared to human studies. Initial evidence of animal methods bias, gathered through researcher surveys and stakeholder workshops, has revealed that it causes unnecessary animal use and is a source of publishing delays and grant rejections for nonanimal research.

The COLAAB has developed recommendations for researchers, publishers, and funders to raise awareness and reduce the incidence and influence of animal methods bias. Mitigating animal methods bias will require multistakeholder interest and effort in this topic, but it overlaps with larger-scale measures needed to overcome the persistence of and reliance on animal use in research and testing. Fully addressing this issue will help reduce unnecessary animal experimentation and accelerate the development and use of nonanimal methods, benefiting both scientific progress and animal welfare.

References

[1] Krebs CE, Lam A, McCarthy J, Constantino H, Sullivan K. A survey to assess animal methods bias in scientific publishing. *ALTEX*. 2023;40(4):665–676. doi:10.14573/altex.2210212.

[2] Krebs CE, Camp C, Constantino H, et al. Proceedings of a workshop to address animal methods bias in scientific publishing. *ALTEX*. 2023;40(4):677–688. doi:10.14573/altex.2210211.

[3] Ingber DE. Is it time for reviewer 3 to request human organ chip experiments instead of animal validation studies? *Adv Sci*. 2020;7(22):2002030. doi:10.1002/advs.202002030.

[4] Krebs CE, Geissler S, Herrmann K, et al. Exploring animal methods bias in biomedical research funding: workshop proceedings and action steps. *NAM J*. 2024:100004. doi:10.1016/j.namjnl.2024.100004.

[5] Trunnell ER, Roe KV. Predominance of animal-based expertise may bias NIH neuroscience grant review: a pilot study with implications for non-animal methodologies. bioRxiv. 2025. doi: 10.1101/2025.02.28.640877.

[6] Chang HY, Jorgenson L, Baines A, et al. *Catalyzing the development and use of novel alternative methods*. National Institutes of Health; 2023:42. https://www.acd.od.nih.gov/documents/presentations/Working_Group_Report.pdf. Accessed 25 July 2024.
[7] Frank J. Technological lock-in, positive institutional feedback, and research on laboratory animals. *Struct Chang Econ Dyn*. 2005;16(4):557–575. doi:10.1016/j.strueco.2004.11.001.
[8] Sandgren EP, Streiffer R, Dykema J, Assad N, Moberg J. Attitudes toward animals, and how species and purpose affect animal research justifiability, among undergraduate students and faculty. *PLOS ONE*. 2020;15(5):e0233204. doi:10.1371/journal.pone.0233204.
[9] Veening-Griffioen DH, Ferreira GS, Boon WPC, et al. Tradition, not science, is the basis of animal model selection in translational and applied research. *ALTEX*. 2021;38(1):49–62. doi:10.14573/altex.2003301.
[10] Del Pace L, Viviani L, Straccia M. Researchers and their experimental models: a pilot survey in the context of the European Union Health and Life Science Research. *Animals*. 2022;12(20):2778. doi:10.3390/ani12202778.
[11] Kooijman M. Why animal studies are still being used in drug development. *Altern Lab Anim*. 2013;41(6):79–81. doi:10.1177/026119291304100627.
[12] Merton RK. The matthew effect in science. *Science*. 1968;159(3810):56–63. doi:10.1126/science.159.3810.56.
[13] Travis GDL, Collins HM. New light on old boys: cognitive and institutional particularism in the peer review system. *ST&HV*. 1991;16(3):322–341. doi:10.1177/016224399101600303.
[14] Langfeldt L. The policy challenges of peer review: managing bias, conflict of interests and interdisciplinary assessments. *Res Eval*. 2006;15(1):31–41. doi:10.3152/147154406781776039.
[15] Clark CJ, Jussim L, Frey K, et al. Prosocial motives underlie scientific censorship by scientists: a perspective and research agenda. *PNAS*. 2023;120(48):e2301642120. doi:10.1073/pnas.2301642120.
[16] Langfeldt L, Reymert I, Svartefoss SM. Distrust in grant peer review—reasons and remedies. *Sci Public Policy*. 2023;51(1):28–41. doi:10.1093/scipol/scad051.
[17] Lee CJ, Sugimoto CR, Zhang G, Cronin B. Bias in peer review. *J Assoc Inf Sci Technol*. 2012;64(1):2–17. doi:10.1002/asi.22784.
[18] Ginther DK, Schaffer WT, Schnell J, et al. Race, ethnicity, and NIH research awards. *Science*. 2011;333(6045):1015–1019. doi:10.1126/science.1196783.
[19] Helmer M, Schottdorf M, Neef A, Battaglia D. Gender bias in scholarly peer review. *eLife*. 2017;6:e21718. doi:10.7554/eLife.21718.
[20] McKenzie ND, Liu R, Chiu AV, et al. Exploring bias in scientific peer review: an ASCO initiative. *JCO Oncol Pract*. 2022;18(12):791–799. doi:10.1200/OP.22.00275.
[21] Peters DP, Ceci SJ. Peer-review practices of psychological journals: the fate of published articles, submitted again. *Behav Brain Sci*. 1982;5(2):187–195. doi:10.1017/S0140525X00011183.
[22] Sarr F, Knight S, Strauss D, et al. Increasing the representation of black, indigenous, and people of colour as students in psychology doctoral programmes. *Can Psychol*. 2022;63(4):479–499. doi:10.1037/cap0000339.
[23] Tomkins A, Zhang M, Heavlin WD. Reviewer bias in single- versus double-blind peer review. *PNAS*. 2017;114(48):12708–12713. doi:10.1073/pnas.1707323114.
[24] Vasconcelos SMR, Sorenson MM, Leta J. A new input indicator for the assessment of science & technology research? *Scientometrics*. 2009;80:217–230. doi:10.1007/s11192-008-2082-z.
[25] Hoppe TA, Litovitz A, Willis KA, et al. Topic choice contributes to the lower rate of NIH awards to African-American/black scientists. *Sci Adv*. 2019;5(10):aww7238. doi:10.1126/sciadv.aaw7238.
[26] Hagelin J, Carlsson HE, Hau J. An overview of surveys on how people view animal experimentation: some factors that may influence the outcome. *Public Underst Sci*. 2003;12(1):67–81. doi:10.1177/0963662503012001247.
[27] Ipsos MORI. Public attitudes to animal research in 2018. 2018:36.
[28] Pifer LK. Exploring the gender gap in young adults' attitudes about animal research. *Soc Anim*. 1996;4(1):37–52. doi:10.1163/156853096x00034.
[29] Strauss M Americans are divided over the use of animals in scientific research. Pew Research Center. 2018. https://www.pewresearch.org/short-reads/2018/08/16/americans-are-divided-over-the-use-of-animals-in-scientific-research/. Accessed 3 September 2024.
[30] Busek M, Aizenshtadt A, Amirola-Martinez M, Delon L, Krauss S. Academic user view: organ-on-a-chip technology. *Biosensors*. 2022;12(2):126. doi:10.3390/bios12020126.
[31] Dickersin K. The existence of publication bias and risk factors for its occurrence. *JAMA*. 1990;263(10):1385–1389. doi:10.1001/jama.1990.03440100097014.

[32] Dickersin K, Min YI. Publication bias: the problem that won't go away. *Ann N Y Acad Sci.* 1993;703(1):135–148. doi:10.1111/j.1749-6632.1993.tb26343.x.
[33] DeVito NJ, Goldacre B. Publication bias. 2019. https://catalogofbias.org/biases/publication-bias/. Accessed 14 February 2025.
[34] Fanelli D. Positive results receive more citations, but only in some disciplines. *Scientometrics.* 2013;94:701–709. doi:10.1007/s11192-012-0757-y.
[35] Greenberg SA. How citation distortions create unfounded authority: analysis of a citation network. *BMJ.* 2009;339:b2680. doi:10.1136/bmj.b2680.
[36] Hammarfelt B. Recognition and reward in the academy: valuing publication oeuvres in biomedicine, economics and history. *Aslib J Inf Manag.* 2017;69(5):607–623. doi:10.1108/AJIM-01-2017-0006.
[37] Rice DB, Raffoul H, Ioannidis JPA, Moher D. Academic criteria for promotion and tenure in biomedical sciences faculties: cross sectional analysis of international sample of universities. *BMJ.* 2020;369:m2081. doi:10.1136/bmj.m2081.
[38] Baddeley A. So where should we publish? *Cortex.* 2001;37(4):598–599. doi:10.1016/S0010-9452(08)70604-X.
[39] Boulbes DR, Costello T, Baggerly K, et al. A survey on data reproducibility and the effect of publication process on the ethical reporting of laboratory research. *Clin Cancer Res.* 2018;24(14):3447–3455. doi:10.1158/1078-0432.CCR-18-0227.
[40] Haven T, Tijdink J, Martinson B, Bouter L, Oort F. Explaining variance in perceived research misbehavior: results from a survey among academic researchers in Amsterdam. *Res Integr Peer Rev.* 2021;6:7. doi:10.1186/s41073-021-00110-w.
[41] Johann D, Neufeld J, Thomas K, Rathmann J, Rauhut H. The impact of researchers' perceived pressure on their publication strategies. *Res Eval.* 2024:rvae011. doi:10.1093/reseval/rvae011.
[42] Smaldino PE, McElreath R. The natural selection of bad science. *R Soc Open Sci.* 2016;3(9):160384. doi:10.1098/rsos.160384.
[43] Tijdink JK, Schipper K, Bouter LM, et al. How do scientists perceive the current publication culture? A qualitative focus group interview study among Dutch biomedical researchers. *BMJ Open.* 2016;6(2):e008681. doi:10.1136/bmjopen-2015-008681.
[44] Rice MJ. The institutional review board is an impediment to human research: the result is more animal-based research. *PEHM.* 2011;6:12. doi:10.1186/1747-5341-6-12.
[45] Whaley P, Hooijmans CR PRIVAT tool. 2022. https://osf.io/bwypt/. Accessed 13 December 2023.
[46] The RIVER working group. Reporting in vitro experiments responsibly–the RIVER Recommendations. MetaArXiv. 2023. doi:10.31222/osf.io/x6aut.
[47] Krebs CE, Camp C, Constantino H, et al. Author guide for addressing animal methods bias in publishing. *Adv Sci.* 2023;10(30):2303226. doi:10.1002/advs.202303226.
[48] Lakens D, Mesquida C, Rasti S, Ditroilo M. The benefits of preregistration and registered reports. *EBT.* 2024;2(1):2376046. doi:10.1080/2833373X.2024.2376046.
[49] Ross-Hellauer T, Horbach SPJM. Additional experiments required: a scoping review of recent evidence on key aspects of open peer review. *Res Eval.* 2024;33:rvae004. doi:10.1093/reseval/rvae004.
[50] McKiernan EC, Bourne PE, Brown CT, et al. How open science helps researchers succeed. Rodgers P, ed. *eLife.* 2016;5:e16800. doi:10.7554/eLife.16800.
[51] Declaration on Research Assessment. San Francisco Declaration on Research Assessment. 2012. https://sfdora.org/read/. Accessed 27 March 2024.
[52] Ross-Hellauer T. What is open peer review? A systematic review. *F1000Res.* 2017;6:588. doi:10.12688/f1000research.11369.2.
[53] Kavanagh O, Krebs CE. Mitigating animal methods bias to reduce animal use and improve biomedical translation. *Sci Prog.* 2024;107(2):00368504241253693. doi:10.1177/00368504241253693.
[54] Bertagnolli MM Statement on catalyzing the development of novel alternative methods. NIH.gov. 2024. https://www.nih.gov/about-nih/who-we-are/nih-director/statements/statement-catalyzing-development-novel-alternatives-methods. Accessed 21 March 2024.
[55] Yee LM, Olufokunbi Sam D CSR advisory council reviewer evaluation working group update. In: *Presented at: Center for Scientific Review Advisory Council Meeting.* 2024. https://public.csr.nih.gov/sites/default/files/2024-08/Update_on_CSRAC_Working_Groups_Reviewer_Evaluation.pdf. Accessed 5 February 2025.
[56] U.S. National Institutes of Health. NIH to prioritize human-based research technologies; 2025. https://www.nih.gov/news-events/news-releases/nih-prioritize-human-based-research-technologies. Accessed 9 June 2026.

[57] U.K. Department for Science, Innovation and Technology. Replacing animals in science: A strategy to support the development, validation and uptake of alternative methods, 2025. https://www.gov.uk/government/publications/replacing-animals-in-science-strategy/replacing-animals-in-science-a-strategy-to-support-the-development-validation-and-uptake-of-alternative-methods. Accessed 1, December 2025.

[58] Kopew JS, Ritskes-Hoitinga M, Singer M, Trunnell ER, Vidaurre G, Krebs CE. Addressing animal methods bias through government-led initiatives. Front Toxicol. 2026;8:1761759. doi: 10.3389/ftox.2026.1761759.

CHAPTER

16

Unlocking funding for human-relevant biomedical research: Navigating the landscape of nonanimal approaches

Angela N. Hvitved[a] *and Francesca Pistollato*[b]

[a]Alternatives Research & Development Foundation (ARDF), Jenkintown, PA, United States

[b]Research and Toxicology, Humane World for Animals, Brussels, Belgium

Abstract

Biomedical research has undergone a paradigm shift in recent years, with a growing emphasis on nonanimal methods to better align scientific endeavors with ethical considerations, cost-effectiveness, and increased translational potential. This chapter serves as a comprehensive guide for researchers seeking funding for basic and applied research activities centered around nonanimal, human-relevant methods in biomedical research. We will explore the historical challenges, ongoing initiatives, and region-specific funding landscapes, with a focus on the European Union and the United States. By delving into the pivotal role of funding in the development and uptake of nonanimal methods, we aim to equip researchers with the knowledge and strategies needed to propel this transformative field forward.

Keywords: Nonanimal research methods; Biomedical research; European Union; NAMs; Research Grants; US Government

Retrospective analysis of the funding landscape for nonanimal research

The use of animals in biomedical research and toxicology studies, and concerns about these approaches, have spurred many research initiatives focused on the development and use of in vitro and in silico human-relevant tools and technologies, either as standalone methods or in combination with other animal or nonanimal approaches. In the context of biomedical research, particularly, the reliance on animals to model human diseases and support new drug discovery and development has been questioned due to the associated

Moving Toward Nonanimal Approaches in Medical Research and Testing
https://doi.org/10.1016/B978-0-443-26514-3.00005-9

high rate of failure for new drugs, especially in areas such as Alzheimer's disease and cancer research.[1–3] To overcome the limitations of animal models or overly simplistic in vitro methods, innovative nonanimal models, such as complex in vitro models (microphysiological systems [MPS], including 3D organoids and organ-on-chip models), in silico models, and artificial intelligence technologies, have been developed to support the study of human disease pathophysiology, drug discovery and development,[4–10] and chemical risk assessment.[11–13] See Part II of this book for deeper discussions of these methods and their applications.

Although insufficient investment toward nonanimal approaches has traditionally been a barrier,[14,15] in recent years, funding opportunities for research activities focused on human-centered methodologies have been launched in the European Union (EU), the United States (US), and other countries. Looking retrospectively, the EU has progressively increased its support for the development and implementation of new approach methodologies (NAMs), from Framework program (FP)5 (1998–2002, €11 million per year), FP6 (2003–2006, €32 million per year), FP7 (2007–2013, €47 million per year), to H2020 (2014–2020, €75 million per year), globally committing over €1 billion to support more than 300 projects over the past two decades.[16] In addition, the industrial sector has furthered these efforts by contributing an additional €150 million.[17]

In March 2024, the European Commission (EC) introduced a European Research Area action plan, advocating for the integration of nonanimal NAMs in scientific research.[18] The purpose of this initiative is to encourage Member States to harmonize their domestic and regional policies with the aim of minimizing reliance on animals in research and experimentation, while expediting the advancement, validation, and adoption of nonanimal methods. This endeavor aligns with the EU's objective outlined in Directive 2010/63/EU on the protection of animals used for scientific purposes[19] to completely substitute animal usage in scientific endeavors, echoing the persistent demands of citizens across all EU Member States[20] for an expedited shift toward nonanimal-based scientific practices.[21–23]

The Netherlands has a clearly defined and transparent initiative in place aimed at transitioning toward nonanimal-based scientific practices.[24] In 2024, the Dutch Growth Fund committee announced its endorsement of a €124.5 million allocation toward the establishment of a Center for Animal-Free Biomedical Translation at the University of Utrecht.[25] Along the same lines, the German government is dedicated to enacting a strategy to decrease animal experimentation and foster research into alternative methods.[26] In 2021, the Flemish Parliament initiated a project aimed at crafting an action plan to diminish the utilization of animals for scientific purposes.[27] In the same year, the Swedish 3Rs Centre introduced an initial strategy to curtail the use of animals in research, testing, and educational settings, as per the directive of the Swedish Government.[28] In addition, in 2020, the Norwegian National Experimental Animals Committee put forth a series of measures aimed at formulating a tangible plan for transitioning toward nonanimal-based scientific approaches.[22,29]

In the US, governmental funding agencies have not historically provided significant funding explicitly directed toward nonanimal methods for the sake of reducing or replacing the use of animals in research. Rather, several federally funded programs have supported the development and implementation of biomedical research approaches that do not use animals, such as MPS and computational models. However, support for nonanimal methods is growing, including for the purpose of complementing or replacing animal methods.

In 2024, the National Institutes of Health (NIH) launched its Complement Animal Research in Experimentation (Complement-ARIE) program, which is aimed at promoting the development and uptake of NAMs for research and regulatory purposes.[30] As the world's largest single public funder of biomedical research,[31] the NIH's increased interest in advancing NAMs suggests a recognition of the potential of NAMs to dramatically change the landscape of biomedical research. However, because the NIH has not historically funded NAMs as a research area, it is challenging to conduct in-depth retrospective analyses of the funding landscape using publicly available data. An analysis presented to an NIH advisory board indicated that NIH investment in "alternative methods" grew from less than 2% of competitive awards in 2001 to 8% in 2021.[32]

State-level public initiatives are also gaining momentum. In 2023, the state of Maryland established a fund to support the development of nonanimal research methods. The first of its kind in the US, the Maryland law requires that animal research facilities in the state contribute to the fund, which provides grants and loans to develop human-relevant alternatives to using animals in medical and product testing.[33,34] Though in its early stages, this effort will support investment in the state's research enterprise while highlighting the need to reduce the use of animals in research. Proponents have indicated an interest in replicating this model in other states as public support for nonanimal methods continues to grow.

The funding landscape for nonanimal methods is evolving quickly in a largely positive direction. Funders appear to recognize the need for increased investment in biomedical research methods that are more accurate and predictive of human biology, while avoiding the ethical concerns inherent in using animals. Although any list of funding initiatives will quickly become outdated, the next section lists examples of major funding opportunities for nonanimal research methods in the EU and the US at the time of writing.

Funding opportunities in Europe

The EC has taken a leading role in promoting human-centered research through targeted calls that support the development of innovative, nonanimal methods. Under the current Horizon Europe program of the EC (running from 2021 to 2027), with a total budget of €95.5 billion, which includes €5.4 billion from the EU recovery program from the COVID-19 crisis, funding calls have supported the development of innovative nonanimal approaches in biomedical research. These initiatives aim to enhance biomedical research's predictive power and translational value by fostering interdisciplinary collaboration and the use of advanced in vitro and in silico models, while prioritizing awareness and capacity-building to facilitate broader adoption of NAMs.

In particular, EC research and innovation actions (supported with about €25 million each) have called for the submission of proposals aimed at developing innovative nonanimal human-based tools and strategies for biomedical research,[35] bioprinting of living cells for regenerative medicine,[36] and accelerating the implementation of NAMs and other innovative nonanimal approaches for the development, testing, and production of health technologies.[37] Similarly, a Horizon Coordination and Support Action call, supported with €2 million, welcomes projects aiming to equip European regulators with advanced knowledge of nonanimal methods and foster collaboration to develop regulatory frameworks using nonanimal data for chemical and pharmaceutical assessment.[38] In addition, a Horizon Innovative Health Initiative Joint Undertaking call, with a budget of about €10–15 million, supports the clinical validation of biomarkers for diagnosis, monitoring disease progression, and treatment response.[39] Funded by Horizon Europe for the 2021–2027 period, the European Partnership for the Assessment of Risks from Chemicals (PARC), led by ANSES,[40] and with a budget of €400 million (50% funded by the EU and 50% by Member States), aims to pioneer next-generation chemical risk assessment methods, with the overarching goal of safeguarding both public health and the environment.[41]

Public and private institutions and organizations based in EU Member States and other non-EU countries, including Norway, Switzerland, and the United Kingdom, broadly support the 3Rs principles and have programs that support researchers working on nonanimal methods development and validation, other 3Rs approaches, and education initiatives (Table 1).[42]

TABLE 1 Funding opportunities supporting NAMs in the EU, Norway, Switzerland, and the United Kingdom[43–56]

Region	Funding program	Funding opportunity title	Funding opportunity description	Link to funding call
EU[i] and [ii]	European Partnership for Alternatives to Animal Testing (EPAA)	Refinement Prize	A €6000 prize is granted every 2 years to a laboratory technician, animal caretaker, or technologist who has demonstrated outstanding achievements in new, novel approaches to advance the implementation and/or awareness-raising of the refinement of animal testing.[43]	https://single-market-economy.ec.europa.eu/calls-expression-interest_en
EU[i] and [ii]	EPAA	3Rs Student Grant	Two full grants of €1000 are available to support students and young scientists with outstanding work in alternative approaches, enabling them to attend a high-profile scientific event.[44]	https://single-market-economy.ec.europa.eu/calls-expression-interest_en
EU[i] and [ii]	EPAA	3Rs Science Prize	A €10,000 prize to promote European research on alternative approaches to animal testing.[45]	https://single-market-economy.ec.europa.eu/calls-expression-interest_en
Norway[i]	Norecopa	3R Prize	A NOK 30,000 prize and diploma to recognize any scientific, technical, or administrative efforts that have contributed to increasing awareness of the principles of the 3Rs.[46]	https://norecopa.no/about/3r-prize/

Region	Funding program	Funding opportunity title	Funding opportunity description	Link to funding call
Germany[i]	Federal Ministry of Education and Research (BMBF)	Alternative methods to animal experiments	Funding opportunities offering €40 million per basic research project; €20 million per project for industrial research; €15 million per project for experimental development; and €7.5 million per study for feasibility studies are aimed at substituting animal experiments with alternative methods, enhancing the appeal of the field of alternative methods for researchers, and supporting validation, adoption, and dissemination of existing alternatives.[47]	https://www.gesundheits-forschung-bmbf.de/de/13910.php and https://www.bmbf.de/SharedDocs/Bekanntmachungen/DE/2024/02/2024-02-07-%c3%84nderungsbekanntmachung-Tierversuch.html
Netherlands[ii]	Willy Van Heumenfonds	Willy Van Heumen Prize	A €15,000 prize is offered every two years to a researcher affiliated with a Dutch organization who is engaged in the advancement of the 3Rs principles.[48]	https://www.proefdierenalt.nl/
Belgium[i]	Belgian Council for Laboratory Animal Science (BCLAS)—scientific organization	Fund in honor of Jean René Maisin	Financial support amounting to €10,000 is available for one or more projects submitted by Belgian institutes that align with the BCLAS mission: to promote and cultivate ethical and responsible care and utilization of laboratory animals, and to demonstrate a commitment to replacing and/or reducing the number of laboratory animals.[49]	https://www.bclas.org/
UK[i]	National Center for the Replacement, Refinement, and Reduction of Animals in Research (NC3Rs)	Project grants	Support for projects aiming to advance the development and commercialization of novel technologies is provided in varying amounts, dependent on the science. Awards are for up to 36 months and are funded at 80% of the full economic cost.[50]	https://nc3rs.org.uk/our-funding-schemes/project-grants

Continued

TABLE 1 Funding opportunities supporting NAMs in the EU, Norway, Switzerland, and the United Kingdom—cont'd

Region	Funding program	Funding opportunity title	Funding opportunity description	Link to funding call
UK[i]	NC3Rs	CRACK IT challenges	Different award schemes support multistakeholder collaborations to solve business and scientific challenges, delivering 3Rs benefits either by improving business processes or developing a commercial product.[51]	https://nc3rs.org.uk/crackit/crack-it-challenges
UK[i]	NC3Rs	International 3Rs Prize	A £28k prize grant and a £2k personal award for a paper that describes outstanding and original work, which has or could have major impacts on the 3Rs in research.[52]	https://nc3rs.org.uk/international-3rs-prize
UK[i]	NC3Rs	PhD studentships	Annual awards for 3–4 years of support, with values of £100k and £135k for non-full economic costs (non-FEC) (i.e., applicants can apply for 100% directly incurred and indirect costs), aimed at embedding the 3Rs in the training of graduate scientists from a broad range of scientific backgrounds.[53]	https://nc3rs.org.uk/our-funding-schemes/phd-studentships
UK[ii]	Replacing Animal Research	Innovation Grant	Awards of £5000 up to £15,000 for 6-month to 2-year pilot projects and proof-of-principle studies that support the replacement of animals in biomedical research.[54]	https://replacinganimalresearch.org.uk/for-scientists/innovation-grants/
Switzerland[ii]	Animal Research Tomorrow	ART Award	Awards of €3000 for 12-month projects focusing on the development and/or testing of novel 3Rs approaches for animal use in scientific research and/or education.[55]	https://animalresearchtomorrow.org/en/award/3rs-award
Switzerland[ii]	Animal Research Tomorrow	Science Communication Awards	Awards of €3000 to support young science communicators in devising tools and strategies aimed at enhancing transparency in animal research.[56]	https://animalresearchtomorrow.org/en/award/scicomm-award

[i]*Public funding source.*
[ii]*Private funding source.*

Funding opportunities in the United States

The US federal funding opportunities that support nonanimal research methods have primarily focused on developing specific technologies, such as organ-on-a-chip models, rather than supporting NAMs as a distinct category of research. NIH's Complement-ARIE initiative marks the largest US federal funding effort specifically directed at developing and implementing NAMs and represents an exciting and encouraging development for US researchers. However, there are still many federal programs that do not have an explicit 3Rs focus but are well suited for supporting biomedical research projects using nonanimal methods. In addition to governmental funding initiatives, several US-based nongovernmental organizations provide grants for activities focused on developing and implementing nonanimal research methods, many with an explicit 3Rs focus. Table 2 lists several active funding opportunities supported by public and private entities based in the US. (Some of these programs are restricted to US-based researchers, while others are open to applicants from any country.)

TABLE 2 Funding opportunities supporting NAM-based research from organizations based in the US (eligibility requirements vary).

Type of funding	Funding entity	Funding opportunity description	Link to program
Public	NIH (Common Fund)	The Complement-ARIE program is a ten-year, $390 million initiative to catalyze the "development, standardization, validation, and use" of human-based NAMs in basic and translational biomedical research.[30] Approximately $40 million is allocated per year for a range of activities, including technology development, data coordination, validation, and training, beginning in FY2025.[57]	https://commonfund.nih.gov/complementarie
Public	NIH (National Center for Advancing Translational Sciences (NCATS))	NCATS funds a range of translational research activities, including the development and testing of MPS models ("tissue chips") and computational methods for biomarker discovery and predictive toxicology, regenerative medicine, and rare disease research. Relevant programs are integrated throughout the NCATS funding portfolio and, although not framed as NAMs research, are well suited to support nonanimal research technologies (e.g.,[58–60])	https://ncats.nih.gov/research/research-activities
Public	National Science Foundation (NSF)	NSF's Engineering of Biomedical Systems program supports projects that use an engineering framework to better understand physiological or pathophysiological processes using a range of technologies, including the development of validated models and advanced biomanufacturing, but the program does not support the development of animal models.[61]	https://www.nsf.gov/funding/opportunities/engineering-biomedical-systems

Continued

TABLE 2 Funding opportunities supporting NAM-based research from organizations based in the US (eligibility requirements vary)—cont'd

Type of funding	Funding entity	Funding opportunity description	Link to program
Private	Alternatives Research & Development Foundation (ARDF)	ARDF's Annual Open program provides one-year grants to develop innovative methods with the potential to replace or reduce the use of animals in research, testing, and education.[62] Its AiR Challenge program focuses on replacement in biomedical research.	https://ardf-online.org/
Private	Association for the Assessment and Accreditation of Laboratory Animal Care International (AAALAC)	The Global 3Rs Awards Program recognizes significant, innovative contributions toward the 3Rs of animal research to advance ethical science by any researcher in academia or industry in any area of biology.[63]	https://www.aaalac.org/awards/global-3rs-awards1/global-3rs-awards/
Private	Johns Hopkins Center for Alternatives to Animal Testing (CAAT)	CAAT's Alan and Helene Goldberg In Vitro Toxicology Grants serve as initial funding opportunities for scientists aiming to pioneer alternatives to the use of animals in biomedical research and safety testing of products.[64]	https://caat.jhsph.edu/alan-and-helene-goldberg-in-vitro-toxicology-grants/
Private	International Foundation for Ethical Research (IFER)	IFER provides graduate fellowships to support student research projects that contribute to the development, acceptance, and adoption of research methods that replace animals in research, testing, or educational contexts.[65]	https://ifer.org/apply-for-grant

Furthermore, many private foundations dedicated to addressing a specific disease or condition also fund research aligned with their organization's mission. These programs are often interested in funding nonanimal approaches that are promising and potentially impactful with respect to their mission. For researchers seeking funding for nonanimal research proposals focused on a particular disease area, such as Alzheimer's disease or different types of cancer, it is worth investigating whether relevant foundations and charities sponsor research grants.

When seeking funding for projects that utilize nonanimal methods, researchers should be open-minded and adaptable in framing their proposals to different types of funders. The range of programs described here shows that, for many valuable funding opportunities, it might be incidental whether the proposed projects use animals or not. Researchers should not unduly limit themselves when considering whether a program is a good fit for their nonanimal research project. Instead, consider whether your project addresses the program's goals and aligns with the funder's mission. The next section describes how to prepare a compelling and competitive application, taking into consideration the different types of funding that might be a good fit for your proposal.

Writing a compelling research proposal to fund nonanimal research

A well-written grant application should clearly present the research project being proposed and be tailored to the funding program to which it is submitted. This section describes how to accomplish these two aims and provides advice for writing a competitive proposal. For most programs, funders use prespecified review criteria to assess applications and decide which proposals to fund. Although review standards can vary significantly between programs, almost all funders want to support projects that seem important and feasible, and they want to feel confident that the awardee is capable of performing the proposed work. The goal of a competitive application is to clearly demonstrate these features to the reviewers and funders.

A competitive research proposal should successfully: (1) explain why the project is interesting and important, with a focus on responding to funder/program objectives; (2) explain the goals or aims of the project; (3) describe the specific activities to be undertaken and how these activities address the project aims; and (4) explain why you, the applicant, are well suited to accomplish the proposed work, given your training and current research environment.[66–68]

Although the specific sections required in a proposal will vary between programs, a strong methodology section is especially crucial for proposals using nonanimal approaches because of the potential for animal methods bias during review (discussed in the next section). The methodology section must provide sufficient detail for the reader to assess the soundness of the approach and provide a compelling explanation as to how the proposed activities will answer the research questions. Consulting with a biostatistician during the design of the study can help ensure that potentially overlooked methodological concerns are raised and addressed during the design phase.[66] This section should also clearly explain the appropriateness of the chosen approach and, for research addressing questions about human health or biology, demonstrate the human-health relevance of the proposed approach.[69] Applicants should discuss potential limitations and challenges of the experimental approach and propose solutions that could address or mitigate them. Finally, for research questions that are predominantly studied using animal models, applicants should consider anticipating and addressing potential reviewer concerns regarding the validity of nonanimal approaches.[70]

Aside from the methodology, the other sections that should be included in the application, and the level of detail expected, will vary depending on the program. For this reason, it is crucial to carefully read the program description, review criteria, and application instructions while preparing the application. Funders use this information to convey to applicants their goals for the program and the types of projects they want to fund to accomplish these goals. If a funder considers a particular topic important enough to mention in the program materials, then it likely merits discussion in a corresponding proposal.

This leads to the second key feature of a competitive application: tailoring your proposal to the funding opportunity. Funders want projects that help them accomplish their organization's mission. The most innovative and scientifically exciting project is unlikely to be compelling to a funder if it is outside the scope of their mission or if it does not clearly align with the goals of their program. A funder whose goal is to advance nonanimal research methods might look for different features in a proposal than a funder that is narrowly focused on a specific disease or condition. Although the same project might be an excellent fit for both programs, it is incumbent on the applicant to clearly demonstrate this fit in the application.

The objectives of the funding program determine how you should frame your project's significance and aims, how the reviewers will assess the proposal, and what types of projects your application will be competing against. It is especially important to ensure that the proposal directly responds to the program description and review criteria. These materials are intended to convey the program's priorities and values to applicants. If a funder uses external reviewers to assess the proposals (the most likely review method), the reviewers use the program's review criteria as the basis for their assessment. Even if a reviewer personally finds a proposal highly compelling, it is their responsibility to evaluate the proposal using the standards set by the funder. This is why it is essential to tailor your application to the funding opportunity being pursued.

A program that aims to advance nonanimal methods might require a section addressing the impact of the project on reducing or replacing animal use (e.g., this has been the case for an EC call[35]) in addition to the more common features of a grant application. If so, it is likely that the review criteria will account for this aspect of the program's mission; for example, reviewers might be asked to assess the project's potential contribution to reducing or replacing the use of animals because it is an explicit goal of the program. In this case, it would be important to have a clear and detailed statement regarding the project's potential role in advancing nonanimal research methods. Some programs ask applicants to estimate the magnitude of animal use that could potentially be avoided or to describe the existing animal-based protocols that the work could supplant. Expectations vary between programs, but carefully reading the application guidelines and review criteria should provide applicants with a clear understanding of what reviewers will look for in their assessments.

Conversely, a grant program that is narrowly focused on a disease or answering a specific biological question might not place a high value on reducing the use of animals. Rather, the funder will be primarily interested in whichever methods and approaches are most likely to successfully address the research question. Indeed, some funders (or reviewers) might not share the goal of aiming to replace animals in research. For these readers, it is important to make a persuasive case regarding the benefits of the proposed nonanimal approach and explain why it is the most effective method to answer the research question. The proposal's significance and impact will probably be framed differently than it would be for a program focused on nonanimal methods. The same project, without any methodological modifications, could be highly competitive in both contexts, but how it is presented and which features are emphasized will depend on the program's priorities.

When in doubt, applicants should consider asking for clarification. Most funders have mechanisms for communicating with potential applicants and responding to specific inquiries. If you have questions about eligibility, program guidelines, budget restrictions, or other key factors, consider contacting the funder. Indeed, some programs encourage potential applicants to discuss their proposals with program staff before submitting an application to ensure that the project is a good fit for the program. Preparing an application is a time-consuming endeavor, and reviewing an application is costly for the program staff and reviewers. It is better for both applicants and funders to avoid the submission of an ineligible or non-compliant proposal.

Finally, applicants should take advantage of the many resources available that provide guidance and advice for writing high-quality grant proposals. For some larger programs, there are funder-specific resources available.[71,72] Other resources provide more general advice.[66–68,73,74] Mentors and colleagues are also excellent resources that should not be overlooked. They can provide feedback on the proposal's scientific strategy and approach, as well as its overall fit with the intended program. It is always good practice to have at least one person provide feedback on an application before it is submitted. In addition to providing scientific feedback on the proposed activities, other readers can flag errors and typos to help ensure that the proposal is polished and professional.

Serving as a reviewer is an excellent way to learn more about the review process and gain valuable experience in understanding the characteristics of a competitive proposal. It can be surprising which features of a proposal might not seem significant as an applicant but, as a reviewer, are much more salient. Reviewing can be an insightful and rewarding way to improve your own grant-writing skills. Some funders have dedicated initiatives for providing review opportunities for less-experienced or early-career researchers, such as NIH's Early Career Reviewer Program.[75] Smaller funders are less likely to have formal initiatives, but many funders are continuously looking to expand their pool of reviewers. Some funders have specific requirements for new reviewers, so it might not be possible for them to accommodate your offer to review, but it is certainly worthwhile to reach out to funders that are well aligned with your area of expertise and express your interest in reviewing.

Animal methods bias in funding

When preparing a proposal for a funding opportunity that is not targeted at nonanimal methods, it can be helpful to anticipate and mitigate the potential risk of bias toward animal-based methods, also called animal methods bias. This bias, often stemming from a preference for familiar methodologies or limited knowledge or expertise regarding emerging methodologies, may occur during the assessment of research proposals by reviewers.[76] As a consequence, reviewers may overlook the suitability and applicability of nonanimal-based approaches, which could inappropriately influence the final scoring of grant proposals. While the prevalence of animal methods bias in the review of grant proposals is not yet known, proactive measures can be taken by researchers to avoid it or mitigate its impact.[69]

Some of the recommendations provided in an author guide for addressing animal methods bias in publishing by the Coalition to Illuminate and Address Animal Methods Bias[70] could also be applicable to researchers preparing grant proposals using nonanimal methods. See Chapter 15 for a deeper discussion of animal methods bias, how it impacts publishing and funding assessments, and recommendations for various stakeholders on mitigating it.

Looking forward and remaining challenges

In recent years, there has been a noticeable increase in funding opportunities for research that does not rely on animal experimentation. This reflects a growing awareness of the ethical concerns surrounding animal testing and a rising demand for alternative methods within the scientific community. However, amidst this positive trend, there remains an issue regarding the ambiguous language used by some funding bodies in their calls for proposals, which frequently emphasize the importance of using innovative human-relevant models while also highlighting the need to improve animal models and/or conduct "reverse translation" research analyses that use data from nonanimal methods to guide the development of new animal models.[77–79]

To truly advance biomedical research, it is crucial not only to develop innovative methods but also to optimize and validate them. Technologies such as organ-on-chip and other MPS models, alongside AI and machine learning (in silico) approaches, hold immense promise for advancing scientific understanding. However, their widespread adoption requires careful optimization and validation to increase confidence in their efficacy. More funding incentives should be directed toward standardizing, benchmarking, and validating innovative nonanimal models and methods. Responding to this need, in mid-2025, the NIH Complement-ARIE Validation and Qualification Network (VQN) was launched as a public-private partnership coordinated by the Foundation for the National Institutes of Health (FNIH) to accelerate the validation and regulatory acceptance of NAMs.[80]

Finally, in order to ensure fair evaluation of proposals that incorporate or focus solely on nonanimal methods, funding bodies should ensure that experts in these approaches are included within study sections and reviewer panels. Their expertise can provide valuable insights into the viability and relevance of nonanimal approaches, thereby promoting a more balanced evaluation process.

Conclusions and final remarks

Around the world, there is growing recognition of the need to shift toward more human-relevant and innovative approaches in biomedical research and safety assessment. (For a deeper discussion on this shift globally and in the US, see Chapters 12 and 13.) In recent years, significant strides have been made in funding human-relevant NAMs, particularly in Europe and the US, signaling a promising trend in research and policy.

In the EU, the EC has taken a proactive role in supporting the development and deployment of human-centered research strategies. Through targeted calls under Horizon Europe, the EU is investing in the creation and implementation of cutting-edge technologies and approaches that not only reduce reliance on animal models but also aim to improve the predictive power and translational success of biomedical research. These initiatives increasingly emphasize the need for robust interdisciplinary collaboration and greater integration of advanced in vitro and in silico models. Importantly, many of these calls also focus on increasing awareness, fostering education, and ensuring capacity-building among researchers and regulators to support the uptake of nonanimal methods.

Beyond the EU, several European countries—including the UK—have also demonstrated a strong commitment to the advancement of the 3Rs principles. Both public and private organizations have played a key role in funding initiatives that seek to modernize research through NAMs, offering opportunities for innovative projects and cross-border collaborations.

In the US, NIH's Complement-ARIE initiative marks a substantial step forward in the federal government's support of NAMs technologies. This program reflects a growing commitment to transforming the biomedical research paradigm and highlights the importance of sustained investment in infrastructure and collaboration to ensure long-term impact.

These transatlantic funding efforts represent a pivotal shift toward more ethical, efficient, and human-relevant science. Continued coordination, sharing of best practices, and strategic investments in capacity-building will be essential to accelerate this transition and fully realize the potential of nonanimal methods in addressing the complex health challenges of the 21st century. Although progress has been uneven, there is good reason to believe that the funding landscape for NAMs research will continue to improve as funders recognize the potential of NAMs technologies and choose to build on past successes.

Researchers seeking funding for nonanimal research projects should ensure that their grant proposals are meticulously drafted, clearly outlining how the research question will be addressed by the proposed approach and providing sufficient detail for reviewers to assess the methodology. Rigorously explaining the methodological approach, the limitations of the approach and potential challenges and mitigation strategies, along with the anticipated short- and long-term impacts of the results, will enhance the proposal's competitiveness and likelihood of success.

In addition to crafting a scientifically sound and feasible project, applicants should tailor their proposals to each funding opportunity. This means carefully reviewing the program guidelines and review criteria to understand the funder's objectives and priorities, and then highlighting responsive features of their proposal. It is important to frame the project so that funders can easily recognize how the proposed work aligns with their funding objectives. Funders want to receive scientifically promising proposals that align with their mission; it is incumbent upon the applicant to demonstrate how the proposed work advances the funder's stated aims.

As funding opportunities for innovative nonanimal approaches continue to expand in both Europe and the US, the research landscape is poised for significant change. Researchers will increasingly benefit from an environment that is more supportive and less biased toward animal experimentation—one in which reliance on outdated animal models gives way to human-relevant tools with greater predictive value. This shift promises not only to reduce the use of animals in science but also to accelerate the development of safer, more effective medical interventions through improved translatability of research findings.

References

[1] Pistollato F, Bernasconi C, McCarthy J, et al. Alzheimer's disease, and breast and prostate cancer research: translational failures and the importance to monitor outputs and impact of funded research. *Anim.* 2020;10(7):1194.

[2] Cummings J, Reiber C, Kumar P. The price of progress: funding and financing Alzheimer's disease drug development. *Alzheimers Dement.* 2018;4:330–343.

[3] Sun D, Gao W, Hu H, Zhou S. Why 90% of clinical drug development fails and how to improve it? *Acta Pharm Sin B.* 2022;12(7):3049–3062.

[4] Ryu S, Chu PH, Malley C, et al. Human pluripotent stem cells for high-throughput drug screening and characterization of small molecules. *Methods Mol Biol.* 2022;2454:811–827.

[5] Pappalardo F, Russo G, Tshinanu FM, Viceconti M. In silico clinical trials: concepts and early adoptions. *Brief Bioinform.* 2019;20(5):1699–1708.

[6] Pappalardo F, Wilkinson J, Busquet F, et al. Toward A regulatory pathway for the use of in silico trials in the CE marking of medical devices. *IEEE J Biomed Health Inform*. 2022;26(11):5282–5286.
[7] Askin S, Burkhalter D, Calado G, El Dakrouni S. Artificial intelligence applied to clinical trials: opportunities and challenges. *Health Technol*. 2023;13(2):203–213.
[8] Paul D, Sanap G, Shenoy S, et al. Artificial intelligence in drug discovery and development. *Drug Discov Today*. 2021;26(1):80–93.
[9] Cova T, Vitorino C, Ferreira M, et al. Artificial intelligence and quantum computing quantum computing (QC) as the next pharma disruptors. In: Heifetz A ed. *Artificial Intelligence in Drug Design*. New York, NY: Springer US; 2022:321–347. (Methods in Molecular Biology). [cited 2024 Jan 18] doi:10.1007/978-1-0716-1787-8_14.
[10] Loewa A, Feng JJ, Hedtrich S. Human disease models in drug development. *Nat Rev Bioeng*. 2023;1(8):545–559.
[11] Schmeisser S, Miccoli A, von Bergen M, et al. New approach methodologies in human regulatory toxicology—not if, but how and when! *Environ Int*. 2023;178:108082.
[12] Hilton GM, Bhuller Y, Doe JE, Wolf DC, Currie RA. A new paradigm for regulatory sciences. *Regul Toxicol Pharmacol*. 2023;145:105524.
[13] Maertens A, Luechtefeld T, Knight J, Hartung T. Alternative methods go green! Green toxicology as a sustainable approach for assessing chemical safety and designing safer chemicals. *ALTEX*. 2024;41(1):3–19.
[14] Craig. Lush Prize. How much money do governments give to animal testing alternatives? 2014. Available from: https://lushprize.org/how-much-money/ [cited 2024 Apr 16].
[15] Langley G. Little funding to develop non-animal testing. *Nature*. 2002;418(6895):273.
[16] New Approach Methodologies: EU-IN Horizon Scanning Report. European Medicines Agency. Amsterdam, The NetherlandsEMA/56850/2025/Rev.1. Available from: https://www.ema.europa.eu/en/documents/report/new-approach-methodologies-eu-horizon-scanning-report_en.pdf. Accessed 17 Apr 2026.
[17] European Commission. Q&A: Citizens' Initiative "Save Cruelty-Free Cosmetics". European Commission; 2023. Available from: https://ec.europa.eu/commission/presscorner/detail/en/qanda_23_3995 [cited 2024 Jan 17].
[18] Desaintes C. Non-animal approaches in biomedical research and testing of pharmaceuticals. 2024. Available from: https://www.inspiring-era.eu/wp-content/uploads/2024/10/EC_Desaintes_NonAnimalApproachesinBiomedicalResearch.pdf. Accessed 26 April 2025.
[19] Directive 2010/63/EU of the European Parliament and of the Council of 22 September 2010 on the protection of animals used for scientific purposes (Text with EEA relevance). Official Journal of the European Union. 2010;L 276:33–79. https://eur-lex.europa.eu/eli/dir/2010/63/oj.
[20] Initiative detail. European Citizens' Initiative; 2021. Available from: https://citizens-initiative.europa.eu/initiatives/details/2021/000006_en [cited 2024 Apr 16].
[21] Public attitudes to animal research in 2018. Ipsos; 2019. Available from: https://www.ipsos.com/en-uk/public-attitudes-animal-research-2018 [cited 2024 Apr 16].
[22] 77% of EU citizens want a transition to non-animal science. Eurogroup for Animals; 2023. Available from: https://www.eurogroupforanimals.org/news/77-eu-citizens-want-transition-non-animal-science [cited 2024 Apr 16].
[23] Animal Research Poll—Eurogroup for Animals—31 March 2023—Savanta. 2023 Available from: https://savanta.com/knowledge-centre/published-polls/animal-research-poll-eurogroup-for-animals-31-march-2023/ [cited 2024 Apr 16].
[24] Ministerie van Landbouw N en V. Home. Ministerie van Landbouw, Natuur en Voedselkwaliteit. Transitie Proefdiervrije Innovatie; 2024. Available from: https://www.transitieproefdiervrijeinnovatie.nl/ [cited 2024 Apr 16].
[25] Dutch National Growth Fund invests €124.5 million in transition to animal-free innovation. RIVM; n.d. Available from: https://www.rivm.nl/en/alternatives-to-animal-testing/dutch-national-growth-fund-invests-eu1245-million-in-transition-to-animal-free-innovation [cited 2024 Apr 16].
[26] Koalitionsvertrag_2021-2025.pdf. n.d. Available from: https://www.spd.de/fileadmin/Dokumente/Koalitionsvertrag/Koalitionsvertrag_2021-2025.pdf [cited 2024 Apr 16].
[27] Schrijver L. Nooijen A. Bastiaanssen V. Actieplan: Gericht op het verminderen van het aantal dierproeven in Vlaanderen. 2023. https://assets.vlaanderen.be/image/upload/v1695300246/Actieplan_Vermindering_Dierproeven_k8inqc.pdf. Accessed 30 Apr 2026.
[28] About the Swedish 3Rs Center. n.d. Available from: https://jordbruksverket.se/languages/english/swedish-board-of-agriculture/animals/the-swedish-3rs-center/about-us [cited 2024 Apr 16].

[29] Forsoksdyrkomitéens-uttalelse-Overgang-til-forskning-uten-forsoksdyr-august-2020.pdf. 2020. Available from: https://www.forsoksdyrkomiteen.no/wp-content/uploads/2020/08/Forsoksdyrkomite%CC%81ens-uttalelse-Overgang-til-forskning-uten-forsoksdyr-august-2020.pdf [cited 2024 Apr 16].

[30] Strategic Planning. NIH Common Fund; n.d. Available from: https://commonfund.nih.gov/complementarie/strategicplanning [cited 2024 Feb 23].

[31] National Institutes of Health (NIH).Direct Economic Contributions; 2023. Available from: https://www.nih.gov/about-nih/what-we-do/impact-nih-research/serving-society/direct-economic-contributions [cited 2024 Sep 6].

[32] Chang H, Jorgenson L. Catalyzing the development and use of novel alternative methods (NAMs) to advance biomedical research. 2023. Available from: https://acd.od.nih.gov/documents/presentations/12142023_Update_Chang_Jorgenson.pdf.

[33] Maryland becomes first state to require animal testing labs to contribute money to non-animal research. The Humane Society of the United States; 2023. Available from: https://www.humanesociety.org/blog/maryland-becomes-first-state-require-animal-testing-labs-contribute-money-non-animal-research [cited 2024 Apr 16].

[34] Legislation. SB0560; n.d. Available from: https://mgaleg.maryland.gov/mgawebsite/Legislation/Details/sb0560?ys=2023RS [cited 2024 Sep 6].

[35] European Commission. Innovative non-animal human-based tools and strategies for biomedical research. EU Funding and Tenders Portal; n.d. Available from: https://ec.europa.eu/info/funding-tenders/opportunities/portal/screen/opportunities/topic-details/horizon-hlth-2024-tool-05-06-two-stage. Accessed 26 April 2025.

[36] European Commission. Bioprinting of living cells for regenerative medicine. EU Funding and Tenders Portal; n.d. Available from: https://ec.europa.eu/info/funding-tenders/opportunities/portal/screen/opportunities/topic-details/horizon-hlth-2024-tool-11-02.

[37] European Commission. Accelerating the implementation of new approach methodologies and other innovative non-animal approaches for the development, testing and production of health technologies. EU Funding and Tenders Portal; n.d. Available from: https://ec.europa.eu/info/funding-tenders/opportunities/portal/screen/opportunities/topic-details/horizon-ju-ihi-2023-05-01. Accessed 26 April 2025.

[38] European Commission. Gaining experience and confidence in new approach methodologies (NAM) for regulatory safety and efficacy testing–coordinated training and experience exchange for regulators. EU Funding and Tenders Portal; n.d. Available from: https://ec.europa.eu/info/funding-tenders/opportunities/portal/screen/opportunities/topic-details/horizon-hlth-2024-ind-06-09. Accessed 26 April 2025.

[39] European Commission. Clinical validation of biomarkers for diagnosis, monitoring disease progression and treatment response. EU Funding and Tenders Portal; n.d. Available from: https://ec.europa.eu/info/funding-tenders/opportunities/portal/screen/opportunities/topic-details/horizon-ju-ihi-2024-07-03-singe-stage. Accessed 26 April 2025.

[40] Anses. Agence nationale de sécurité sanitaire de l'alimentation, de l'environnement et du travail. ANSES; n.d. Available from: https://www.anses.fr/en/content/home [cited 2024 Sep 11].

[41] Partnership for the assessment of risks from chemicals. PARC; n.d. Available from: https://www.eu-parc.eu.

[42] Funding opportunities to support 3Rs research. RE-Place; n.d. Available from: https://www.re-place.be/news/funding-opportunities-support-3rs-research [cited 2024 Apr 15].

[43] Refinement Prize 2023: call for submissions. European Commission; 2023. Available from: https://single-market-economy.ec.europa.eu/calls-expression-interest/refinement-prize-2023-call-submissions_en [cited 2024 Apr 16].

[44] 3RS Student Grants 2025. Call for submission. 3RS Student Grants 2025; 2025. Available from: https://single-market-economy.ec.europa.eu/calls-expression-interest/3rs-student-grants-2025-call-submissions_en [cited 2024 Apr 16].

[45] European Commission. 3Rs Awards; n.d. Available from: https://single-market-economy.ec.europa.eu/sectors/chemicals/european-partnership-alternative-approaches-animal-testing/3rs-awards_en [cited 2024 Apr 16].

[46] Noreocopa. 3R Prize; n.d. Available from: https://norecopa.no/about/3r-prize/ [cited 2024 Sep 11].

[47] Internetredaktion RBL. Federal ministry of education and research—BMBF Gesundheitsforschung. Startseite—BMBF Gesundheitsforschung; n.d. Available from: https://gesundheitsforschung-bmbf.de/en/index.php [cited 2024 Apr 16].

[48] Stichting Stimuleringsfonds. Alternatieven voor Proefdieren; n.d. Available from: https://www.proefdierenalt.nl/ [cited 2024 Apr 16].

[49] BCLAS. Homepage. BCLAS; n.d. Available from: https://www.bclas.org/ [cited 2024 Apr 16].

[50] Project grants. NC3Rs; n.d. Available from: https://nc3rs.org.uk/our-funding-schemes/project-grants [cited 2024 Apr 16].
[51] CRACK IT Challenges. Innovation platform. CRACK IT Challenges; n.d. Available from: https://nc3rs.org.uk/crackit/crack-it-challenges [cited 2024 Apr 16].
[52] International 3Rs prize. NC3Rs; n.d. Available from: https://nc3rs.org.uk/international-3rs-prize [cited 2024 Apr 16].
[53] PhD studentships. NC3Rs; n.d. Available from: https://nc3rs.org.uk/our-funding-schemes/phd-studentships [cited 2024 Apr 16].
[54] FRAME. Innovation grant scheme. FRAME; n.d. Available from: https://frame.org.uk/what-we-do/innovation-grants/ [cited 2024 Apr 16].
[55] ART 3R Awards. n.d. Available from: https://animalresearchtomorrow.org/en/3rs-award [cited 2024 Apr 16].
[56] ART SciComm Awards. n.d. Available from: https://animalresearchtomorrow.org/en/scicomm-award [cited 2024 Apr 16].
[57] Complement animal research in experimentation (Complement-ARIE). A Common Fund Proposal; 2024. Available from: https://dpcpsi.nih.gov/sites/default/files/2024-01/1-1PM-OSC-Concept-Complement-ARIE-Rutter-Woychik-onepager-508.pdf.
[58] RFA-TR-24-025. Tissue chips in space 2.0: translational multi-organ tissue chip systems for drug efficacy, toxicity testing, and personalized medicine in human health, aging and associated diseases (UG3/UH3 clinical trial not allowed). RFA-TR-24-025; 2024. Available from: https://grants.nih.gov/grants/guide/rfa-files/RFA-TR-24-025.html [cited 2024 Sep 6].
[59] PAR-23-267. Industrialization and translation of extracellular vesicles for use in regenerative medicine (U43/U44 clinical trials not allowed). PAR-23-267; 2023. Available from: https://grants.nih.gov/grants/guide/pa-files/PAR-23-267.html.
[60] RFA-TR-23-017. Miniaturization and automation of tissue chip systems (MATChS) (U43/U44 clinical trial not allowed). RFA-TR-23-017; 2023. Available from: https://grants.nih.gov/grants/guide/rfa-files/RFA-TR-23-017.html [cited 2024 Sep 6].
[61] Engineering of Biomedical Systems. U.S. National Science Foundation (NSF); 2023. Available from: https://new.nsf.gov/funding/opportunities/engineering-biomedical-systems [cited 2024 Sep 6].
[62] Annual Open Grants. Alternatives Research and Development Foundation; n.d. Available from: https://www.ardf-online.org/ardf-grants.html [cited 2024 Apr 16].
[63] AAALAC. Global 3Rs awards. AAALAC; n.d. Available from: https://www.aaalac.org/awards/global-3rs-awards/ [cited 2024 Apr 16].
[64] Alan and Helene Goldberg In Vitro Toxicology Grants. The Johns Hopkins Center for alternatives to animal testing; n.d. Available from: https://caat.jhsph.edu/alan-and-helene-goldberg-in-vitro-toxicology-grants/ [cited 2024 Apr 16].
[65] International Foundation for Ethical Research. Home—IFER; n.d. Available from: https://ifer.org/ [cited 2024 Apr 16].
[66] Al-Riyami A. How to prepare a research proposal. *Oman Med J*. 2008;23(2):66–69.
[67] Research.com. How to write a research proposal in 2024: structure, examples & common mistakes. Research.com; 2022. Available from: https://research.com/research/how-to-write-a-research-proposal [cited 2024 Apr 16].
[68] Giddings M. *Four Steps to Funding: Avoid Rejection and Get Your Grant Funded on the Next Try with This Simple Four Step Formula*, Paperback 1.2 ed. Marketing Your Science LLC; 2011.
[69] Krebs CE, Geissler S, Herrmann K, et al. Exploring animal methods bias in biomedical research funding: workshop proceedings and action steps. *NAM J*. 2025;1:100004. Available from: https://www.sciencedirect.com/science/article/pii/S3050620424000046. ISSN 3050-6204. doi:10.1016/j.namjnl.2024.100004.
[70] Krebs CE, Camp C, Constantino H, et al. Author guide for addressing animal methods bias in publishing. *Adv Sci*. 2023. doi:10.1002/advs.202303226.
[71] Common mistakes to avoid when applying for Horizon Europe funding. European Commission; n.d. Available from: https://rea.ec.europa.eu/news/common-mistakes-avoid-when-applying-horizon-europe-funding-2023-02-09_en [cited 2024 Sep 6].
[72] Brock MV, Bouvet M. Writing a successful NIH mentored career development grant (K Award). *Ann Surg*. 2010;251(6):1013–1017.

[73] Brownson RC, Colditz GA, Dobbins M, et al. Concocting that magic elixir: successful grant application writing in dissemination and implementation research. *Clin Transl Sci*. 2015;8(6):710–716.

[74] Smith D, Chowdhury AS, Oxford JT. Overview on grant writing for graduate student research. *Curr Protoc*. 2022;2(9):e545.

[75] NIH Center for Scientific Review. Early career reviewer (ECR) program; n.d. Available from: https://public.csr.nih.gov/ForReviewers/BecomeAReviewer/ECR. Accessed 14 April 2025.

[76] Krebs CE, Lam A, McCarthy J, Constantino H, Sullivan K. A survey to assess animal methods bias in scientific publishing. *ALTEX*. 2023;40(4):665–676. doi:10.14573/altex.2210212.

[77] Marshall LJ, Constantino H, Seidle T. Phase-in to phase-out—targeted, inclusive strategies are needed to enable full replacement of animal use in the European Union. *Anim*. 2022;12(7):863.

[78] Constantino H, Pistollato F, Seidle T. Transitioning biomedical research toward human-centric methodologies: systems-based strategies. *Drug Discov Today*. 2024;29(4):103947.

[79] UK Medical Research Council. JPND Research and Innovation Strategy. 2019. https://www.neurodegenerationresearch.eu/wp-content/uploads/2019/04/Full-JPND-Research-and-Innovation-Strategy-3.04.pdf.

[80] National Institutes of Health. Foundation for NIH Seeks New Approach Methodology Concepts for the Complement-ARIE Validation and Qualification Network. 2025. https://commonfund.nih.gov/complementarie/news/foundation-nih-seeks-new-approach-methodology-concepts-complement-arie. Accessed 30 Apr 2026.

CHAPTER

17

Building confidence in new approach methodologies through reporting, validation, and benchmarking

Kristifor Sunderic

National Center for Advancing Translational Sciences, National Institutes of Health, Rockville, MD, United States

Abstract

This chapter explores building trust and confidence in new approach methodologies (NAMs) through standardization, validation, benchmarking, and transparent reporting. It discusses the limitations of animal models in terms of predictivity and reproducibility, and the drive to develop human-relevant NAMs for safety and efficacy testing. It then outlines validation frameworks and benchmarking methods to compare NAM performance against traditional models. The different regulatory approaches in the United States and European Union are compared, illustrating how legislation, policy, and collaboration influence NAM adoption. Lastly, it discusses key initiatives such as the NIH Complement-ARIE program and public–private partnerships, and their roles in accelerating regulatory acceptance of NAMs.

Keywords: New approach methodologies (NAMs); National Institutes of Health (NIH); Complement Animal Research in Experimentation (Complement-ARIE); Validation; Benchmarking; Animal models

Introduction

The high failure rate in preclinical and clinical drug development, coupled with advances in biomedical engineering, computational approaches, and omics technologies, has led to a growing need to pivot to human-based new approach methodologies (NAMs) to study human health and disease mechanisms. NAMs are a broad set of human-based methods, including in vitro, in chemico, or in silico methods, that can be used to assess safety and efficacy independently of or in complement with animal models. Enabling widespread adoption of NAMs requires building trust and confidence in these new methods among end users: researchers, regulatory agencies, industry, and the public. This chapter explores how standardization, validation, benchmarking, and reporting practices can strengthen

https://doi.org/10.1016/B978-0-443-26514-3.00001-1

confidence in these methods and enable wider adoption of NAMs. Rigor, reproducibility, and good reporting practices are essential, along with lessons learned from animal research. NAMs also need a flexible validation framework to establish the fitness of a NAM for a specific context of use. Finally, this chapter discusses a roadmap for increasing the use of NAMs, compares the United States and EU regulatory environments, and highlights the progress made so far, along with remaining challenges.

Rigor and reproducibility: Learning from animal research pitfalls and reporting best practices

One of the strongest arguments for enabling wider adoption of NAMs is the limitations of animal research. A wide body of preclinical research has shown that animal studies can suffer from flawed experimental design, high variability, poor reproducibility, poor translation to human physiology, and incomplete reporting.[1,2]

Less than 10% of drugs entering clinical trials result in an approved medicine, largely due to species-specific differences.[3,4] These differences mean that findings from animal models do not always translate to humans.[5] About 56% of drug candidates fail due to a lack of efficacy in humans, and 28% fail due to toxic effects.[6] Additionally, the accumulation of external inputs, stressors, and life experiences in humans uniquely contributes to how diseases manifest and how they respond to treatments in different populations.[7] These translational failures underscore the fundamental physiological differences between humans and other species, leading to wasted resources and ethical concerns over animal use. Pivoting away from animal model results to human cell-based models, which have more predictive safety and better reflect the diversity of human genetics, may lead to improved treatments in real-world applications.

However, more confidence is needed for NAMs to fully deliver on safety and accuracy. To that end, the NAMs community must prioritize rigorous and reproducible study design. Cell-based assays must use well-characterized cell lines or tissue models, standardized culture conditions, and robust experimental protocols.[8] And researchers must design NAM experiments with the same rigor expected in clinical research, including randomization of test samples, robust statistical analysis, blinding of analysis to reduce bias, and meeting predetermined criteria for data inclusion.[1] Applying these best practices to NAMs will help prevent avoidable errors and artifacts.

Another key aspect is the thorough reporting of methods and results. Guidelines such as ARRIVE (Animal Research: Reporting of In Vivo Experiments) were developed to improve transparency among animal studies by ensuring researchers provide all essential details of their experimental design and execution.[9] The ARRIVE checklist prompts authors to report items such as animal strain, sex, randomization method, sample size rationale, handling of data, and more, to enable others to replicate or evaluate the work.[1,9] Standards like the Organisation for Economic Co-operation and Development's (OECD) Good In Vitro Method Practices (GIVIMP) can also provide a framework for consistent documentation and quality control in in vitro assays.[10] Over 1000 journals and funding agencies have adopted such guidelines, but compliance is not always enforced.[1,11] There are also several other checklists and guidelines for assessing animal studies, demonstrating a need to standardize tools and reporting. However, when guidelines are followed, studies show improved reproducibility and fewer false results.[1] Poorly reported studies not only waste resources but also fail to ethically amortize animal suffering by generating reliable data.

Researchers can also demonstrate trust in NAMs by acknowledging what a method can and cannot do for its intended use. This is known as **fitness for purpose**—the suitability of a method or tool for its intended use and ensuring it can generate reliable and relevant information to support regulatory decisions. A NAM does not need to model an entire organism to be useful; it only needs to reliably model the specific biological mechanism or endpoint for which it is intended. Reporting should then focus on the method's biological relevance. For example, if a liver-on-a-chip model is proposed to model drug-induced injury, reporting should include data demonstrating that the chip emulates the structural and functional characteristics of the human liver, including testing its performance on compounds with known liver toxicity profiles. Fitness for purpose is a critical aspect of acceptance in regulatory decision-making.

It is worth mentioning that reproducibility itself is an ongoing challenge across scientific disciplines,[12–14] in both animal and in vitro studies. However, agencies are taking steps to address this issue. In April 2025, the NIH Director, Dr. Jay Bhattacharya, announced a renewed focus on addressing reproducibility, including the establishment of a new scientific journal dedicated to replication research.[15]

NAMs must meet or exceed the rigor and transparency standards of animal studies. Robust experimental design, thorough reporting, and open data practices can mitigate reproducibility issues and increase confidence that NAM results are dependable. Regulatory agencies and the wider scientific community will better trust NAM data as credible and reliable, accelerating wider adoption. Standardization of methods and rigorous reporting are, therefore, pillars in building confidence in nonanimal approaches.

Validation and benchmarking of NAMs: Regulatory requirements versus research use

A new method must work reliably and relevantly for its intended purpose; i.e., it must be fit for purpose. This is the domain of validation and benchmarking. Validation asks: does the method accurately measure or predict what it is supposed to with sufficient consistency? Benchmarking asks: how does the new method's performance compare to existing accepted methods for the endpoint in question?

Four interdependent pillars have been described for fit-for-purpose validation: biological relevance, technical characterization, data integrity, and information transparency.[16,17] More specifically, a valid NAM should: (1) have a sound biological basis linking it to the real outcome, (2) be technically well characterized and controlled, (3) generate reliable data with integrity, and (4) be transparently reported and open to outside evaluation. Previously, validation of alternative methods was often a lengthy, one-size-fits-all procedure. A new in vitro test might undergo a multistep evaluation: intra-laboratory optimization, a multilab ring trial to demonstrate reproducibility, and then an independent peer review of the data supported by a validation body like the US Interagency Coordinating Committee on the Validation of Alternative Methods (ICCVAM) or the EU's European Center for Validation of Alternative

Methods (ECVAM).[18] Only then could it be considered for regulatory acceptance. However, years and even decades could pass before it became an internationally accepted OECD test guideline. While this ensured rigor, it was slow and method-centric, with each assay validated in isolation. Today, both regulators and scientists recognize the need for more nimble, **context-of-use**-driven validation approaches; that is, based on what the NAM is intended to do and how it will be used in decision-making.[17] Rather than asking, "Is this method valid for all purposes?" we ask, "Is this method valid for the specific purpose and context we need?" Validation should be a robust yet flexible process where studies can be tailored to the intended context of use.

Benchmarking is the comparison of a NAM's performance to existing methods or benchmarks. A new toxicology test, for example, would traditionally be benchmarked by comparing its results on a list of chemicals or agents to the results of the established animal test for those same reagents. For example, a new skin sensitization assay might be evaluated on a panel of chemicals with known outcomes in the animal test[19] measuring sensitivity, specificity, and overall performance. If the NAM achieved a similar performance to the animal test, it would be considered "validated."

However, ideal benchmarking is not always about reproducing the animal test results, especially if the animal test itself is an imperfect predictor of human outcomes.[20] For example, rodent toxicity tests are reported to predict only 40%–65% of human toxicities correctly,[21–26] even though they are treated as a gold standard largely due to convention and lack of alternatives. A NAM does not need to replicate the exact results of an established method, only support a scientifically rigorous decision.[18] The true benchmark is human risk prediction and protection, not the animal test.

Therefore modern benchmarking practices may incorporate comparisons against both animal data and any available human data. Where human data on the effect are available, they can reveal the animal test's limitations and show that a NAM might actually outperform the animal test in human relevance.[27–30] In cases where no direct human data exist, benchmarking might rely on demonstrating that the NAM can capture known mechanisms of toxicity, such as genetic mutation. This is known as mechanistic validation, which ties the NAM to key events in an adverse outcome pathway. If the key mechanisms are reproduced, it suggests the NAM is relevant even if long-term human outcomes are not immediately available for comparison.[17]

Lastly, it is important to distinguish regulatory validation requirements from NAM use in basic biomedical research. In the regulatory area, there are formal expectations and guidance documents outlining validation criteria. This can include demonstrating the method's relevance and reliability through intra- and inter-lab studies, as well as an independent peer review of the validation study, among other criteria.[31] Agencies like the Food and Drug Administration (FDA), Environmental Protection Agency (EPA), and European counterparts generally will not replace a required animal test with a NAM until such evidence exists and, preferably, an OECD or other official test guideline is established. Regulators participate in validation efforts through bodies like ICCVAM, ECVAM, the Japanese Center for the Validation of Alternative Methods, and the International Cooperation on Alternative Test Methods (ICATM) network.[18,32] Regulatory validation also increasingly involves qualification programs. Rather than expecting a NAM to replace an existing test outright, regulators may "qualify" it for a specific use. The FDA's Drug Development Tool (DDT) qualification program and the newer Innovative Science and Technology Approaches for New Drugs (ISTAND) program are examples.[18] Under these programs, a tool is evaluated for a defined context of use and, if it meets criteria, the FDA issues a qualification determination. Once qualified, the tool can be used by sponsors in regulatory submissions for that purpose without needing to re-prove its validity each time.[18] This effectively serves the same role as validation but in a more use-case-specific and streamlined way.

By contrast, in academic or exploratory biomedical research, the "validation" of a new model is far less formal. Researchers are not bound by regulatory validation requirements to use a NAM model to study disease mechanisms. Instead, the model gains credibility if the researcher demonstrates certain hallmarks (e.g., that an organoid expresses key neuronal markers). Publication in peer-reviewed journals serves as a kind of validation, and there is more freedom to use a novel model with the understanding that it is exploratory.

Roadmap for adoption: Timeline for increasing use and ensuring safety

Table 1 provides a comparative overview summarizing key aspects of regulatory support in the United States and the European Union toward broader NAMs adoption and reductions in animal testing.

TABLE 1 Comparing US and EU regulatory landscapes.

Domain	United States	European Union
Legal requirements for animal tests	Historically, US law required animal data for drug approvals, but the 2022 FDA Modernization Act 2.0 removed the mandate for animal studies in new drug applications and explicitly allows validated nonanimal methods to be used. In chemical regulation, the Toxic Substances Control Act encourages minimizing animal testing. The EPA intends to reduce animal toxicity testing, though timelines are being reevaluated. Generally, agencies have discretion to accept a NAM in lieu of an animal test if it is sound and fit for purpose.	Since 2013, it has been illegal to test cosmetic ingredients on animals or to sell cosmetics tested on animals. For chemicals and pharmaceuticals, EU laws require animal data for many endpoints unless an alternative is accepted. The Registration, Evaluation, Authorization, and Restriction of Chemicals (REACH) regulation follows a "last resort" principle, where animal testing is permitted only when no other alternatives exist. In practice, many tests still use animals due to the lack of fully validated alternatives.[20] Notably, EU hazard classification is based on traditional animal test outcomes, but in 2021, the EU Parliament overwhelmingly passed a resolution to phase out animal testing and seek replacements in all scientific areas.[33]

Continued

TABLE 1 Comparing US and EU regulatory landscapes—cont'd

Domain	United States	European Union
Validation and acceptance frameworks	The United States relies on interagency, science-driven processes. ICCVAM coordinates the validation and evaluation of new methods across 17 federal agencies and issues recommendations that agencies can adopt.[17] In addition, agencies develop their own roadmaps for NAM integration, such as the FDA's dedicated NAMs initiative to spur alternative method use across all its centers, as well as its qualification pathways, DDT, and ISTAND.[2] The EPA and OSHA participate in OECD efforts to create international test guidelines for NAMs. Regulatory acceptance can be decentralized, with each agency or even subcenter having its own criteria, but there is an increasing push for harmonization. The 2023 NIH ACD Working Group on NAMs recommended more interagency coordination and investment to speed acceptance,[34] leading to the Complement-ARIE program.[35]	The EU has a centralized process via the EU Reference Laboratory for Alternatives to Animal Testing, which evaluates new methods and can propose them for adoption as an OECD Test Guideline or EU regulatory test methods. The EU's REACH regulation explicitly supports alternative methods. Its Annex XI allows waiving animal tests if "scientifically acceptable" alternatives or weight-of-evidence approaches are available.[36] The challenge is often proving to regulators that a NAM is scientifically acceptable to avoid animal tests.[20] The EU is active in the OECD Test Guidelines Programme and ICATM to jointly validate methods.[32] In terms of guidance, the European Medicines Agency aligns with guidelines from the International Council for Harmonisation, which allow alternative methods if justified. The EU is developing a roadmap to accelerate the transition to NAMs, including increased funding and proposed revisions to regulatory requirements.
Progress and case studies	The United States has notable success stories. In vitro ocular irritation assays have been accepted by the FDA in place of the classic rabbit eye test for certain product evaluations, essentially eliminating rabbit eye testing for many ophthalmic safety screens.[18] The agency's ISTAND program accepted its first organ-on-a-chip into the qualification pipeline.[37] The EPA has approved alternative tests for skin sensitization, including the direct peptide reactivity assay. In drug development, computational and cell-based NAMs have aided in decisions, for example, by avoiding a clinical trial on a potentially carcinogenic drug by detecting the risk via in vitro and in silico methods.[18]	The EU's cosmetics animal testing ban led to the development of many alternative assays that are now standard. There are now OECD Test Guidelines for reconstructed human skin models to test irritation instead of rabbit skin, and for various cell assays and ex vivo methods for eye irritation.[20] Integrated testing strategies using multiple NAMs have been validated for skin sensitization, showing equal or better performance than mouse models, which are accepted under REACH.[20] Europe has also launched collaborative public–private partnerships to develop and evaluate NAM approaches for systemic toxicity, which have produced successful case studies.[20] Some EU member states are increasing funding for NAM research centers, and the Horizon Europe research program prioritizes projects on NAM validation.

Domain	United States	European Union
Challenges	While US law now permits alternatives, implementation is the next challenge. FDA and EPA staff need training and experience to evaluate NAM-derived data. Guidance documents are being updated, but some are still implicitly oriented toward animal data. Regulatory guidelines often specify animal test designs, for example, but these are being revised. There's also an issue of infrastructure: ensuring labs are equipped and personnel are skilled to perform validated NAM tests. In the short term, conducting NAM and animal tests in parallel may be needed to build familiarity. Meanwhile, continued demonstration of NAM successes and publication of predictive performance are needed to build confidence. Funding is also an ongoing challenge. Significant investment is needed to develop complex NAMs like multiorgan systems and to conduct large validation studies.	The EU faces legislative and inertia challenges. Some regulations explicitly demand animal data or have criteria tied to animal test outcomes, such as in chemicals regulation. Changing those laws is not trivial. There can be a conservative attitude among regulators. Even though REACH outlines the use of animals only as a last resort, authorities sometimes hesitate to accept novel data in place of a long-established animal test due to fear of unknown risks or legal challenges. The hazard-based regulatory mindset needs to shift to a mechanism and risk-based approach using NAM data. However, this requires consensus on new safety factors or prediction models. Europe's broad stakeholder environment also means more coordination is needed among industry, regulators, and NGOs. Public sentiment in Europe is creating political pressure to shift away from animal testing. Like in the United States, investment is also key and remains an ongoing challenge.[20]

A roadmap for increasing NAM utilization

The FDA has outlined an approach to reduce animal testing in preclinical safety studies with validated NAMs.[2]

Phase 1: Foundation Building (next ~5 years)

In the near term, agencies and stakeholders will focus on investing heavily in NAM research and validation studies. Rapid development of combinatorial NAMs is being targeted by the NIH's Complement-ARIE program.[38] Agencies should also publish clear guidance on NAM data for regulatory submissions, something the FDA is already engaged in. Training programs should be expanded for regulators, industry scientists, and academics to familiarize them with NAM technologies. By the end of this period, safety testing will be largely NAM-driven due to regulatory pressure and the wide availability of OECD-approved tests. Validated NAMs are available for many toxicity endpoints, regulators are accepting NAM data on a case-by-case basis, and a NAM-savvy workforce is emerging.

Phase 2: Integration and expansion (next ~5–10 years)

In the next phase, NAMs are largely integrated into regulatory frameworks. Validated NAMs cover nearly all critical areas, and FDA requirements default to being NAM-based. A significant portion of routine toxicology tests is done with NAMs. Animal testing is no longer required if an approved NAM or a battery of NAMs can be used instead, but may be used in rare cases where one does not exist. Still, animal studies are the exception rather than the norm, and soon, conventional animal testing will not be required for safety.

Phase 3: Full transition (~10+ years)

Finally, animal testing is dramatically minimized to only necessary cases. Validated NAMs exist for all major regulatory endpoints, either as individual methods or integrated testing strategies. Here, human-relevant NAM data are the default for safety and risk assessment, and any proposal to use animal testing must be scientifically justified. When animal testing is performed, it is highly limited and highly specialized. International agencies have shared OECD test guidelines and standards covering NAMs, ensuring that a test performed in one country is accepted worldwide. Human-relevant, multiorgan microphysiological systems that simulate whole-body interactions are used for personalized medicine and predicting disease outcomes, and education and training have fully incorporated NAM technologies.

Of course, regulators and the public will not accept transitioning away from animal testing if they believe there is a risk to safety and public health, though safety evaluation will, in fact, improve by using more human-relevant tools. To mitigate concerns, there are strategies to help maintain confidence as we transition away from animal testing:

1 **Next-generation risk assessment (NGRA)**

NGRA is a human-relevant risk assessment framework for evaluating the safety of chemicals and other products.[39] It incorporates NAMs into risk assessment to ensure that the use of a product does not cause harm to humans, with little or no new animal testing. NGRA outlines several principles for risk assessment, one of which is the use of tiered testing strategies. This principle states that the total amount of resources allocated to any risk assessment should be no less and no more than what is required to provide adequate precision to reach a conclusion. For example, a chemical might be run through a battery of NAMs covering various toxicological endpoints. If results indicate acceptably low risk, then no animal testing is necessary. Otherwise, the chemical may be advanced to a higher tier of testing, including population-specific exposures or some animal testing. Another principle is characterizing sources of uncertainty. Uncertainty has always been a feature of risk assessment and can arise from limitations in knowledge or imperfections in instruments, models, or techniques. These uncertainties should be expected but must always be well documented to support a decision. Depending on the uncertainties, regulators can build in conservative assumptions to ensure safety margins, so even if risk is under predicted, that risk is accounted for in the margin. While originally intended for assessing cosmetics, NGRA has found applications in other areas of safety decision-making, such as occupational health.[40]

2 **Redundant parallel testing**

Agencies often use NAMs alongside existing animal tests to cross-verify results. Overlapping evidence pointing to the same conclusion indicating that NAMs work at least as well as the animal tests they are being compared to. The NAM is not "missing" a hazard that an animal test would have detected. Over time, as NAMs consistently capture relevant effects, parallel animal testing can be discontinued.

3 **Continuous surveillance and feedback**
Through NAM's use, agencies will implement monitoring systems to track their outcomes. Predictive data will be continuously collected and compared to real-world outcomes to further improve them. Unexpected adverse effects are fed back to refine the models. This is already done with animal-tested products, where unexpected toxicities in humans force adjustments to testing protocols. A database comparing NAMs and real-world outcomes can track performance and update models as new data emerge.

4 **Public engagement, communication, and transparency**
Maintaining public trust will require proactive communication from agencies about how NAMs help protect public health and safety. This includes publicizing when NAMs are as predictive as or more predictive than animal testing and explaining the validation process. Being transparent about the careful evaluation of submissions and the reasoning of review panels will help demonstrate responsible due diligence.

5 **Maintaining momentum through advocacy**
Finally, advocacy organizations like the Physicians Committee for Responsible Medicine act as partners and watchdogs, maintaining pressure on regulators and industry to stick to reduction goals. The Physicians Committee supports NAM researchers, organizes training, and has contributed to papers on NAM validation. The organization has been instrumental in advocating for legislative change by presenting evidence to lawmakers that alternatives can work. The Physicians Committee also commissioned a public opinion survey showing that over 85% of Americans across the political spectrum support phasing out animal testing in favor of modern methods,[2] representing broad societal support for this change. Such advocacy ensures ongoing challenges are continuously addressed.

Conclusion

Building trust in nonanimal models depends on making them more human-relevant and more reproducible. Here, we explored how standardization, validation, benchmarking, and reporting contribute to that trust. By enforcing rigorous experimental design and transparency, NAM data become more robust and credible. Through validation and benchmarking efforts, NAMs are shown to be fit for purpose, reliably answering the questions they are meant to, and meeting or exceeding the performance of older methods. By comparing new methods against the best available benchmarks, we focus on predictive power for human and environmental outcomes.

Policy changes and new research initiatives across the globe represent the high-level commitment to reducing animal testing. Collaborative initiatives and working groups like ICCVAM in the US and ECVAM in the EU have created frameworks to evaluate and integrate these tools. A growing list of success stories, such as cosmetics without animal testing, vaccines tested on human cell platforms, and chemicals screened via high-throughput in vitro assays, offers proof that NAMs can work in the real world. Even pharmaceutical research, traditionally dependent on animal models, is increasingly incorporating human organoid and in silico models early in development to derisk candidates before any animal or human is involved, saving time, money, and lives. Continued international collaboration will further our progress.

In the United States, significant policy momentum has been built in departing from animal testing. The 2022 FDA Modernization Act 2.0 allows nonanimal test methods for drug approval.[41] This led the NIH to launch the Complement-ARIE program in 2024, designed to speed the development, standardization, validation, and use of human-based NAMs.[42] It targets key scientific, infrastructure, and adoption gaps through multiple coordinated components. A dedicated Technology Development Center will create NAM models in the areas of greatest need, and a Validation and Qualification Network, established as a public–private partnership, will set validation standards and streamline evaluation and regulatory acceptance of new methods. In mid-2025, the NIH announced it would no longer release funding opportunities exclusively seeking animal studies.[43] These policies will further drive the adoption and use of NAMs by encouraging researchers to integrate in vitro platforms, organ-on-a-chip systems, and in silico models into their experiments.

There are still many mountains to climb to establish full confidence in NAMs. Improving the biological relevance of NAMs is an ongoing challenge, but continued advances in stem cell biology, 3D cultures, and computational modeling are helping bring these technologies to their full potential. We must continue refining criteria for validation to ensure they are flexible and adaptable.

Fitness for purpose and context of use should guide our expectations of NAMs. We know that no model is perfect, whether animal or nonanimal. A mouse is not a model human, and a liver organoid is not a fully grown liver. However, each can be fit for a particular purpose. A mouse may mimic physiology to an extent, while a liver organoid can mimic specific human liver functions. We can tailor our toolset to answer the question at hand, and confidence in NAMs will grow when the best tool is used for each question. Additionally, we can pursue a more holistic goal of improving human safety science alongside reducing animal testing.[20]

The next generation of scientists will be developing many of these methods in labs around the world, pushing the boundaries of what NAMs can do. It is essential that they receive training not just in science and technology but also in good study design and validation, as discussed herein.

Building trust in NAMs will require innovation, cultural change, and policy reform, but it will eventually lead to a paradigm of safety and efficacy testing that protects humans and animals alike. We are gradually shifting from a time when animal testing was synonymous with human safety to a future where the best test is the one that predicts human response most accurately. The coming decades will solidify this paradigm, and from that vantage point, we will look back and marvel at the progress made for humans and animals alike.

Acknowledgments

The author is deeply grateful to Dan Tagle, Margaret Ochocinska, Victoria Ledbetter, Geetha Senthil, Andrew Wright, Chariz Johnstone, and Christine Happel for their invaluable support in writing this chapter.

References

[1] National Institutes of Health. *Fostering Rigorous Research: Lessons Learned from Nonhuman Primate Models and Charting the Path Forward*. Bethesda, MD: NIH; 2020. Available at: https://osp.od.nih.gov/wp-content/uploads/NHP_Workshop_Report_FINAL_20200218.pdf. Accessed Aug 14, 2026.

[2] Food and Drug Administration. Roadmap to reducing animal testing in preclinical safety studies; 2025. Available at: https://www.fda.gov/media/186092/download. Accessed Aug 14, 2026.

[3] Sun D, Gao W, Hu H, Zhou S. Why 90% of clinical drug development fails and how to improve it? *Acta Pharm Sin B*. 2022;12(7):3049–3062.

[4] Wong CH, Siah KW, Lo AW. Estimation of clinical trial success rates and related parameters. *Biostatistics*. 2019;20(2):273–286.

[5] Bell J. Aspirin killed the cat: animal research models do not always apply to humans. *Expert Opin Drug Metab Toxicol*. 2019;15(9):683–685.

[6] Arrowsmith J, Miller P. Phase II and Phase III attrition rates 2011–2012. *Nat Rev Drug Discov*. 2013;12(569). https://www.nature.com/articles/nrd4090.

[7] Zeggini E, Gloyn A, Barton A, Wain L. Translational genomics and precision medicine: moving from the lab to the clinic. *Science*. 2019;365(6460):1409–1413.

[8] Pamies D, Ekert J, Zurich M-G, et al. Recommendations on fit-for-purpose criteria to establish quality management for microphysiological systems and for monitoring their reproducibility. *Stem Cell Rep*. 2024;19:604–617.

[9] National Centre for the Replacement Refinement and Reduction of Animals in Research. ARRIVE guidelines. [Online]. Available: https://nc3rs.org.uk/arrive-guidelines. [Accessed 2025].

[10] Organization for Economic Cooperation and Devleopment. *No. 286: Guidance Document on Good in Vitro Method Practices (GIVIMP)*. OECD Publishing; 2018.

[11] Song J, Solmi M, Carvalho AF, Shin JI, Ioannidis JP. Twelve years after ARRIVE guidelines: animal research has not yet arrived at high standards. *Lab Anim*. 2024;58(2):109–115.

[12] Ioannidis JPA. Why most published research findings are false. *PLoS Med*. 2005;2(8):e124. doi:10.1371/journal.pmed.0020124.

[13] Begley CG, Ellis LM. Raise standards for preclinical cancer research. *Nature*. 2012;483:531–533.

[14] Open Science Collaboration. Estimating the reproducibility of psychological science. *Science*. 2015;349:aac4716. doi:10.1126/science.aac4716.

[15] C. o. S. S. Associations. NIH director Bhattacharya gives first public remarks at NIH council of councils. Consortium of Social Science Associations; 2025. [Online]. Available: https://cossa.org/nih-director-bhattacharya-gives-first-public-remarks-at-nih-council-of-councils/. [Accessed 2025].

[16] van der Zalm AJ, Barroso J, Browne P, et al. A framework for establishing scientific confidence in new approach methodologies. *Arch Toxicol*. 2022;96(11):2865–2879.

[17] Interagency coordinating committee on the validation of alternative methods, validation, qualification, and regulatory acceptance of new approach methodologies, 2024. Available at: https://ntp.niehs.nih.gov/sites/default/files/2024-03/VWG_Report_27Feb2024_FD_508.pdf. Accessed Aug 14, 2026.

[18] Nelson CP, Brown P, Fitzpatrick S, et al. Advancing alternative methods to reduce animal testing. *Science*. 2024;386(6723):724–726.

[19] OECD. *Guideline No. 497: Defined Approaches on Skin Sensitisation*. Paris: OECD Publishing; 2025.

[20] Sewell F, Alexander-White C, Brescia S, et al. New approach methodologies (NAMs) identifying and overcoming hurdles to accelerated adoption. *Toxicol Res*. 2024;13:1–9.

[21] Olson H, Betton G, Robinson D, et al. Concordance of the toxicity of pharmaceuticals in humans and in animals. *Regul Toxicol Pharmacol*. 2000;32(1):56–67.

[22] Bailey J, Thew M, Balls M. An analysis of the use of animal models in predicting human toxicology and drug safety. *Altern Lab Anim*. 2014;42(3):181–199.

[23] Monticello T, Jones T, Dambach D, et al. Current nonclinical testing paradigm enables safe entry to first-in-human clinical trials: the IQ consortium nonclinical to clinical translational database. *Toxicol Appl Pharmacol*. 2017;334:100–109.

[24] Clark M, Steger-Hartmann T. A big data approach to the concordance of the toxicity of pharmaceuticals in animals and humans. *Regul Toxicol Pharmacol*. 2018;96:94–105.

[25] Atkins J, George G, Hess K, et al. Pre-clinical animal models are poor predictors of human toxicities in phase 1 oncology clinical trials. *Br J Cancer*. 2020;123(10):1496–1501.
[26] Rooney J, Choksi N, Ceger P, et al. Analysis of variability in the rabbit skin irritation assay. *Regul Toxicol Pharmacol*. 2021;122:104920. doi:10.1016/j.yrtph.2021.104920.
[27] Eskes C, Cole T, Hoffman S, et al. The ECVAM international validation study on in vitro tests for acute skin irritation: selection of test chemicals. *Altern Lab Anim*. 2007;35(6):603–619.
[28] Spielmanh H, Hoffmann S, Liebsch M, et al. The ECVAM international validation study on in vitro tests for acute skin irritation: report on the validity of the EPISKIN and EpiDerm assays and on the skin integrity funtion test. *Altern Lab Anim*. 2007;35(6):559–601.
[29] Alepee N, Adriaens E, Abo T, et al. Development of a defined approach for eye irritation or serious eye damage for neat liquids based on cosmetics Europe analysis of in vitro RhCE and BCOP test methods. *Toxicol Vitro*. 2019;57:154–163.
[30] Strickland J, Truax J, Corvaro M, et al. Application of defined approaches for skin sensitization to agrochemical products. *Front Toxicol*. 2022;4:852856. doi:10.3389/ftox.2022.852856.
[31] OECD. *Guidance Document on the Validation and International Acceptance of New or Updated Test Methods for Hazard Assessment*. Paris: OECD Publishing; 2005.
[32] US Food and Drug Administration. Advancing new alternative methodologies at FDA; 2021. Available at: https://www.fda.gov/media/144891/download. Accessed Aug 14, 2026.
[33] Parliament E. European Parliament resolution of 16 September 2021 on plans and actions to accelerate the transition to innovation without the use of animals in research, regulatory testing and education; 2022. Available at: https://eur-lex.europa.eu/legal-content/EN/TXT/PDF/?uri=oj:JOC_2022_117_R_0009. Accessed Aug 14, 2026.
[34] *Catalyzing the Development and Use of Novel Alternative Methods*. National Institutes of Health, Bethesda, Maryland, 2023. https://acd.od.nih.gov/documents/presentations/Working_Group_Report.pdf.
[35] National Institutes of Health. Complement Animal Research In Experimentation (Complement-ARIE) Program; 2024. [Online]. Available: https://commonfund.nih.gov/complementarie. [Accessed 2025].
[36] Off J Eur Union. EC. REGULATION (EC) no 1907/2006 of the European Parliament and of the council of 18 December 2006 concerning the registration, evaluation, authorisation and restriction of chemicals (REACH), establishing a European chemicals agency. Off J Eur Union; 2006, pp. 396. Available at: https://eur-lex.europa.eu/eli/reg/2006/1907/oj/eng. Accessed Aug 14, 2026.
[37] FDA. FDA's ISTAND Pilot Program accepts a submission of first organ-on-a-chip technology designed to predict human drug-induced liver injury (Dili); 2024. [Online]. Available: https://www.fda.gov/drugs/drug-alerts-and-statements/fdas-istand-pilot-program-accepts-submission-first-organ-chip-technology-designed-predict-human-drug. [Accessed 2025].
[38] Sunderick K, Wright AM, Kleinstreuer N, et al. Complement-ARIE: catalyzing the development and adoption of new approach methodologies. *NAM J*. 2025;1:100026. doi:10.1016/j.namjnl.2025.100026.
[39] Dent M, Amaral RT, Silva PAD, et al. Principles underpinning the use of new methodologies in the risk assessment of cosmetic ingredients. *Comput Toxicol*. 2018;7:20–26.
[40] Leso V, Nowack B, Karakoltzidis A, et al. Next generation risk assessment and new approach methodologies for safe and sustainable by design chemicals and materials: perspectives and challenges for occupational health. *Toxicology*. 2025;517:154211. doi:10.1016/j.tox.2025.154211.
[41] Adashi EY, O'Mahony DP, Cohen IG. The FDA Modernization Act 2.0: drug Testing in Animals is Rendered Optional. *Am J Med*. 2023;136(9):853–854.
[42] National Institutes of Health. Partnership aims to increase adoption and implementation of new approach methodologies; 2024. [Online]. Available: https://commonfund.nih.gov/complementarie/news/partnership-aims-increase-adoption-and-implementation-new-approach. [Accessed 2025].
[43] N. I. o. H. G. &. Funding. How DOES the NIH initiative to prioritize human-based research affect research proposing the use of laboratory animals? National Institutes of Health; 2025. [Online]. Available: https://grants.nih.gov/news-events/nih-extramural-nexus-news/2025/07/how-does-the-nih-initiative-to-prioritize-human-based-research-affect-research-proposing-the-use-of-laboratory-animals. [Accessed 18 November 2025].

CHAPTER

18

Phasing out animal tests: An analysis of the successful transition from rabbits to human-relevant methods in Europe

Justine C. Watkins[a], *Thomas Hartung*[b,c], *Ingrid Visseren-Hamakers*[d], *Dilyana Filipova*[e], *Howard White*[f], *Luísa Ferreira Bastos*[g], *Love Hansell*[h], *and Merel Ritskes-Hoitinga*[i]

[a]Utrecht University, Utrecht, Netherlands
[b]Johns Hopkins University, Baltimore, MD, United States
[c]CAAT-Europe, University of Konstanz, Konstanz, Germany
[d]Radboud University, Nijmegen, Netherlands
[e]Doctors Against Animal Experiments, Cologne, Germany
[f]Research and Evaluation Centre, London, United Kingdom
[g]Eurogroup for Animals, Brussels, Belgium
[h]Institute for Management Research, Radboud University, Nijmegen, Netherlands
[i]Faculty of Veterinary Medicine, Institute for Risk Assessment Sciences (IRAS), Utrecht University, Utrecht, Netherlands

Abstract

Efforts to replace animal testing with human-relevant methods have advanced. Pyrogenicity quality control testing illustrates the progress of this transition well. The rabbit pyrogen test (RPT), once the gold standard, has limitations that prompted the development of in vitro alternatives like the monocyte activation test (MAT). While MAT is validated and accepted in the EU, the United States, Russia, and India, only the European Pharmacopoeia has fully phased out RPT, making MAT compendial. In contrast, RPT remains the gold standard in the United States, reflecting regulatory inertia and global inconsistencies.

Moving Toward Nonanimal Approaches in Medical Research and Testing
https://doi.org/10.1016/B978-0-443-26514-3.00017-5

This chapter analyzes the historical, scientific, and regulatory evolution of pyrogen testing, comparing Europe and the United States. It finds the United States lags in phasing out RPT, and that even in Europe, the transition took nearly 40 years. Proactive regulation and simultaneous method phase-in/out are essential to accelerate future transitions.

Keywords: Pyrogenicity; Endotoxin; European pharmacopeia; United States pharmacopeia; Nonanimal replacement; Rabbit pyrogen test (RPT); Monocyte activation test (MAT)

Introduction

As scientists, manufacturers, and regulators seek to develop more ethical, human-relevant, and accurate methods to predict human health risks, alternatives to animal testing have become especially valuable.[1–5] In Europe alone, regulatory testing consumed, on average, just under 1 million animals per year between 2018 and 2022, 58% of which were used for the quality control (QC) of products.[6] In general, quality assessment (potency, purity, safety, and efficacy testing) of biologically based products, like vaccines, is conducted to ensure products meet QC specifications defined within the product's market authorization.[3] Pharmacopoeias such as the European Pharmacopoeia (Ph. Eur.), have significant influence in outlining the standard test methods manufacturers use to receive market approval and/or complete residual batch safety testing for their products.[7–9] Ph. Eur. alone encompasses 3000 reference standards with over 31 signatory states and 31 observer states, representing all continents.[9] Within QC testing, there are significant opportunities to reduce animal numbers, but so far, there are limited examples of alternatives that have been officially validated and accepted.[3,4,10,11]

Pyrogenicity testing for QC offers enlightening insights. This test is used to determine the presence and/or amount of (non-)endotoxin pyrogens in a biological sample to ensure the product does not cause the undesired activation of the host's immune response.[3,10,12] For decades, the gold standard for pyrogen testing was the rabbit pyrogen test (RPT), a pass/fail test that detects fever reactions in rabbits.[10,13,14] However, RPT is unethical because it uses animals (when alternatives exist), cannot be used to test for pyrogens in a variety of products (i.e., radiopharmaceuticals, steroids, etc.), and, as summarized in this chapter, is scientifically inaccurate.[12,15]

While there have been significant advancements in scientific innovation, legislation, and regulatory acceptance of nonanimal alternatives for pyrogenicity, the shift to their application has been slow, and RPT still remains required in many regions and is embedded in global pharmacopoeias and regulatory texts.[9,16] Even in Europe, where the monocyte activation test (MAT), a human-based alternative, has been a compendial method for many years, there was continued use of the RPT for QC testing until at least 2020.[17] However, as of July 2025, RPT will be phased out of the Ph. Eur.,[18] representing a monumental success within this field.

To understand the success of MAT in breaking boundaries where other nonanimal alternatives have not, we can analyze the successes and failures of this transition. This analysis aims to answer the following questions: What are the scientific, regulatory, and historical changes that contributed to the replacement of RPT by the alternative MAT, and how does this transition compare between and within Europe and the United States? What are the historical, regulatory, and scientific differences between Europe and the United States that led to the different stages in the MAT transition? And what are the historical, regulatory, and scientific changes that influenced the phase-in of MAT and the phase-out of RPT in the European Pharmacopoeia?

We completed a literature review and interviewed experts to develop a historical timeline of social, scientific, technological, and regulatory changes that occurred in the United States and European contexts. We then performed a content analysis, comparing relevant US and EU legislations, as well as an analysis of relevant changes to European Pharmacopoeial text related to MAT and RPT over time. We discuss the outcomes of these analyses using an adapted framework from Smith and Raven[19] on the socio-technical regime. (According to Smith and Raven, the sociotechnical regime can be described as a set of technologies, practices, rules, and cultural norms that shape society and address the needs of a specific sector.) The adapted framework is used to trace the social, scientific, regulatory, and (partial) economic influences garnered from this case study to outline suggestions for future transitions to nonanimal alternatives in regulatory safety assessment.

Background

Regulatory background

Pharmacopoeias set quality standards for medicines by establishing reference standards for the assessment of medicinal products, which are then elaborated in a set of monographs and general texts.[7–9,20] Pharmacopoeias are established through collaboration between manufacturers, regulators, and international stakeholders.[8] The World Health Organization (WHO) has global oversight of over 60 pharmacopoeias and develops international guidance to help harmonize global pharmacopoeial standards.[8] The European Pharmacopoeia Convention (EPC) is responsible for the elaboration of Ph. Eur., and the European Directorate for the Quality of Medicines and Healthcare (EDQM) publishes it.[7,20] In the United States, the US Pharmacopoeia (USP) takes on these roles, but unlike the Ph. Eur., which is a facet of the Council of Europe via EDQM, the USP is a nonprofit company.[8,21]

Pharmacopoeial text is embedded in national legislation, making certain monographs legally binding. For instance, in the EU, compliance with Ph. Eur. is necessary to obtain market approval in EU Member States.[8] In addition, in Europe, there are national pharmacopoeias that apply only to single signatory states, such as in France, Germany, and Switzerland.[5,7] Some countries, like Sweden, Finland, and the Netherlands, have discontinued the use of their national pharmacopoeia in favor of the Ph. Eur.[5,7] Generally, the Ph. Eur. takes precedence and should become part of the official standards in each European country. [5,7]

Given that the Ph. Eur. plays a leading role in the transition to nonanimal methods for pyrogenicity testing, analyzing their pharmacopoeial text is critical to understanding which changes may be needed to move away from RPT and can serve as an example for other pharmacopoeias like the USP. The Ph. Eur. has gradually replaced RPT with nonanimal alternatives—in 2010, it officially included MAT as a compendial method to RPT.[10,20,22,23] In 2016 and 2020, the MAT monograph (2.6.30) was further revised, and in the latter year, EDQM added another nonanimal method called recombinant factor C (rFC).[9,16,22] In 2024, the MAT

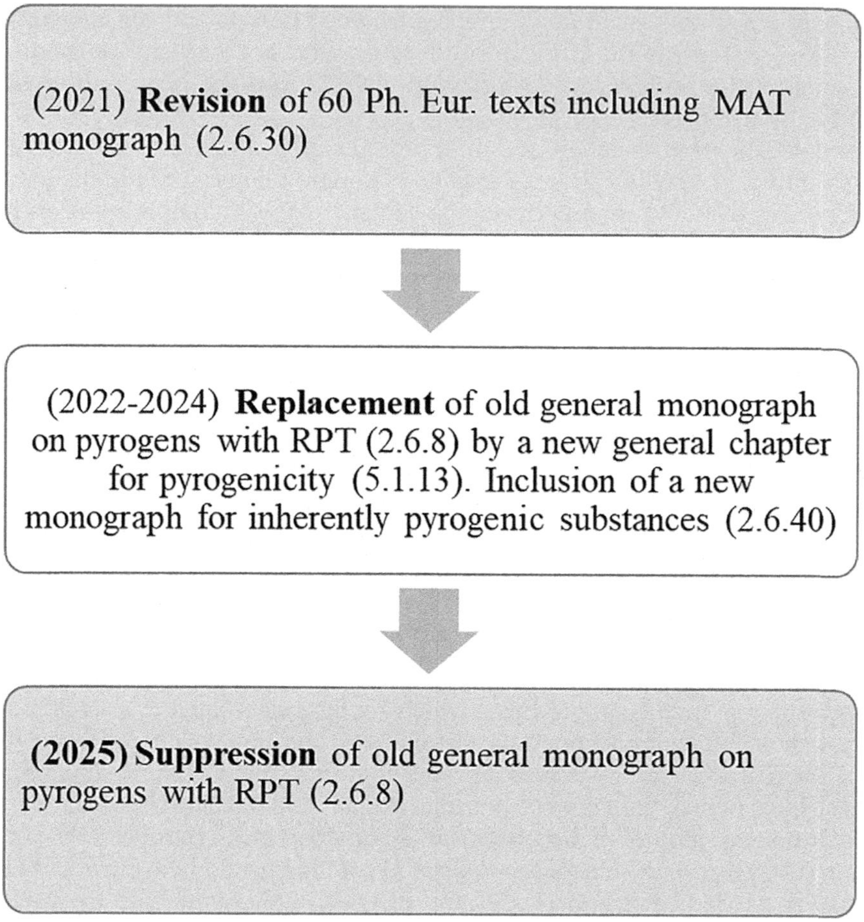

FIG. 1 Overview of the EDQM's strategy to phase out RPT. The original phase-out strategy predicted an end date in 2026; this has since been modified to July 2025.

monograph was further revised, and they developed a new monograph (2.6.40) specifically for vaccines that normally contain pyrogens.[24] That same year, during its 179th meeting, the EPC made a landmark decision to remove the RPT completely, revising 60 texts mentioning RPT and introducing a new general monograph on pyrogen testing (5.1.13), officially ending the use of the RPT in the Ph. Eur. by 2025. An overview of EDQM's phase-out strategy can be seen in Fig. 1.

In addition to pharmacopoeial text, regulations and directives also impact the selection of testing methods. In the European Union, Directive 2001/83/EC established Europe-wide regulations on medicinal products for human use.[25] In 2010, this key regulation was further impacted by EU Directive 2010/63/EU, which discusses the protection of animals used for scientific purposes and establishes a legal obligation for practitioners to use alternatives to animal testing when feasible (i.e., when there is a validated and accepted alternative).[6] In the

United States, the Code of Federal Regulations (CFR) codifies "the general and permanent rules published in the Federal Register by the executive departments and agencies of the Federal Government".[26] General regulations for biologics, including specifications regarding pyrogenicity standards, are described in Title 21 Subchapter F of the CFR 96.[27] While these regulations are more a formality, as manufacturers mostly refer to pharmacopoeias, the CFR and EU Directives may be referred to by regulators to satisfy legal disputes or make regulatory decisions. Thus, analyzing these texts is important to understand if and how pyrogenicity and RPT are embedded in regulation.

Scientific background

Endotoxin and non-endotoxin pyrogens

An "endotoxin" is a toxin that is part of the bacterial membrane, only found in Gram-negative bacteria.[28–30] This was later identified as lipopolysaccharide (LPS), which can cause immune responses in the host, such as fever reactions.[31] Non-endotoxin pyrogens (NEPs), which are attributed to Gram-positive bacteria, have also become recognized as important pyrogens, attributed to components like teichoic and lipoteichoic acid present in their cell wall.[12,32,33]

Pyrogenicity testing

The regulatory methods used to test pyrogenicity are RPT, bacterial endotoxin tests (BET)—which include the limulus amebocyte lysate (LAL) and rFC—and the MAT. The most common and widely accepted method for testing (non-)endotoxin pyrogens is RPT. This test generally requires the intravenous injection of a sample into approximately 3–5 rabbits.[12,15,34] While constrained, their temperatures are recorded via a rectal thermometer for 180 min; the sample passes if the summed temperature change does not exceed a predetermined range.[12,15,34] While this method is globally recognized, it suffers from numerous limitations as further outlined in Table 1.

The BET is a (partial) alternative to RPT and outlines a set of assays used to detect EPs. BETs include LAL and rFC assays.[10,12,15] The LAL assay became popularized after the discovery that LPS reacts with the immune cells in *Limulus polyphemus* (horseshoe crab) blood, causing a clotting reaction.[40] The LAL is a simple, globally accepted in vitro assay and is the preferred method for testing endotoxin pyrogens.[12,22] However, LAL impacts animal welfare and biodiversity because it requires the extraction of 20–25% of hemolymph from horseshoe crabs, which kills them and has contributed to the endangerment of their populations in Japan and the United States.[12,23,40] A nonanimal alternative to the LAL is the rFC, which is an in vitro assay that uses a genetically engineered version of the factor C protein to detect LPS by activating a cascade that cleaves a detectable substrate.[3,10,12,15,43] Both BET assays suffer from a number of limitations outlined further in Table 1. Most notably, BET does not detect NEPs, produces false positive reactions to glucans, and does not reflect the potency differences of different LPS in mammals.[10,12,14,15,35]

In contrast, MAT can fully replace RPT because it can detect both NEPs and endotoxin pyrogens.[10–12,14,22,46] In MAT, human monocytes, which can be collected from various sources (i.e., cell lines, human whole blood, cryopreserved blood, etc.), are assessed in vitro for pyrogenicity.[10,11,38,39,46] Five versions of MAT have been validated and accepted by the EU Reference Laboratory for Alternatives to Animal Testing (EURL ECVAM).[47] Furthermore, MAT can be

TABLE 1 Evaluation of the benefits and limitations of four (regulatory) methods to test pyrogenicity. Some assays (*) are BET assays and thus only detect bacterial endotoxins from Gram-negative bacteria.

Test Method	Description	Benefits	Limitations
Rabbit pyrogen test (RPT)	In vivo test: Intravenous injection of a sample into 3–5 rabbits, after which the change in temperature is recorded and compared to a prespecified range	• Detects all endotoxins • Internationally accepted and harmonized test method • Well established, easy to interpret	• In vivo test using rabbits • High variability; the sensitivity of the test depends on the strain of the rabbit • Detection limit is > 50 pg/mL; humans have a lower fever threshold • Cannot be used to determine the maximum acceptable endotoxin concentration • Cannot be used to detect pyrogens in various products, i.e., cytokine-inducing drugs, steroids, radiopharmaceuticals, etc. • Qualitative test; cannot be used to determine the maximum acceptable endotoxin concentration • Time constraints (>3 days); costly
Limulus amebocyte lysate (LAL) Test*	In vitro test: Detects LPS by causing a clotting reaction using horseshoe crab blood	• Simple in vitro assay • Internationally accepted and harmonized test method • Time-efficient (within hours); cost-effective • Higher sensitivity for endotoxin than RPT	• Suffers from sustainability issues, namely the depletion of horseshoe crabs in the wild • Uses animal material • Does not detect non-endotoxin pyrogens • Interference from sample preparation
Monocyte activation test (MAT)	In vitro test: Detects pyrogenic cytokines (e.g., IL-1β, TNF-α, IL-6) by activating human monocytes, which can be from various sources, i.e., cell lines, peripheral blood mononuclear cells, or whole human blood	• Human relevant; uses human material • Detects all endotoxins • Validated alternative to RPT; compendial in Ph. Eur. • Higher sensitivity than RPT; detection limit is > 10 pg/mL • Time-efficient (>1 day) compared to RPT • Higher accuracy than LAL • Broad application; no cell preparation	• Not internationally harmonized • Time constraints (greater than 1 day) and more costly than LAL • Learning curve; more complex assay than rFC and LAL • Access to human monocytes may be limited
Recombinant factor C (rFC) assay*	In vitro test: Synthetic protein-based assay. Comparable to or better than LAL	• Overcomes ethical and sustainability issues of LALComparable or better performance than LAL	• Not internationally harmonizedSimilar limitations to LAL: does not detect NEPs

See references [2,3,10–12,15,16,23,34–46]

used for a broader range of products than RPT. For instance, it does not react with aluminum hydroxide, which is found in a variety of vaccines.[12,48] However, MAT also suffers from some limitations (Table 1).

Summary of methodology

This chapter employed a multimethod qualitative approach to explore the transition from RPT to MAT (and other nonanimal alternatives), focusing on regulatory, scientific, and historical dimensions. Three main methods were used: a literature review, semi-structured expert interviews, and document analysis. Methodological details can be found in the Appendix to this chapter.

Literature review

A comprehensive literature review was conducted to establish background knowledge on pyrogenicity testing, including its historical evolution, regulatory development, and scientific foundations. Sources included academic literature, government publications, project meetings, webinars, conference materials, and gray literature retrieved through the Overton database. The Overton search strategy and inclusion criteria are detailed in Appendix Parts A and B.

Semistructured interviews

Five semistructured interviews were carried out with experts across regulatory agencies, national control laboratories, academia, and consultancy agencies. Interviews were between 60 to 120 mins. Stakeholders self-identified their sectoral affiliation, and interview questions were tailored accordingly. Interview insights were used to deepen the historical and regulatory analysis, enrich interpretations of pharmacopoeial documents, and explore barriers and drivers in adopting MAT. Details are provided in Appendix Part C.

Document and content analysis

Details are provided in Appendix Parts D through H. The chapter examines relevant regulatory and pharmacopoeial documents from EDQM, the European Union, and the United States. A selection process combined Overton-based search results, expert recommendations, and specific exclusion criteria (e.g., geography, language, document accessibility). Documents analyzed included EU Directive 2001/83/EC,[25] US Code of Federal Regulations (CFR) Title 21 Subchapter F,[27] and select European Pharmacopoeia (Ph. Eur.) monographs (2.6.8, 2.6.30, and 2.6.40).[24,49–52] The USP was excluded due to the absence of MAT as a compendial method.

Two types of content analyses were conducted:

1. Comparative analysis of EU and US regulations to assess textual support for nonanimal testing. A binary scoring system was used to determine whether documents met predefined criteria related to in vitro methods, with a focus on explicit or implicit references to in vivo (especially the RPT) or in vitro tests.

2. Pharmacopoeial text analysis, applying a similar binary framework to evaluate the presence and prominence of in vivo versus in vitro methods across the three Ph. Eur. monographs.

Criteria for both analyses were derived from themes identified in the literature and the documents reviewed. Specific search terms and criteria used for document coding are available in the appendix.

Theoretical framework

To interpret findings in a broader socio-regulatory context, the study adapted a framework (Fig. 2) from Smith and Raven,[19] which considers science, technology, market, policy, and historical-social context as key factors influencing the uptake of nonanimal methods. This framework was applied to map interactions among these dimensions and to visualize the enabling conditions for the adoption of MAT in Europe versus the United States.

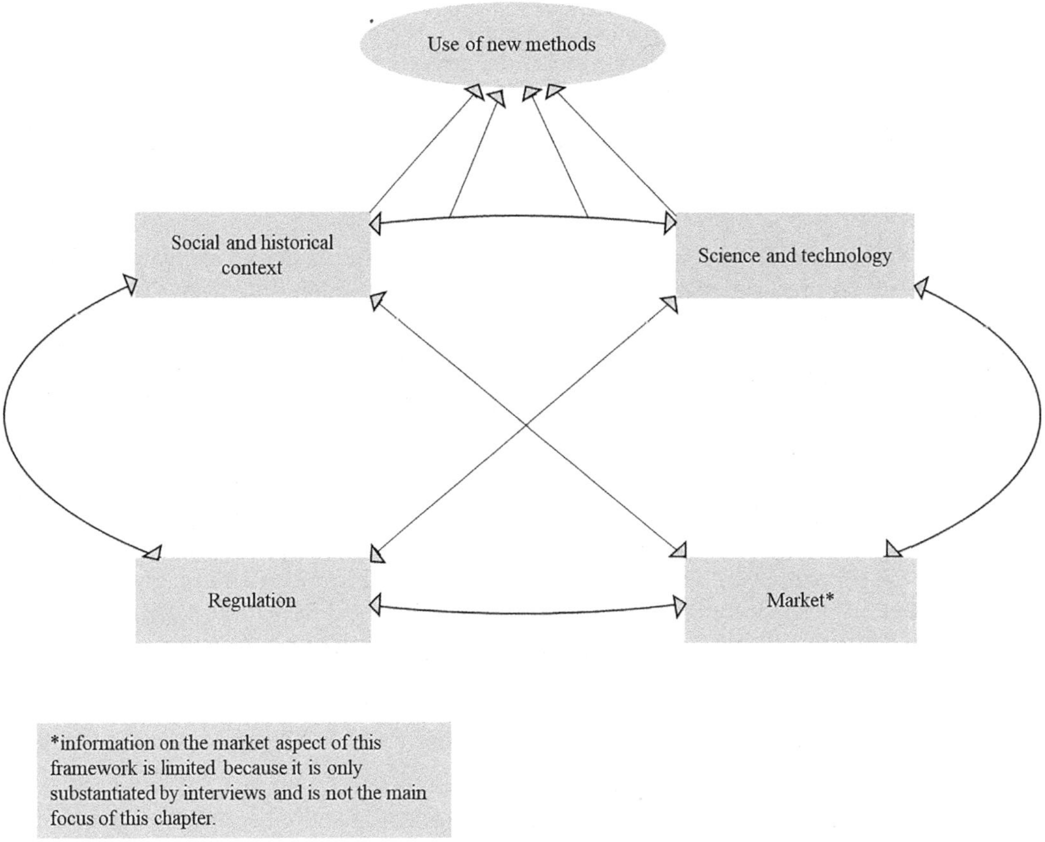

FIG. 2 Theoretical framework to analyze multifactorial elements in nonanimal method implementation with examples from the pyrogenicity case study. Visualization of the impact of social and historical context, science and technology, policy, and market on the implementation of new methods.

Findings

Historical overview of pyrogenicity testing in Europe and the United States

Richard Pfeiffer, a German scientist, first described endotoxin in 1892.[44] The scientific discovery of pyrogens, then called *water* or *injection fever*, began in the late 19th to early 20th centuries after scientific achievements in infusion and injection were popularized by Paul Ehrlich. This was especially catalyzed by the invention of Salvarsan, an anti-syphilis drug (**Interview #5**). This time period marked the first instances of widespread injection technology, which led to the discovery that different batches of biological products had the potential to induce fever in patients (**Interview #5**). In 1912, researchers Hort and Penfold were responsible for establishing RPT as a scientific method to standardize "the pyrogenetic content of a vaccine," citing that rabbits were a better model than previously used guinea pigs because of their ability to have 'relatively stable' temperature measurements.[53] This landmark study set the tone for the next century, and in that same year, the British Pharmacopoeia incorporated RPT into its official text.[10,12] Decades later, in 1942, the USP followed suit, incorporating RPT into USP 'Chapter < 151 > Pyrogenicity Test (RPT)'.[12,15,37] USP's decision followed the events of World War II, where many injured soldiers had transfusions requiring the use of intravenous technology.[15] Notably, the Ph. Eur. would only be established over a decade later in the 1960s,[54] officially incorporating RPT in Monograph 2.6.8 on pyrogenicity in 1971.[16,20] The delayed incorporation of Monograph 2.6.8 in the Ph. Eur. can be attributed to the fact that the predecessors of the European Union only started to become established in the late 1940s, with cooperation between European countries steadily growing over the next four decades.[54]

In 1951, Lüderitz and Westphal's landmark study[29] enabled the further development of novel methods to detect pyrogens. This study characterized the polysaccharide structure of endotoxins, which was, at this point, widely unknown. In addition, this study called attention to differences between Gram-negative and Gram-positive bacteria, which importantly distinguished the BET from RPT and MAT (**Interview #5**). Between the 1950s and 1980s, a series of studies found that horseshoe crab blood coagulated in the presence of endotoxin and could also be used to characterize pyrogens.[55] In the United States, the FDA allowed the horseshoe crab blood-based assay to be used by manufacturers, most notably Baxter Travenol, to validate its use.[42] At this stage, technology was rapidly developing, and the need to standardize methods became even more pertinent. By the 1980s, LAL was internationally recognized and accepted as a compendial method for testing bacterial endotoxins in USP (USP Chapter < 85 >) and the second edition of the Ph. Eur. (Monograph 2.6.14).[56,57]

During this same period, the scientific basis for the MAT was being explored. The landmark study by Dinarello[58] was crucial in MAT development because it translated into their finding of a key fever-inducing reaction in humans. Dinarello[59] used human peripheral blood mononuclear cell (PBMC) activation to show the presence of pyrogens; however, they used the RPT to measure the release of endogenous pyrogen. A few years later, Poole et al.[60] became the first to introduce MAT in PBMC samples to measure pyrogen content. Different groups then aimed to replace PBMC with monocytic cell lines, but these turned out to be more difficult to handle.[12] It is important to note that, at this time, the term "MAT" was still not used. Hartung and Wendel further optimized MAT by using human whole blood, making it the third version of MAT that was developed at the time.[38]

By 2003, five versions of MAT assays (In Vitro Pyrogen Test using Human Cryopreserved Whole Blood/IL-1 Beta, In Vitro Pyrogen Test using Human Whole Blood/IL-6, In Vitro Pyrogen Test using MonoMac 6 cells/IL-6, In Vitro Pyrogen Test using Human Whole Blood/IL-1 Beta, and In Vitro Pyrogen Test using Peripheral Blood Mononuclear Cells/IL-6)[10] successfully underwent formal validation under the EURL ECVAM. The process from submission to acceptance into the pharmacopoeia took approximately 9–10 years, and in 2010, MAT was officially added to the Ph. Eur.[10] Later, in 2017, MAT monograph 2.6.30 was revised to improve accessibility,[10] and in 2020, the rFC assay was incorporated into the European Pharmacopoeia in Monograph 2.6.32.

Only in 2021 did the EPC decide to completely remove RPT from Ph. Eur.,[61] which most likely stemmed from many factors. First, Europe already had a broad regulatory emphasis on promoting the reduction, refinement, and replacement (3Rs) of animal testing, consistent with several European policies (i.e., European Convention for the Protection of Vertebrate Animals Used for Experimental and Other Scientific Purposes; EU Directive 2010/63/EU) seeking to phase out animal testing.[20,62] Second, in 2012, the EDQM conducted a survey with Contract Research Organizations and vaccine manufacturers to determine cases of contamination over the last 10 years. This survey highlighted issues with existing (animal) methods.[41] Third, in 2018, the European Partnership for Alternative Approaches to Animal Testing (EPAA) also conducted a survey, which found that there was little incentive for manufacturers to transition to MAT due to "regulatory burden," highlighting a discrepancy between EU Directive 2010/63/EU (Directive 2010/63/EU is European legislation that regulates the use of animals in scientific research, aiming to improve animal welfare and promote alternatives in line with the 3Rs principle—Replacement, Reduction, Refinement) and pharmacopoeial monographs. Thus, a phase-out of RPT in the legal texts appeared imperative to realize the complete ending of RPT in Europe.[20] Lastly, the public, scientific, and regulatory support for MAT could have acted as an additional driver.[16,20] Effective July 2025, RPT has been phased out of Ph. Eur. and replaced by a new monograph on pyrogenicity (5.1.13).[18]

In contrast, in the United States, the adoption of the Monocyte Activation Test (MAT) into the USP has lagged behind Europe. Although the *USP < 1225 > on validating alternative methods* was introduced in 1989, it was not until 2017 that the USP explicitly permitted in vitro alternatives.[10] This followed a key FDA guidance document in 2012, which identified MAT and rFC as alternatives to RPT.[63] Interestingly, in 2001, the FDA was the first regulatory body to accept the use of MAT for regulatory purposes for a specific pharmaceutical product from Novartis® (**Interview #1**). Further progress includes the USP's decision to add a chapter on rFC, which is expected to be published in November 2025.[64] In addition, the FDA Modernization Act 2.0, passed in 2022, removed the requirement of animal testing for market authorization.[65] While these developments reflect a shifting regulatory landscape, MAT remains non-compendial to RPT in the United States, meaning products require product-specific validation or waivers to use the MAT. This regulatory barrier may discourage widespread adoption of MAT by US manufacturers, in contrast to the European context. A general overview of major historical, social, and scientific changes that occurred in Europe and the United States is outlined in Fig. 3.

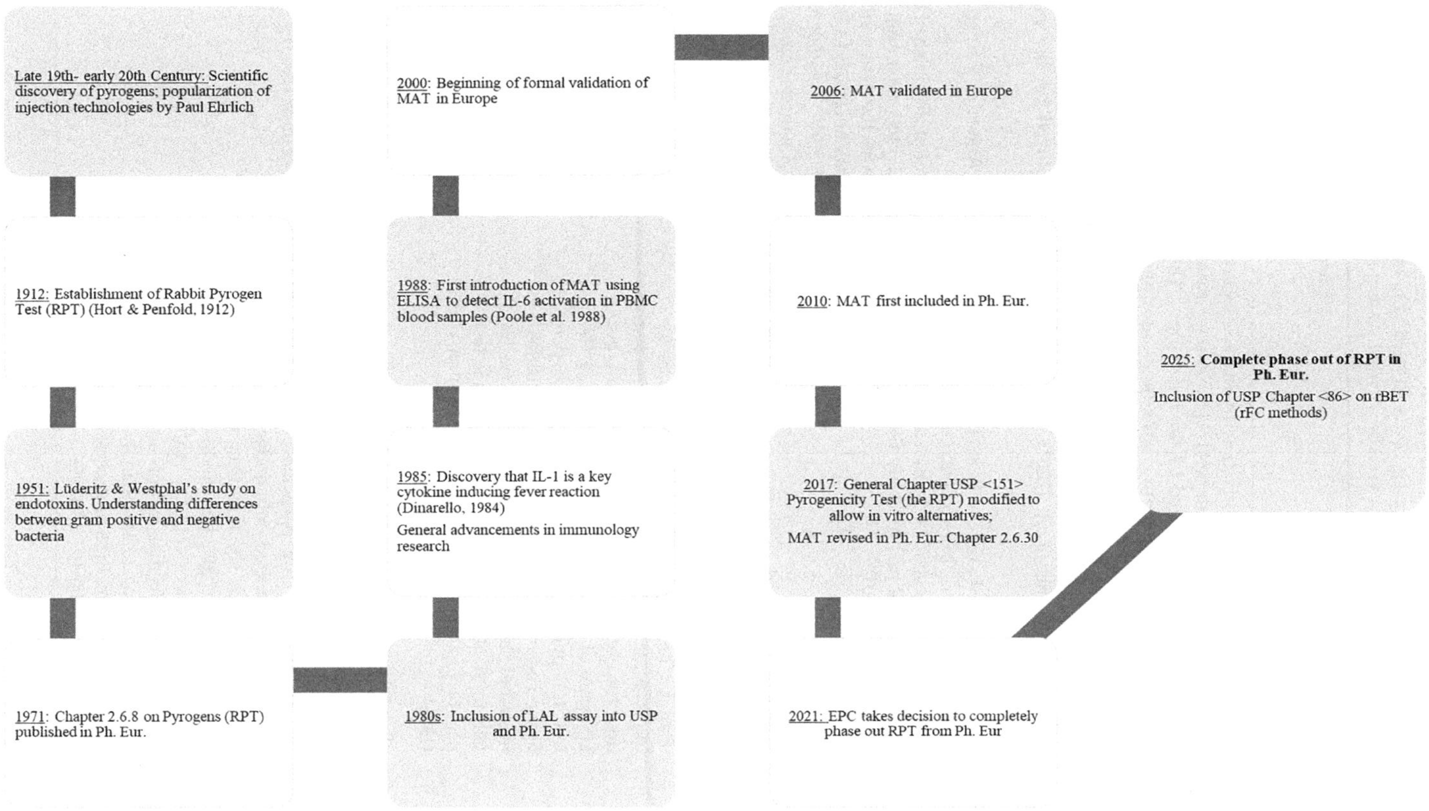

FIG. 3 Timeline of relevant events in the pyrogenicity case study. Historically relevant social, scientific, and regulatory events leading to MAT incorporation in the EU and RPT phase-out in Europe.

TABLE 2 Specific text related to pyrogenicity testing in Directive 2001/83/EC and CFR Subchapter F and their relative change over time. Text from 2022 for Directive 2001/83/EC has remained consistent since its last revision in 2004. Specific changes related to in vivo or in vitro testing for pyrogenicity are underlined.

Regulatory provision (Regulation, Annex/ Subchapter)	Past text (year of version)	Current text (year of version)
EU Directive 2001/83/ EC Article 8	"[…] To obtain an authorization to place a medicinal product on the market […] the application shall be accompanied by the following particulars and documents, submitted in accordance with Annex I […] (h) Description of the control methods employed by the manufacturer (including) […] tests for the presence of pyrogenic substances […]" **(2001)**	Complete deletion of the term *"pyrogen"* from the Directive text. **(2022)** "The (market authorization) application shall be accompanied by the following particulars and documents, submitted in accordance with Annex I: […] (h) Description of the control methods employed by the manufacturer." **(2022)**
EU Directive 2001/83/ EC Annex I, Part 2, F, 1.4, 1	"[…] particulars of safety tests, such as sterility, bacterial endotoxin, pyrogenicity and local tolerance in animals shall be included in the analytical particulars […] to verify the quality of the product." **(2001)**	Replaced by Annex I, Part 1, 3.2.2.5 **(2003)** "For the control of the finished medicinal product […] the maximum acceptable deviation in the active substance content […] shall not exceed ± 5% at the time of manufacture. Detailed information on the […] methods of analysis and their validation shall be provided."—Annex I, Part 1, 3.2.2.5 **(2022)**
US CFR Subchapter F 610.13 Purity	"[…] any product intended for use by injection shall be tested for pyrogenic substances by intravenous injection into rabbits […] the test dose for each rabbit shall be at least 3 mL per kilogram of body weight of the rabbit […] The test for pyrogenic substances shall be performed according to the requirements specified in United States Pharmacopeia XX […] If the lot fails to meet the test requirements […] the test may be repeated once using five other rabbits" **(1996)**	No significant change **(2023)**
US CFR Subchapter F 600.3 Definitions	"Purity includes but is not limited to relative freedom from […] pyrogenic substances" **(1996)**	No significant change **(2023)**
US CFR Subchapter F 610.9 Equivalent methods and processes	"Modification of any particular test method […] shall be permitted only under the following conditions: (a) The manufacturer presents evidence, […] that the modification will provide assurances of the safety, purity, potency, and effectiveness of the biological product equal to or greater than the assurances provided by the method or process specified in the general standards … " **(1996)**	No significant change **(2023)**

TABLE 3 Binary analysis of European and US regulations.

Criteria	Past EU Directive 2001/83/EC	Current EU Directive 2001/83/EC	Past US CFR Subchapter F	Current US CFR Subchapter F
Does **not** mention RPT explicitly	Yes	Yes	No	No
Does **not** refer implicitly or explicitly to any in vivo method in the text	No	Yes	No	No
Mentions MAT in text	No	No	No	No
Refers to any in vitro alternative in the text	No	No	No	No
Only refers to MAT in the text	No	No	No	No
Only refers to an in vitro alternative in the text	No	No	No	No
Nonanimal alternatives to (in vivo) test methods are allowed **and** explicitly expressed in the text section	Yes	Yes	Yes	Yes

Comparative analysis of US and European regulations on pyrogenicity

Comparing the two regulatory texts—Title 21 Subchapter F of the US CFR and EU Directive 2001/83/EC—the latter has undergone important changes between its first introduction in 2001 and its most updated revision in 2004, while the former has not undergone significant revision since its introduction in 1996 (Table 2). The EU Directive opted to remove the term "pyrogen" from its text in 2004, instead keeping the text vague and placing responsibility on the manufacturer to determine the methods they use for QC tests. In contrast, the CFR specifically mentions pyrogenicity as a required standard test (Table 3).

Even though CFR Section 610.9 of Subchapter F specifically states that alternative methods are permissible, RPT is still embedded in CFR Section 610.13 (Table 3). When we evaluate the EU Directive for its alignment with supporting nonanimal alternatives, we see there is only one distinct difference between the 2001 and 2004 text; more specifically, except for not mentioning in vivo methods, the EU Directive remains vague on the use of (nonanimal) alternatives for purity testing.

Analysis of the European pharmacopoeia transition to MAT

Analyzing the text within different versions of Monograph 2.6.8 (referencing RPT) and Monograph 2.6.30 (on MAT) explicates how the EPC has tackled the phase-out of RPT. As suggested by the National Center for the Replacement, Refinement, and Reduction of Animals in Research,[43] the use of or reference to in vivo terms or test methods can act as a barrier to the use of in vitro alternatives, suggesting that the language of documents like pharmacopoeias can impact the choice of test methods and regulatory opinions. The EPC has made significant changes to the MAT monograph between 2017 and 2024 by removing references to RPT altogether (Table 4).

Although there were no significant changes to the RPT monograph between 2016 and 2020, the new general monograph (5.1.13) will completely replace the RPT monograph and will only focus on in vitro alternatives.[18] Both new monographs on MAT, which are effective July 2025, completely remove the word "rabbit" from their text apart from one instance referring to historical data:

> "If available, historic batches found to be contaminated with non-endotoxin contaminants that caused positive responses in the rabbit pyrogen test or adverse drug reactions in humans should also be included."[24]

In addition, even though both reviewed versions of the RPT monograph refer to MAT, it is important to note that it is only referenced in the final sentence of the general monograph:

> "Wherever possible, and after product-specific validation, the pyrogen test is replaced by the monocyte-activation test (2.6.30)."[49,50]

This indicates that although MAT has been integrated into the Ph. Eur. for many years (since 2010), RPT was still referenced as the standard test method in the text. Referencing and including MAT in a separate monograph was not enough because the general monograph on pyrogens (2.6.8) still upheld RPT as the gold standard. The EPC's decision to remove the general monograph 2.6.8, develop a new general monograph on pyrogens (5.1.13), include and revise monographs on MAT by removing any in vivo-related terminology, and revise all other monographs referring to 2.6.8, offers great insight into the regulatory needs for test method phase-out.

TABLE 4 Binary analysis of European pharmacopoeial text related to RPT and MAT between 2016 and 2024.

Criteria for animal-friendly chapters in the Ph. Eur.	2.6.8 (2016)	2.6.30 (2017)	2.6.8 (2020)	2.6.30 (2024)	2.6.40 (2024)
Does not refer to RPT or in vivo testing (excluding historical analysis)	No	No	No	Yes	Yes
Refers to MAT or in vitro-related methods	Yes	Yes	Yes	Yes	Yes
Refers to MAT or in vitro-related methods more than RPT or in vivo testing	No	Yes	No	Yes	Yes
Refers primarily (>50%) to RPT only in the context of in vitro or other related terms	No	Yes	Yes	Yes	Yes
Clearly indicates that the focus of the chapter is on pyrogenicity and/or endotoxin testing	Yes	Yes	Yes	Yes	Yes

Discussion

Comparing US and EU transitions to MAT

When comparing Europe to the United States in the pyrogenicity transition, it is clear that the United States is far behind (Table 5).

Europe has integrated MAT into the pharmacopoeia, making it a compendial method. EU legislation has become more neutral toward animal testing and, in theory, necessitates the replacement of animal studies where possible.[66] Generally, the relevant text of EU Directive 2001/83/EC [67] is more neutral than the US CFR because it does not refer to any in vivo methods nor specify pyrogenicity as a mandatory test. The neutral language of the Directive likely has a positive effect because it creates room for deviations from standard methods and does not automatically signal to regulators that there is a singular gold standard, but instead encourages case-by-case review (**Interview #5**).

In general, the United States is thought to be 10–15 years behind Europe in this transition (**Interview #1**). The reasons for this can be attributed to many factors. First, according to the USP, the United States lacks appropriate data to transition to MAT. Even though data is available from the Ph. Eur., in the United States, they require additional product-specific validation to implement MAT into the US regulatory framework—which differs from the European approach.[68] However, the USP is laying the foundations for a MAT monograph and encourages companies to share their data.[69]

In addition, both the Ph. Eur. (1971) and EU Directive 2001/83/EC are much younger than texts from the USP (1942) and US CFR (1996). The Ph. Eur. only incorporated RPT in the 1980s, after LAL methods were known and accepted. In the United States, RPT was embedded in the USP since the 1940s, four decades before LAL was recognized. This could explain why the US CFR, but not Directive 2001/83/EC, explicitly refers to RPT. Even though Directive 2001/83/EC originally stated animals should be used, the text was adapted only two years later, citing such changes were due to scientific and technical progress, changing definitions, and a focus on harmonization to allow for the assessment of products placed on the market. In contrast, instead of overhauling older legislative text, the FDA appears to develop new text that allows flexibility through waivers (i.e., 21 U.S. CFR 610.9) and provides guidance for industry.[68] In general, legislature referencing RPT (and in vivo tests) may be more resistant to change in the United States than in the EU because US regulatory texts are more longstanding, which in turn embed RPT into legislation, creating regulatory inertia.[70]

However, inertia is not the only barrier. There have been opportunities in the United States to further this transition, but attempts have been thwarted. Even though the FDA was the first

TABLE 5 Comparing differences in MAT transition in Europe and the United States.

Europe	United States
• Current version of EU Directive 2001/83/EC does not refer to the RPT or the use of animals in textCurrent version of EU Directive 2001/83/EC does not refer to the testing of pyrogens	• Current version of the US CFR refers specifically to the RPT in text • Current version of US CFR refers to the testing of pyrogens as a standard and required test
• MAT has been compendial to RPT for over a decade	• MAT is not compendial to RPT
• RPT is undergoing phase-out from the European Pharmacopoeia	• RPT is not undergoing phase-out from the USP

regulatory body to accept MAT for regulatory use (**Interview #1**), in 2018, the FDA rejected the opportunity to fund an application to formally validate MAT.[10] After a change in the organization, the new Microbiology Working Group is investigating ways to incorporate alternatives like the MAT into the USP (**Interview #1**). This is further demonstrated by USP's move to include rFC in a compendial monograph and their organization of a workshop on endotoxins in May 2024.[71] Overall, this demonstrates that the USP and relevant FDA authorities have, at different times, shown both interest in and resistance to the formal incorporation of MAT methods into pharmacopoeial and regulatory texts, with emerging advancements favoring in vitro methods.

In addition, in the United States, LAL may have acted as a barrier to MAT uptake. Here, unlike in Europe, there is an economic interest in the LAL assay (**Interviews #1 and #2**). An estimated 4–5 companies have an FDA license to produce this assay, representing a billion-dollar industry[14] (**Interviews #1 and #2**). The global pyrogen testing market was estimated to be worth $927 million in 2020 and is poised to reach $1689 million by 2025.[72,73] Decreased (economic) interest in the rFC could act as a barrier to other alternatives in pyrogen testing, namely MAT, because these test methods and transitions are interrelated. In transformative governance literature, transitions can be interrelated such that they can suffer from similar underlying hindrances to their acceleration.[74] Applied to the MAT transition, focus on maintaining LAL for economic gain could reveal underlying sentiments such as an overall lack of interest in animal ethics. This same relationship could also explain why the past USP committee was more resistant to alternatives in the pyrogenicity sector (**Interview #1**). In general, this could indicate economic interests—such as the large US horseshoe crab (*Limulus*) industry—have played a role in the current status of MAT in the United States. However, further investigation would need to be completed to substantiate this hypothesis, and generally, the emerging advancements in the United States indicate a changing tide, which could be positive for the status of MAT.

In contrast, in Europe, MAT has been formally validated by EURL ECVAM and was accepted by the Ph. Eur. in 2010, with increasing initiatives to further prioritize MAT over the next decades.[10,16,20] Importantly, the prioritization of MAT in Europe can be attributed to Directive 2010/63/EC, which mandates animal alternatives when methodologically feasible. While MAT was already compendial to RPT in the Ph. Eur. by January 2010—seven months before Directive 2010/63/EC went into effect—the new directive created a legal obligation for European organizations to stop the use of animals for pyrogen testing since BET and MAT methods were already compendial. Glaringly, this legal incentive is missing in US legislation, even with the passing of the FDA Modernization Act 2.0, because it does not *obligate* manufacturers to default to nonanimal methods, making it less stringent than Directive 2010/83/EC.

Overall, when comparing the MAT transition in Europe and the United States, the following factors differentiate the two regions: longstanding regulatory inertia and differing economic and regulatory incentives. In the United States, deeply ingrained RPT legislation, hesitation toward investment in nonanimal methods, the interrelated nature of the LAL market, a lack of regulatory incentive to use nonanimal methods, and lack of formal validation of MAT are some of the reasons explaining why the United States lags behind Europe. Still, it is likely other factors are at play, and further investigation would be necessary to unearth these interaction effects.

Analyzing MAT transition in the European Pharmacopoeia

Even though Europe has successfully decreased rabbit use for pyrogenicity,[17,75] the process to implement MAT and phase out RPT took nearly 40 years from the first introduction of

MAT in 1988. Understanding why this transition was so slow is critical. Although this chapter cannot comprehensively address all the drivers and barriers to MAT implementation, it reveals the interplay between key scientific advancements, historical events, economic influences, and regulatory changes.

First, the use of MAT in Europe, even after its implementation into Ph. Eur. in 2010, was relatively low among manufacturers, which could be in part because MAT was not a fully fledged monograph until 2017 (**Interview #1**). There was also an increased focus on the inability of RPT to properly detect contamination. In 2012, a survey conducted by EDQM revealed gaps in compendial methods and the poor quality of in vivo data to predict contamination over a ten-year period.[41,76] This created further incentive to move toward more scientifically appropriate methods.[41] In 2018, the EPAA conducted a survey revealing that there was little incentive among stakeholders to replace RPT due to discrepancies between EU Directive 2010/63/EC and Ph. Eur. monographs, which prompted the EDQM's decision to remove RPT to ensure the exclusive use of BET or MAT methods.[20]

The initial low use of MAT among European stakeholders could be attributed to the lower availability of MAT test kits and the existence of patented technology (**Interview #2**). It is likely that the development of a wide variety of MAT kits using various monocyte sources was prompted by the standardization of reagents, international recognition of MAT, and the expiration of specific patents that existed in MAT technology earlier, among other influences (**Interview #2**). The global MAT market is growing: in 2022, it was projected to be worth \$60.7 million, and it is projected to increase to \$305.3 million by 2031.[73]

Acknowledging the regulatory, scientific, economic, social, and technical opportunities and hindrances to the pyrogenicity transition reveals that it was likely slow because the original phase-in of MAT was not enough to discourage the use of RPT. Incorporation of clear, specific monographs on MAT, increased availability of testing kits, awareness of the scientific limitations by regulators, standardization of reagents, increased demand for nonanimal alternatives by EU legislation, coupled with the phase-out of RPT, have all finally aligned to make MAT a true replacement for RPT in the European context. Thus, this case study demonstrates that transitions to nonanimal methods are multifaceted and dynamic, relying not only on the phase-in of nonanimal approaches but also on the phase-out of archaic in vivo methods.

Transitioning to nonanimal methods in the broader context: lessons learned from the pyrogenicity case-study

The transition to MAT provides valuable insights into the complexities that interact and shape transitions to nonanimal methods. Social and historical aspects especially appear to impact or catalyze legislative action and scientific research, as depicted in Fig. 4. The implementation, use, and/or propagation of a new method appear directly driven by regulation and the market, where regulations describe what is legally possible or necessary, while the market reflects what is available. Fig. 4 provides a simplified visualization of the major interactions explored in this research.

In general, impacting several key domains may accelerate transitions to new methods. In the regulatory context, a lack of (global) harmony within regulation acts as a fundamental barrier (**Interviews #2 and #4**). For instance, if a product is tested to support global markets

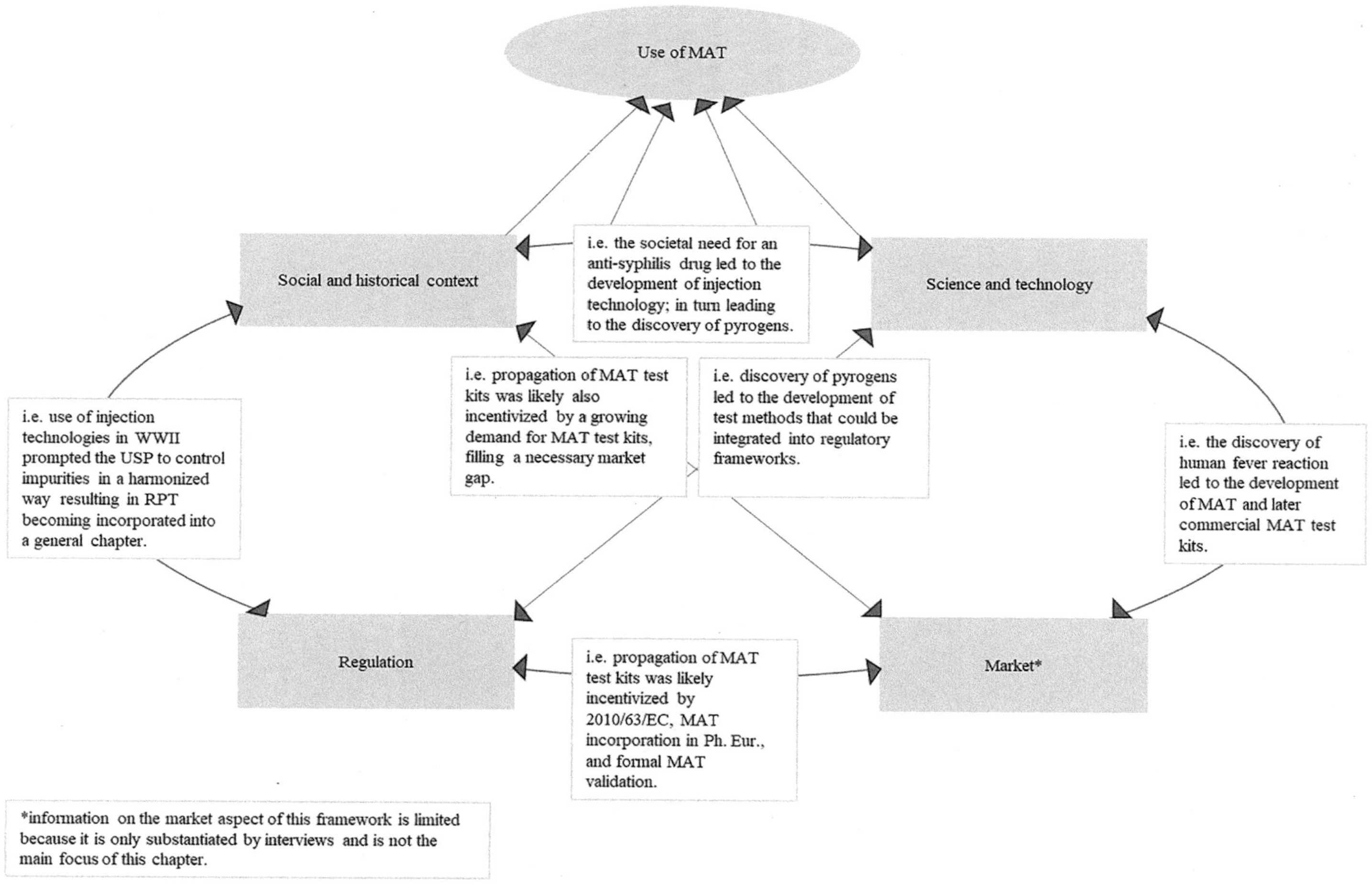

FIG. 4 Theoretical framework to analyze multifactorial elements in nonanimal method implementation, filled in with examples from present research.

(i.e., vaccines) and one region requires RPT, then companies will likely still use RPT because the priority is to deliver products to patients in all regions (**Interview #2**). To overcome this, the development of international working groups and general forums to discuss with diverse stakeholders should be established (**Interviews #2 and #4**). In addition, there should be increased emphasis on animals as a last resort (**Interviews #4**); e.g., the FDA should adopt language similar to Directive 2010/63/EC. There should also be increased enforcement of these principles by authorities (**Interviews #1 and #4**). For instance, in 2018, the European Commission enforced Directive 2010/63/EC by sending a letter to a German organization stating they were no longer allowed to grant general permissions for RPT use unless submitted dossiers established a five-year plan to phase out RPT.[22] For successful transitions, authorities must enforce rules, and they should establish consequences when these rules are not followed (**Interview #4**). Therefore, it is advised that relevant European and US authorities should disincentivize animal use through increased surveillance and repercussions for using animal methods where (validated and accepted) alternatives exist.

Finally, based on this case study, there was a clear pattern of regulation being reactionary. To move transitions further and faster, proactive regulatory change could decrease regulatory inertia. For instance, the RPT phase-out was found to be important to the phase-in of MAT only decades after MAT implementation into the Ph. Eur. Thus, initiating or developing an effective phase-in strategy should be done in parallel with phase-out strategies, such that all relevant text is reviewed, edited, and implemented at once rather than in isolated spurts.

In the scientific-technical realm, data transparency and availability act as barriers (**Interview #4**). Lack of data is cited as a barrier to the transition to MAT in the United States.[69] To overcome these barriers, safe spaces for data exchange (i.e., regulatory sandboxes) should be developed and promoted as collaborative efforts between relevant stakeholders (**Interview #2 and #4**). In addition, a lack of knowledge of new methods can be limiting.[70,72] For instance, the interpretation of results of a MAT assay versus an RPT or BET test requires more specific knowledge and training (**Interview #1 and #2**). Thus, the creation of learning centers, webinars, and/or investment in education for relevant manufacturers and legislators on new methods is an important step and could be developed by national and international organizations like EDQM, USP, and WHO.

Finally, in the market realm, the availability and supply of specific technology can act as a barrier (**Interview #1**). To overcome this, there should be greater incentives for industry and academia to develop nonanimal methods and for industry to propose new approaches (**Interview #4**). For instance, authorities should allocate more funding toward organizations developing nonanimal methods or create incentives for innovative nonanimal methods. The involvement of regulators from the beginning should, in turn, increase the acceptance of new methods and provide some guarantees that the new test method will become validated and eventually accepted.

Conclusions

The MAT transition shows a clear divergence between Europe and the United States. Europe has successfully integrated nonanimal methods like MAT into its pharmacopoeia and legislation, promoting a shift away from animal testing. EU Directive 2001/83/EC is more neutral than the US CFR, as it does not explicate specific in vivo methods and allows greater flexibility for the adoption of new technologies. For this particular case study, the United States lags behind by approximately 10–15 years, generally due to the entrenched use of the RPT and a slower legislative response to advancements in alternative methods.

Despite Europe's success with MAT, the transition was slow, taking nearly four decades from MAT's introduction. Initially, in 2010, MAT use was low due to its lack of a comprehensive monograph until 2017. The decision to phase-out RPT unfortunately took another four years, meaning RPT remained embedded in the Ph. Eur. and in use by manufacturers, revealing discrepancies between EU directives and existing monographs (as further explicated by EPAA's survey in 2018). Therefore, the phase-in and use of MAT in favor of RPT were facilitated by increased regulatory incentives (i.e., 2010/63/EC, revisions of Monograph 2.6.30, etc.), increased attention to regulatory discrepancies leading to the phase-out of RPT, validation and acceptance of MAT technology, and an increased availability and market for MAT test kits, among other factors.

Overall, the transition to MAT underscores several factors that influence the adoption of nonanimal methods:

1. **Proactive Regulatory Action:** Regulatory bodies should proactively plan and implement phase-in and phase-out strategies in parallel to minimize inertia. Reactive regulation should be minimized, and instead, regulation should be developed on a scientific and practical basis (i.e., 2012 EDQM survey, 2018 EPAA survey).
2. **Regulatory Flexibility for Nonanimal Alternatives**: There should be flexibility in regulation to allow for alternative methods. Keeping the text vague may enable manufacturers to use nonanimal methods and may discourage regulators from defaulting to requesting in vivo methods in regulatory decision-making.
3. **Regulatory Incentive to Develop Nonanimal Alternatives**: The use of animal methods should be explicitly referred to as a last resort and implemented into regulation to encourage the development of nonanimal methods. When nonanimal methods are available, this regulation should be enforced by relevant authorities. Eventually, such incentives should have global reach and be encouraged through international collaborations, which can be facilitated by organizations like the WHO. Global regulatory alignment is crucial, and discrepancies between regions can hinder adoption, as seen with manufacturers continuing RPT use for products marketed globally.

Future research should investigate the economic impacts of transitioning from RPT to MAT, examine, through surveys and interviews with stakeholders, resistance (or lack thereof) to alternative methods, explore nonanimal alternatives, and study the interrelation between different transitions in nonanimal testing methods (i.e., MAT and rFC) to identify common barriers and opportunities that could inform broader regulatory transitions. By addressing these areas, future research can provide a clearer framework for faster (global) adoption and use of nonanimal methods.

Acknowledgments

Justine Watkins would like to extend her heartfelt gratitude to the individuals who contributed their valuable time and insights to this chapter. Special thanks go to Dr. Thomas Hartung, Dr. Ingo Spreitzer, and Thierry Decelle for their enlightening interviews, which provided crucial perspectives on pyrogenicity testing. Their willingness to share their expertise and experiences greatly enhanced the depth and quality of this work. She is also grateful to the other contributors and interviewees whose contributions, though not all mentioned by name, were equally significant in shaping the arguments and findings presented here.

This research is part of the SAFE project (NWA.1395.20.004), supported by the Dutch Research Council.

Appendix

Part A List of relevant webinars and trainings

- Humane Society International (HSI); Webinar: Transition to nonanimal based vaccine batch release testing. Online. 27 March 2024.
- 2024 EDQM virtual training program: Ph. Eur. Texts Related to Biologicals and Microbiology Chapters. Online. February 1, 2024.
- EDQM Training: Modules 1–4. Online. January 30–February 1, 2024.
- USP Webinar: USP Modern Methods for Endotoxins and Pyrogens Testing Virtual Workshop, 14–May 15, 2024.

Part B Terms used to find relevant text in Overton

To identify text related to pyrogenicity, the term "pyro" was used to filter relevant text from Subchapter F in the CFR and EU Directive 2001/83/EC. In addition, as the term "purity" encompasses "pyrogenicity" in the official definition of the word (see CFR 600.3 Definitions), "purity" was also used as a relevant search term in the US CFR. If the term pyrogenicity was no longer used in later versions of the text, sections of the text that originally included the term were referred to, and the new text was analyzed. Text discussing containers, water, bleeding of animals, etc., was not included because of its relation to BET, which is less relevant for the purposes of this study. In addition, only text discussing standards for biological products was further reviewed.

In the sections deemed relevant, the terms "rabbit" and "animal" were used to determine if in vivo or RPT test methods were referenced. Other terms that explicitly referred to in vivo testing (i.e., in vivo, rodent) did not appear in relevant texts. After manual review, text was highlighted, copied, and checked in all subsequent versions of the regulation. Any changes to the text or sections were documented, including the year of change and specific changes to the text. A subset of these changes is presented in the section Comparative analysis of US and European regulations on pyrogenicity. In addition, to account for regulatory changes in legislation referencing pyrogenicity, current and archived text from EU Directive 2001/83/EC and CFR Subchapter F is listed in the chapter.

Part C Semistructured interviews

Five semi-structured expert interviews were conducted, ranging between 60 and 120 min. Experts represented different stakeholders, such as regulators, national release authorities (or national control laboratories), academics, legal practitioners, consultants, and experts in animal alternatives for pyrogenicity. All experts were requested to self-identify with a stakeholder category of their choosing. A list of questions was prepared for each expert to facilitate the conversation. Questions were tailored to each expert's stakeholder category and thus differed between experts. Interviews were used to inform the historical overview, literature review, and enrich the analysis of historical, scientific, regulatory, and pharmacopoeial texts. Furthermore, interviews offered unique insights into the opportunities and barriers to nonanimal methods and were used to further develop the themes explored in the document analysis.

Interviewee Number	Name	List of Prepared Questions
1	Dr. Ingo Spreitzer	• Can you briefly describe your background, current role, and general relationship to pyrogenicity? • What are the main barriers to ending RPT once and for all in the EU or US? • What do you consider the most influential text for batch safety testing in the EU or US? • Based on data obtained from the Alures database I saw that 11/28 EU countries + Norway continue to use RPT, Germany being one of them, could you speak on why this may be?
2	Thierry DECELLE	• Can you briefly describe your background, current role, and general relationship to pyrogenicity? • What are the main barriers to ending RPT once and for all in the EU or US? • What do you consider the most influential text for batch safety testing in the EU or US? • Based on data obtained from the Alures database I saw that 11/28 EU countries + Norway continue to use RPT, Germany being one of them, could you speak on why this may be?
3	P.T. Theunissen, PhD, ERT	• What is the current structure of the regulatory system? • What is EMAs position within the regulatory system? More specifically "who" or "what" controls which tests are required by law within the EU (esp. in relation to biologics)? • If I was a manufacturer of a vaccine (tetanus for instance) what am I most likely to or advised to follow in regard to vaccine/biological quality and safety assessment? • How does EMA cooperate with EDQM? NCAs? ICH? WHO? other international bodies like FDA? • What new initiatives are EMA involved in relating to 3Rs, NAMs or non-animal alternatives this upcoming year?
4	Julia Pochat	• Can you tell me a little about yourself such as your general background and your position within Eurogroup for animals? • What is your specific involvement within the pharmaceutical legislation? • What are some common trends that you see? For instance, is 3Rs more readily discussed than reduction and animal testing as a last resort? • Based ongoing changes in pharmaceutical legislation at the EU level, what is your opinion on the ability for these new revisions to impact animal numbers? For instance, does the April 26th Reform of the EU pharmaceutical legislation go far enough? • What are some current/ongoing revisions in pharmaceuticals that you know about, that were not in my email headed by EU bodies i.e. EMA, EDQM, EU commission, DG SANTE? • Where are some possible opportunities for intervention? What are major barriers? • Do you have general knowledge or affiliation with the EU PHR. or EDQM? If so, do you have an idea of what would be considered "influential text" for quality control? • Are you aware of/ have knowledge of pyrogenicity testing? IF So, are you aware of ongoing work to remove RPT from pharmacopeial text? What do you think this will achieve?
5	Thomas Hartung	• On the removal of RPT: What do you think this will achieve? • Do you have any relevant text that can help me understand the nature of pyrogens? • What is the history of pyrogenicity testing? • What do you think are reasons for rabbits still being used for EU purposes/EU legislation? • What do you think would be emergent themes in legislative or pharmacopeial text regarding pyrogenicity testing? • *Do you have general knowledge or affiliation with the USP or Eur Ph.? If so, what would be considered "influential text" for quality control?* Which "influential text" are legally binding?

Expert #	Name	Affiliation	Stakeholder Category
1	Dr. Ingo Spreitzer	Paul-Ehrlich-Institut (PEI)	National Release Authority
2	Thierry Decelle	DCL Solutions	Consultant
3	P.T. Theunissen, PhD, ERT	CBG-meb	Regulatory
4	Julia Pochat	Eurogroup for Animals	–
5	Thomas Hartung	Johns Hopkins University	Academia

Part D Identifying relevant EU and US regulations and pharmacopoeial texts

To determine relevant regulatory and pharmacopoeial text, a combination of approaches was utilized, including a comprehensive search in Overton (https://www.overton.io/) and expert-based recommendations via interviews and webinars. To search for relevant policy-related literature in Overton, we used the search term "rabbit pyrogen," which returned 77 results. Duplicates were removed and text was excluded if outside of geographic range of interest (Europe and United States), was not published in English, did not regard biologicals i.e. medical devices, or if text was not accessible. A total of 33 sources remained after exclusion criteria were applied. Based on the review of these resources and information obtained via experts, a selection of relevant regulatory and pharmacopeial text was chosen for further analysis see table below:

Jurisdiction	Key Regulation	Key Pharmacopeial Text
European	European Directive 2001/83/EC on medicinal products for human use	2.6.8 General Chapter on Pyrogens (RPT) 2.6.14 BET (LAL assay) 2.6.30 MAT 2.6.32 BET (rFC assay) NEW: 2.6.40 MAT for inherently pyrogenic substances. DRAFT: 5.1.13 General Chapter on Pyrogens
United States	United States Code of Federal Regulations (CFR) Subchapter F, 600-680 on Biologics.	USP General Chapter <1225> Validating Alternatives USP Chapter <151> Pyrogenicity Test USP Chapter <85> BET

Relevant European and US regulatory and pharmacopoeial text on pyrogenicity testing

The USP was not further reviewed, as this research seeks to study the successful implementation of MAT into pharmacopoeial text, and MAT is not compendial in the USP. Furthermore, as BET assays are not a full replacement for RPT, like MAT, any text referencing BET was not reviewed. Only text regarding RPT, MAT, and general pyrogen monographs was analyzed further. Monograph 5.1.13 of Ph. Eur. was not further analyzed because, at the time of analysis, this monograph was still in draft format. Access was granted to two versions of both Ph. Eur. Monographs 2.6.8 and 2.6.30 from 2016 and 2020, and 2017 and 2024, respectively. Monograph 2.6.40 only had one version from 2024. CFR documents ranged from 1996 to 2024, while text from Directive 2001/83/EC ranged from 2001 to 2024.

Content analyses

Two types of content analyses were conducted:

1. A comparative analysis of US CFR Subchapter F, 600-680 (1996), and EU Directive 2001/83/EC, and
2. An analysis of Ph. Eur. Chapters 2.6.8(Pharmacopoeia, 2017a, 2016), 2.6.30(Pharmacopoeia, 2017b, 2024), and 2.6.40(European Pharmacopoeia. 2.6.40 Monocyte-Activation Test For Vaccines Containing Inherently Pyrogenic Components, 2024).

The comparative analyses

To identify relevant text and sections from the relevant text, specific terms were chosen and used to filter relevant text from Title 21, Subchapter F of the U.S. CFR and EU Directive 2001/83/EC. Both current and archived texts of the relevant legislation were reviewed. The relevant texts reviewed were: EU Directive 2001/83/EC: Article 8, Annex I (Part 2, F, 1.4, 1) and Annex I (Part 1, 3, 2.2, 5); and U.S. CFR Subchapter F sections 610.13, 600.3, and 610.9.

To analyze the relevant text, a list of nine criteria was developed by identifying themes in (policy) documents that explicitly discussed the replacement of animal models. All text used to derive these criteria was from documents identified in Overton.

Using these criteria, text was ranked on a binary basis ("yes" or "no"), expressing whether a criterion was or was not met. For instance, regulatory text that would be considered highly animal-friendly and in support of in vitro test methods would score "yes" for all criteria. A binary scale was used to allow for a comparison between European and US regulations.

Part F: The terms of interest used to analyze pharmacopoeial text

Search terms were identified and found in pharmacopoeial text using the "command F" function of a MacBook Air (M1, 2020):

> ("rabbit" or "in vivo"), ("alternative" or "replace"), ("monocyte-activation test" or "in vitro" or "MAT" or "human"), ("2.6.30"), ("pyrogen"), ("gram-positive"), ("gram-negative"), ("endotoxin"), ("non-endotoxin"), and ("2.6.8").

Part G: Analysis of European pharmacopoeial text

To analyze Monographs 2.6.8, 2.6.30, and 2.6.40, all text containing relevant search terms listed in Part F above was extracted and further analyzed using a criteria list. Criteria were developed by identifying themes in Monographs 2.6.8, 2.6.30, and 2.6.40. The following criteria were then used to determine if the past or current text demonstrated support for animal-free innovation in pyrogenicity testing:
Using these criteria, Monographs 2.8.6, 2.6.30, and 2.6.40 were analyzed on a binary scale following a similar methodology as described above in Section D: The Comparative Analyses.

1. Does not refer to RPT or in vivo testing (excluding historical analysis).
2. Refers to MAT or in vitro-related methods.
3. Refers to MAT or in vitro-related methods more than RPT or in vivo testing.
4. Refers primarily (>50%) to RPT only in the context of in vitro/alternative-related terms.
5. Clearly indicates that the focus of the monograph is pyrogenicity and/or endotoxin testing.

Part E List of sources collected from Overton with exclusion and other selection criteria applied

Type of document	Country or region	Citation
Report	Europe	• European Commission. (2020, February 7). COMMISSION STAFF WORKING DOCUMENT Accompanying the document REPORT FROM THE COMMISSION TO THE EUROPEAN PARLIAMENT AND THE COUNCIL: 2019 report on the statistics on the use of animals for scientific purposes in the Member States of the European Union in 2015–2017. https://op.europa.eu/en/publication-detail/-/publication/d9b6a1c1-4a57-11ea-9b32-01aa75ed71a1 • European Commission. (2021, July 19). COMMISSION STAFF WORKING DOCUMENT: Summary Report on the statistics on the use of animals for scientific purposes in the Member States of the European Union and Norway in 2018. https://op.europa.eu/en/publication-detail/-/publication/1b3ff348-91e4-11eb-b85c-01aa75ed71a1 • European Commission. (2023, March 31). COMMISSION STAFF WORKING DOCUMENT Summary Report on the statistics on the use of animals for scientific purposes in the Member States of the European Union and Norway in 2020. https://op.europa.eu/en/publication-detail/-/publication/4b7c1c1b-b1b7-11ec-83e1-01aa75ed71a1 • Joint Research Centre. (2020, April 28). EURL ECVAM status report on the development, validation and regulatory acceptance of alternative methods and approaches (2019). https://publications.jrc.ec.europa.eu/repository/handle/JRC121939 • European Commission. (2021, March 25). Towards replacement of animals for scientific purposes. https://op.europa.eu/en/publication-detail/-/publication/98c8474a-3edc-11eb-9e17-01aa75ed71a1 • Institute for Health and Consumer Protection. (2003, July 16). Institute for Health and Consumer Protection Activity Report for 2002. https://publications.jrc.ec.europa.eu/repository/handle/JRC23202 • Institute for Health and Consumer Protection. (2002, September 24). Institute for Health and Consumer Protection Activity Report for 2001. https://publications.jrc.ec.europa.eu/repository/handle/JRC23201
	Europe (Switzerland)	• Stiftung Forschung 3R Fondation Recherches 3R. (2016, January 19). Replacement, reduction, and refinement of animal testing in the quality control of human vaccines: brief description of ongoing projects. https://www.forschung3r.ch/en/projects • Stiftung Forschung 3R Fondation Recherches 3R. (2016, January 8). Replacement, Reduction, and Refinement of Animal Testing in the Quality Control of Human Vaccines. https://www.forschung3r.ch/en/projects

Continued

Part E List of sources collected from Overton with exclusion and other selection criteria applied—cont'd

Type of document	Country or region	Citation
Guidance and Guidelines	Europe	• European Medicines Agency. (2007, December 21). Draft note for guidance on plasma derived medicinal products (CPMP/BWP/269/05 rev. 3) - Addendum on the replacement of rabbit pyrogen testing by an alternative test for plasma derived medicinal products (EMEA/CHMP/BWP/452140/2006). https://www.ema.europa.eu/en/documents/scientific-guideline/draft-note-guidance-plasma-derived-medicinal-products-cpmp/bwp/269/05-rev3_en.pdf • European Medicines Agency. (2011, July 28). Guideline on plasma-derived medicinal products. https://www.ema.europa.eu/en/documents/scientific-guideline/guideline-plasma-derived-medicinal-products_en.pdf • European Chemicals Agency. (2022, December 22). Guidance on information requirements and chemical safety assessment: Appendix R7-1 for nanomaterials applicable to Chapter R7a Endpoint specific guidance: Version 4.0. https://echa.europa.eu/documents/10162/13643/r7a-nanomaterials_en.pdf
	United States	• US Food and Drug Administration. (2000, April 24). Guidance for Industry. https://www.fda.gov/media/72712/download • US Centers for Disease Control and Prevention. (2014, October 23). Guidelines for environmental infection control in health-care facilities; recommendations of CDC and Healthcare Infection Control Practices Advisory Committee (HICPAC). https://www.cdc.gov/infectioncontrol/guidelines/environmental/index.html • US Centers for Disease Control and Prevention. (2017, February 23). Guidelines for environmental infection control in health-care facilities; recommendations of CDC and the Healthcare Infection Control Practices Advisory Committee (HICPAC). https://www.cdc.gov/infectioncontrol/guidelines/environmental/index.html • US Food and Drug Administration. (2012, June 22). Pyrogen and Endotoxins Testing: Questions and Answers. https://www.fda.gov/media/81338/download • US Food and Drug Administration. (2018, July 20). CMC Information for Human Gene Therapy IND Applications. https://www.fda.gov/media/113469/download • US Food and Drug Administration. (2020, July 13). Setting Endotoxin Limits During Development of Investigational Oncology. https://www.fda.gov/media/140074/download • US Food and Drug Administration. (2008, April 8). CMC for Human Gene Therapy INDs. https://www.fda.gov/media/81771/download
(Draft) Reflection Papers	Europe	• European Commission. (2016, November 17). Draft reflection paper providing an overview of the current regulatory testing requirements for medicinal products for human use and opportunities for implementation of the 3Rs - First version. https://www.ema.europa.eu/en/documents/scientific-guideline/draft-reflection-paper-providing-overview-current-regulatory-testing-requirements-medicinal_en.pdf • European Commission. (2018, October 30). Reflection paper providing an overview of the current regulatory testing requirements for medicinal products for human use and opportunities for implementation of the 3Rs - First version. https://www.ema.europa.eu/en/documents/scientific-guideline/reflection-paper-providing-overview-current-regulatory-testing-requirements-medicinal-products_en.pdf

Government Hearings	United States	• US Senate Committee on Appropriations. (2003, May 8). S.Hrg. 108-155 — Agriculture, rural development, and related agencies appropriations for fiscal year 2004. https://www.congress.gov/congressional-report/108th-congress/senate-report/ • US Senate Committee on Appropriations. (2003, July 16). S.Hrg. 108-629 — Agriculture, rural development, and related agencies appropriations for fiscal year 2005. https://www.congress.gov/congressional-report/108th-congress/senate-report/ • US Senate Committee on Appropriations. (2003, October 27). Agriculture, rural development, and related agencies appropriations for fiscal year 2004: Hearings before a subcommittee of the Committee on Appropriations, United States Senate, One Hundred Eighth Congress, first session, on H.R. 2673/S. 1427, an act making appropriations for agriculture, rural development, Food and Drug Administration, and related agencies programs for the fiscal year ending September 30, 2004. https://www.congress.gov/congressional-report/108th-congress/senate-report/ • US Senate Committee on Appropriations. (2004, February 6). Senate Hearing, 108th Congress — Agriculture, rural development, and related agencies appropriations for fiscal year 2005. https://www.congress.gov/congressional-report/108th-congress/senate-report/ • US Senate Committee on Appropriations. (2004, October 5). Agriculture, rural development, and related agencies appropriations for fiscal year 2005: Hearings before a subcommittee of the Committee on Appropriations, United States Senate, One Hundred Eighth Congress, second session, on H.R. 4766/S. 2803, an act making appropriations for agriculture, rural development, Food and Drug Administration, and related agencies programs for the fiscal year ending September 30, 2005. https://www.congress.gov/congressional-report/108th-congress/senate-report/
Other	Europe	• Various Authors. (2023, April 24). Non-technical summaries granted in 2022. https://www.example.com/document_1 • Various Authors. (2020, March 16). Non-technical summaries granted in 2018. https://www.example.com/document_2 • Various Authors. (2017, December 21). Non-tech summaries 2016: projects on regulatory uses. https://www.example.com/document_3 • Various Authors. (2016, June 21). Non-tech summaries 2015: projects on regulatory use. https://www.example.com/document_4 • Various Authors. (2022, April 20). The Research Modernisation Deal: A Strategy to Ending Animal Experiments. https://www.example.com/document_6
	United States	• US Food and Drug Administration. (2008, April 7). Companion Document. https://www.fda.gov/media/72710/download • US Food and Drug Administration. (2008, February 15). Coronary and Carotid Embolic Protection Devices - [510(k)] Submissions. https://www.fda.gov/media/71769/download

Part H. Framework for discussing the pyrogenicity case study in the broader context

To discuss the transition from in vivo methods to nonanimal methods in regulatory testing in the broader context, the following framework was adapted from Smith and Raven (2012).

Their original framework describes six aspects (science, technology, industry, market, culture, and policy) of the sociotechnical regime that interact, allowing us to analyze the underlying (in)direct factors that stimulate or create change in a transition. We adapted this framework to analyze how science, technology, policy, and market impact the implementation or use of new (nonanimal) methods, using pyrogenicity as a case study (Fig. 4). Furthermore, this adapted framework integrates social and historical context as a specific aspect relevant to method implementation. The use of this framework allows us to visualize how these aspects interact more generally, leading to the successful utilization of nonanimal methods in regulatory research.

AI Disclosure

During the preparation of this work, the author(s) used Perplexity to assist in developing the abstract and conclusion sections. All other sections of the text were not written with AI assistance. AI was also used as a brainstorming partner in the initial phase of chapter development. After using this tool/service, the author(s) reviewed and edited the content as needed and take(s) full responsibility for the content of the publication.

References

[1] Carmicahel PL, Baltazar MJ, Cable S, et al. Ready for Regulatory Use: nAMs and NGRA for Chemical Safety Assurance. *Altex*. 2022;39(3):359–366. doi:10.14573/altex.2204281.

[2] Lilley E, Isbrucker R, Ragan I, Homes AM. Integrating 3Rs approaches in WHO guidelines for the batch release testing of biologicals. *Biologicals*. 2021;74:24–27. doi:10.1016/j.biologicals.2021.10.002.

[3] van den Biggelaar RHGA, Hoefnagel MHN, Vandebriel RJ, et al. Overcoming scientific barriers in the transition from in vivo to nonanimal batch testing of human and veterinary vaccines. *Expert Rev Vaccines*. 2021;20(10):1221–1233. doi:10.1080/14760584.2021.1977628.

[4] Meigs L. Animal testing and its alternatives – the most important omics is economics. *Altex*. 2018;35(3):275–305. doi:10.14573/altex.1807041.

[5] Milne C (2020). "Overview of the EDQM Biological Standardisation Programme" *in European Pharmacopoeia Training Session on Biologicals*. https://www.edqm.eu/documents/52006/244695/european-pharmacopoeia-overview-of-biologicals-standardisation-programme-bsp-by-catherine-milne.pdf/074e138d-badd-a37a-d15a-7651799429a9?t=1635322040897.

[6] European Commission. *COMMISSION STAFF WORKING DOCUMENT: union overview on the implementation of Directive 2010/63/EU on the protection of animals used for scientific purposes in the Member States of the European Union in 2018–2022*. Available at: https://transcience.ovh/wp-content/uploads/2025/02/Second-EU-Implementation-Report-SWD2024183-final.pdf. Accessed Aug 15, 2026.

[7] EDQM. (2024). *European regulations for medicines: Place & role of the EDQM & the European Pharmacopoeia (Ph. Eur.) - 2024 Training*. Available at: https://www.edqm.eu/documents/52006/115295/european-regulations-for-medicines-cathie-vielle-october2017.pdf/2e91dd83-548f-0cc5-21d1-68febab5f5ad. Accessed Aug 15, 2026.

[8] Rustogi S, Sharma SK, Chauhan J, et al. Pharmacopoeia roles and responses: a systemic resilience approach to COVID-19 pandemic. *Saudi Pharm J*. 2022;30(5):613–618. doi:10.1016/j.jsps.2022.02.009.

[9] Gwenaël C. Oral presentation "Towards animal free pyrogen tests in the Ph.Eur: latest progress". In: 11th Pharmalab 2023 Conference. 20–22 November 2023, Dusseldorf. n.d.

[10] Hartung T. Pyrogen Testing Revisited on Occasion of the 25th Anniversary of the Whole Blood Monocyte Activation Test. *Altex*. 2021;38(1):3–19. doi:10.14573/altex.2101051.

[11] Hartung T. The human whole blood pyrogen test – lessons learned in twenty years. *ALTEX*. 2015;32(2):79–100. doi:10.14573/altex.1503241.

[12] Hasiwa N, Daneshian M, Bruegger P, et al. T4 Report evidence for the detection of non endotoxin pyrogens by the whole blood monocyte activation test. *Altex*. 2013;30(2):169–208. doi:10.14573/altex.2013.2.169.

[13] Vydelingum S, et al. (2024) *USP Modern Methods for Endotoxins and Pyrogens Testing Virtual Workshop.*

[14] Hartung T (2024) *USP Modern Methods for Endotoxins and Pyrogens Testing Virtual Workshop.*

[15] Spreitzer, Ingo. *Evolution and Characteristics of the Monocyte Activation Test (MAT).* Springer Science and Business Media LLC; 2019:523–535. doi:10.1007/978-3-030-17148-3_14.

[16] Charton E. *The Eur. Ph. strategy for the replacement of the Rabbit Pyrogenicity Test.* [Webinar]. Transition to non-animal based vaccine batch release testing. Policy and regulations theoretical aspects and case studies; 2024. https://www.afsacollaboration.org/wp-content/uploads/2024/01/6.-E-Charton-Towards-animal-free-pyrogen-tests-in-the-Ph-Eur-HSI-March-2024-002.pptx-Lecture-seule.pdf.

[17] European Commission (2022), ALURES – animal Use Reporting – EU System EU Statistics Database on the Use of Animals for Scientific Purposes Under Directive 2010/63/EU. https://webgate.ec.europa.eu/envdataportal/content/alures/section2_number-of-uses.html.

[18] EDQM (2024). *Ph. Eur. bids adieu to rabbit pyrogen test in its monographs - European Directorate for the Quality of Medicines & HealthCare - EDQM.* https://www.edqm.eu/en/-/ph.-eur.-bids-adieu-to-rabbit-pyrogen-test-in-its-monographs.

[19] Smith A, Raven R. What is protective space? Reconsidering niches in transitions to sustainability. *Res Policy.* 2012;41(6):1025–1036. doi:10.1016/j.respol.2011.12.012.

[20] Charton E (2022) *European Pharmacopoeia Approach to Testing for Pyrogenicity.* Available at: https://www.americanpharmaceuticalreview.com/Featured-Articles/583968-European-Pharmacopoeia-Approach-to-Testing-for-Pyrogenicity/. Accessed June 4, 2025.

[21] Devine J (2024). "USP Welcome and Opening Remarks" in *USP Modern Methods for Endotoxins and Pyrogens Testing Virtual Workshop.* https://www.usp.org/sites/default/files/usp/document/events-and-training/2024-endotoxin-and-pyrogen-workshop-agenda-speakers-titles%20v1.pdf.

[22] Spreitzer I (2024) *The current status and implementation of MAT. USP Modern Methods for Endotoxins and Pyrogens Testing Virtual Workshop.*

[23] Zelter E (2023) *Moves to Replace Animal-Based Detection of Pyrogens By In-Vitro Testing. American Pharmaceutical Review.* Available at: https://www.americanpharmaceuticalreview.com/Featured-Articles/595243-Moves-to-Replace-Animal-Based-Detection-of-Pyrogens-By-In-Vitro-Testing/ Accessed June 4, 2025.

[24] *European Pharmacopoeia (2024) 2.6.40 Monocyte-Activation Test For Vaccines Containing Inherently Pyrogenic Components.* Available at: https://www.walderwyss.com/en/news/2024-07-05_addendum-11-5-of-the-european-pharmacopoeia-in-force. Accessed Aug 25, 2026.

[25] *European Directive 2001/83/EC on the EC code relating to medicinal products for human use - EU monitor* (2001). Available at: https://eur-lex.europa.eu/eli/dir/2001/83/oj Accessed June 4, 2025.

[26] National Archives (2023). *About the Code of Federal Regulations.* Available at: https://www.archives.gov/federal-register/cfr/about.html Accessed June 4, 2025.

[27] Code of Federal Regulations. Title 21 Food and Drugs; Subchapter 7 Biologics. Available at: https://www.ecfr.gov/current/title-21/chapter-I/subchapter-F. Accessed June 4, 2025.

[28] Caroff M, Novikov A. Lipopolysaccharides: structure, function and bacterial identification. *OCL.* 2020;27. doi:10.1051/ocl/2020025.

[29] Lüderitz O, Tanamoto K, Glanos C, et al. Lipopolysaccharides: structural principles and biologic activities. *Rev Infect Dis.* 1984;6(4):428–431. doi:10.1093/clinids/6.4.428.

[30] Rietschel ET, Kirikae T, Schade FU, et al. Bacterial endotoxin: molecular relationships of structure to activity and function. *FASEB J.* 1994;8(2):217–225. doi:10.1096/fasebj.8.2.8119492.

[31] Westphal O, Lüderitz O, Bister F. Über die Extraktion von Bakterien mit Phenol/Wasser. *Zeitschrift Für Naturforschung B.* 1952;7(3):148–155. doi:10.1515/znb-1952-0303.

[32] Morath S, Geyer A, Hartung T. Structure-function relationship of cytokine induction by lipoteichoic acid from Staphylococcus aureus. *J Exp Med.* 2001;193(3):393–397. doi:10.1084/jem.193.3.393.

[33] Rockel C, Hartung T. Systematic Review of Membrane Components of Gram-Positive Bacteria Responsible as Pyrogens for Inducing Human Monocyte/Macrophage Cytokine Release. *Front Pharmacol.* 2012;3. doi:10.3389/fphar.2012.00056.

[34] Hoffmann S, Lüderitz-Püchel U, Montag T, et al. Optimisation of pyrogen testing in parenterals according to different pharmacopoeias by probabilistic modelling. *J Endotoxin Res.* 2005;11(1):25–31. doi:10.1179/096805105225006678.

[35] Dehus O, Hartung T, Hermann C. Endotoxin evaluation of eleven lipopolysaccharides by whole blood assay does not always correlate with Limulus amebocyte lysate assay. *J Endotoxin Res*. 2006;12(3):171–180. doi:10.1177/09680519060120030401.
[36] Ding JL, Ho B. A new era in pyrogen testing. *Trends Biotechnol*. 2001;19(8):277–281. doi:10.1016/S0167-7799(01)01694-8.
[37] Dubczak J, et al. (2020) *American Pharmaceutical Review*. Available at: https://www.americanpharmaceuticalreview.com/Featured-Articles/571720-Past-Present-and-Future-of-Endotoxin-Testing/ Accessed June 4, 2025.
[38] Hartung T, Wendel A. Detection of pyrogens using human whole blood. *In Vitro Mol Toxicol*. 1996;9(4):353–359.
[39] Hartung T, et al. Novel Pyrogen Tests Based on the Human Fever Reaction. *Altern Lab Anim*. 2001;29(2):99–123. doi:10.1177/026119290102900203.
[40] Levin J, Bang FB. Clottable protein in Limulus; its localization and kinetics of its coagulation by endotoxin. *Thromb Diath Haemorrh*. 2018;19(1):186–197. doi:10.1055/s-0038-1651195.
[41] Mallet L (2024). *"NGS for adventitious viruses: ongoing initiatives and upcoming events " in Webinar: Transition to nonanimal based vaccine batch release testing. Policy and regulations theoretical aspects and case studies."* https://www.afsacollaboration.org/sciencex_event/transition-to-non-animal-based-vaccine-batch-release-testing/.
[42] Mascoli CC, Weary ME. Applications and advantages of the Limulus amebocyte lysate (LAL) pyrogen test for parenteral injectable products. *Prog Clin Biol Res*. 1979;29:387–402.
[43] National Centre for the Replacement, Refinement and Reduction of Animals in Research (NC3Rs). (2023). Review of animal testing requirements in WHO guidelines and recommendations for biologics. https://nc3rs.org.uk/our-portfolio/review-animal-use-requirements-who-biologics-guidelines.
[44] Rietschel ET, Cavaillon JM. Richard Pfeiffer and Alexandre Besredka: creators of the concept of endotoxin and anti-endotoxin. *Microbes Infect*. 2003;5(15):1407–1414. doi:10.1016/j.micinf.2003.10.003.
[45] Tindall B, Demircioglu D, Uhlig T. Recombinant bacterial endotoxin testing: a proven solution. *BioTechniques*. 2021;70(5):290–300. doi:10.2144/btn-2020-0165.
[46] da Silva CC, Presgrave O, Hartung T, de Moraes AM, Delgado IF. Applicability of the Monocyte Activation Test (MAT) for hyperimmune sera in the routine of the quality control laboratory: comparison with the Rabbit Pyrogen Test (RPT). *Toxicol In Vitro*. 2016;32:70–75. doi:10.1016/j.tiv.2015.12.004.
[47] European Commission, Joint Research Centre. (2022). *Tracking System for Alternative methods towards Regulatory acceptance (TSAR)*. https://tsar.jrc.ec.europa.eu/.
[48] Fennrich S, Wendel A, Hartung T. New Applications of the Human Whole Blood Pyrogen Assay (PyroCheck). *ALTEX*. 1999;16(3):146.
[49] European Pharmacopoeia. *Chapter 2.6.8 Pyrogens*. Strasbourg: Council of Europe; 2016.
[50] European Pharmacopoeia. *Chapter 2.6.8 Pyrogens*. Strasbourg: Council of Europe; 2017.
[51] European Pharmacopoeia. *Chapter 2.6.30 Monocyte-activation Test*. Strasbourg: Council of Europe; 2017.
[52] European Pharmacopoeia. *Chapter 2.6.30 Monocyte-activation Test*. Strasbourg: Council of Europe; 2024.
[53] Hort EC, Penfold WJ. Microorganisms and their relation to fever. *J Hyg*. 1912;12(3):361–390. doi:10.1017/S0022172400005052.
[54] EDQM. (2024) "History" in *European Directorate for the Quality of Medicines & HealthCare*. https://www.edqm.eu/en/edqm/about/history.
[55] Ozato K, Tsujimura H, Tamura T. Toll-like receptor signaling and regulation of cytokine gene expression in the immune system. *BioTechniques*. 2002;33(4):S66–S75. doi:10.2144/oct0208.
[56] Bacterial Endotoxins/Pyrogens (2018). *FDA*. https://www.fda.gov/inspections-compliance-enforcement-and-criminal-investigations/inspection-technical-guides/bacterial-endotoxinspyrogens.
[57] Charton E (2021) *Using recombinant factor C for bacterial endotoxin testing in the Ph. Eur. how far have we come, how far have we to go?*. https://www.edqm.eu/documents/52006/115334/recombinant-factor-c-emmanuelle-charton.pdf/ea9de4ea-f831-d948-b834-986c32352533.
[58] Epstein FH, Dinarello CA. Interleukin-1 and the Pathogenesis of the Acute-Phase Response. *N Engl J Med*. 1984;311(22):1413–1418. doi:10.1056/nejm198411293112205.
[59] Dinarello CA. Interleukin-1. *Clin Infec Dis*. 1984;6(1):51–95. doi:10.1093/clinids/6.1.51.
[60] Poole S, Thorpe R, Meager A, Gearing AJ. Assay of pyrogenic contamination in pharmaceuticals by cytokine release from monocytes. *Dev Biol Stand*. 1988;69:121–123.
[61] EDQM (2021). *European Pharmacopoeia to put an end to the rabbit pyrogen test*. https://www.edqm.eu/en/-/european-pharmacopoeia-to-put-an-end-to-the-rabbit-pyrogen-test.

[62] Hartung T. Comparative analysis of the revised directive 2010/63/EU for the protection of laboratory animals with its predecessor 86/609/EEC - A t4 report. *Altex*. 2010;27(4):285–303. doi:10.14573/altex.2010.4.285.

[63] U.S. Food and Drug Administration. (2012). Pyrogen and Endotoxins Testing: questions and Answers. Available at: https://www.fda.gov/media/83477/download. Accessed June 4, 2025.

[64] Eckford C. Widening adoption of non-animal regents for endotoxin testing. *Eur Pharm Rev*. 2024. https://www.europeanpharmaceuticalreview.com/news/227037/widening-adoption-of-animal-free-reagents-for-endotoxin-testing/.

[65] U.S. Congress. (2022). *FDA Modernization Act 2.0, S. 5002, 117th Cong*. https://www.congress.gov/bill/117th-congress/senate-bill/5002.

[66] European Parliament, & Council of the European Union. Directive 2010/63/EU of the European Parliament and of the Council of 22 September 2010 on the protection of animals used for scientific purposes. *Off J Eur Union*. 2010;276:33–79. Available at https://eur-lex.europa.eu/legal-content/EN/TXT/?uri=CELEX:32010L0063.

[67] European Parliament, & Council of the European Union. Directive 2001/83/EC of the European Parliament and of the Council of 6 November 2001 on the Community code relating to medicinal products for human use. *Off J Eur Communities*. 2001;311:67–128. https://eur-lex.europa.eu/legal-content/EN/TXT/?uri=CELEX:32001L0083.

[68] U.S. Food and Drug Administration. (2000). *Guidance for Industry on Bioavailability and Bioequivalence Studies for Orally Administered Drug Products—general Considerations*. https://www.federalregister.gov/documents/2000/10/27/00-27602/guidance-for-industry-on-bioavailability-and-bioequivalence-studies-for-orally-administered-drug.

[69] Schweitzer M (2024) *USP Modern Methods for Endotoxins and Pyrogens Testing Virtual Workshop*.

[70] Eyal-Cohen M. Unintended legislative inertia. *SSRN Electron J*. 2019. doi:10.2139/ssrn.3574126.

[71] U.S. Pharmacopeia. *Expert Committee Approves Endotoxin Testing Using Non-animal Derived Reagents*. U.S. Pharmacopeia; 2024. https://www.usp.org/news/expert-committee-approves-endotoxin-testing-using-non-animal-derived-reagents.

[72] Marketsandmarkets (2024) *Pyrogen Testing Market Size, Share, Trends and Revenue Forecast*. Available at: https://www.marketsandmarkets.com/Market-Reports/pyrogen-testing-market-255266155.html Accessed June 4, 2025.

[73] Transparency Market Research. *Monocyte Activation Test Market*. Global Industry Analysis, Size, Share, Growth, Trends, and Forecast. 2023-2031. (2023). Available at: https://www.transparencymarketresearch.com/monocyte-activation-test-market.html. Accessed June 4, 2025.

[74] Visseren-Hamakers IJ, et al. How to Save a Million Species? Transformative Governance through Prioritization. In: Visseren-Hamakers IJ, Kok M, eds. *Transforming Biodiversity Governance*. Cambridge University Press (CUP; 2022:67–90. doi:10.1017/9781108856348.005.

[75] Busquet F, Kleensang A, Rovida C, et al. New European Union statistics on laboratory animal use - What really counts! *Altex*. 2020;37(2):167–186. doi:10.14573/altex.2003241.

[76] Cirefice G (2024) "Eur. Activities and Perspectives on Next Generation Sequencing" in *USP Biologics Stakeholder Forum*. https://www.usp.org/sites/default/files/usp/document/events-and-training/04-edqum-pharmacopeia-europe-activities-in-the-field-of-ngs-hts.pdf.

CHAPTER

19

Safety within REACH? a transition analysis for the European Commission roadmap to animal-free safety assessment of industrial chemicals

Love Hansell[a], *Luísa Ferreira Bastos*[b], *Kelsey Beltz*[c], *Merel Ritskes-Hoitinga*[d,e], *Thomas Hartung*[f,g], *Julia Pochat*[b], *Anne Van Veen*[d], *I.J. Visseren-Hamakers*[a], *Justine Watkins*[d] and *Howard White*[h]

[a]Institute for Management Research, Radboud University, Nijmegen, Netherlands
[b]Eurogroup for Animals, Brussels, Belgium
[c]The Good Lobby, Brussels, Belgium
[d]Faculty of Veterinary Medicine, Institute for Risk Assessment Sciences (IRAS), Utrecht University, Utrecht, Netherlands
[e]Department of Clinical Medicine, Aarhus University, Aarhus, Denmark
[f]CAAT-Europe, University of Konstanz, Konstanz, Germany
[g]Johns Hopkins University, Baltimore, MD, United States
[h]Research and Evaluation Centre, London, United Kingdom

Abstract

The European Union (EU) is formally committed to the full replacement of animal-based models in safety assessment, but in practice, the transition to an animal-free safety assessment system faces significant hurdles. As the European Commission (EC) develops its roadmap to transition to the animal-free safety assessment of chemicals, the legislative landscape of chemical safety assessment in Europe is changing. Registration, Evaluation, Authorisation and Restriction of Chemicals (REACH) and Classification, Labelling and Packaging (CLP) are being revised, and it is more important than ever to discuss the hurdles to the transition to animal-free safety assessment. This chapter presents a transition analysis of barriers to this transition and categorizes them according to the main underlying and direct causes, allowing for a more nuanced discussion about

Moving Toward Nonanimal Approaches in Medical Research and Testing
https://doi.org/10.1016/B978-0-443-26514-3.00010-2

facilitating change. The EC and the European Chemicals Agency have multiple avenues in the short and long term to facilitate the transition, providing better safety for its citizens while reducing animal testing.

Keywords: Transition to nonanimal methods; Animal-free safety assessment of chemicals; New Approach Methodologies; Transformative Governance

Introduction

The European Union (EU) is officially committed both to the full replacement of animal-based experimental models and to the "last resort" principle, which states that animal testing should "only be considered where a nonanimal alternative is unavailable."[1] The transition to animal-free safety assessment can therefore be seen as a necessity recognized by the EU, despite its slow pace, as recognized by European Parliament (EP) resolutions and European Citizens' Initiatives (ECIs), which call for reform and bans on animal testing in multiple sectors, including cosmetics.[2–5] In response to the ECI "Save Cruelty Free Cosmetics—Commit to a Europe Without Animal Testing,"[3] the European Commission (EC) committed to developing a roadmap[2] to accelerate the transition "with the aim to ultimately move to an animal-free regulatory system under chemicals legislation."[6]

This gives pause to consider the challenges such a roadmap faces in its implementation. The aim of this chapter is to examine the existing governance system of industrial chemical safety assessment in the EU. Based on the transition and transformation literature, we present a transition analysis aiming to distill the underlying causes of the barriers that hinder the transition to animal-free safety assessment.

Background

The EU uses around 8.5 million nonhuman animals annually for research, testing, and education.[7] Of these, 13% were used for regulatory testing (animals used to satisfy regulatory requirements for producing and marketing various products) in 2022.[8] The ethics of inflicting harm and death on our fellow coinhabitants of Earth has long been questioned and is part of a broader change in how humans think about nonhuman animals. Some polls show opposition from the general public to subjecting millions of individual animals to experiments, but more research is needed.[9] Aside from the deeply problematic ethics of animal testing, its scientific relevance is also questioned. The relevance of animal studies for human health (translatability) is often overestimated[10–12] (see Chapter 3 for a deeper discussion); a substance could be safe for an animal but not for humans (and vice versa). Moreover, replicating animal studies is difficult due to poor reporting of testing protocols[13–16] and the fact that animal data is highly variable.[11,17] Despite this, animal testing is still seen as the "gold standard" in safety assessment.[12,18]

In efforts to address the shortcomings of animal testing, new approach methodologies (NAMs) have been developed, defined as "any technology, methodology, approach, or combination that can provide information on chemical hazard and risk assessment, and avoid the use of animals."[19] These methods include computer modeling and artificial intelligence (AI) (in silico), the use of extracted or cultured living cells outside the body of an organism in a controlled environment (in vitro),[20–22] the use of chemical substances without organic material (in chemico),[23] and other approaches such as systematic reviews.[24]

Many NAMs have already been shown to deliver data as accurate as or more accurate than animal-based models and have the ability to replace animal testing in safety assessments.[20,25–28] NAMs have the potential to generate more reproducible results, which are more reflective of human biology and therefore more relevant for human health.[20,27] Despite demonstrating scientific relevance for regulatory acceptance, many NAMs have yet to be validated and applied in existing safety assessment regimes, while animal studies remain the norm. However, even if a NAM is validated, its uptake into regulatory safety assessments and its replacement of an existing animal-based model is very slow. This is exemplified in the case of the monocyte activation test for pyrogen testing (see Chapter 18), which, despite being validated, still has its animal-model counterpart being used abundantly.[29]

The toxic ignorance of chemical safety in the EU

There are around 100,000 industrial chemicals on the European market; 70,000 of them are very poorly understood, and only 500 are considered sufficiently well understood in terms of human and environmental risks.[30] In addition, chemical exposure chiefly involves mixtures of chemicals, which significantly increases the amount of testing needed to assess the safety of substances.[31] Considering this, the statement "We know almost nothing about the impacts of chemicals on human health and the environment" rings eerily true.[32] This large data gap means that the time and money needed to assess one chemical—let alone 70,000 (plus the mixtures of these)—while relying on animal testing, is an immense obstacle to effective safety assessment of chemicals on the market and of new chemicals in the future. One cancer bioassay can cost up to €700,000 and take up to five years to conduct.[33] Thus, in terms of both costs and time, using animal testing to establish the safety of both new chemicals and chemicals already on the market is simply unfeasible. If animal testing is seen as the gold standard, it is clear that all that glitters is not gold.

There is a growing realization that a new regulatory system is needed, with calls for a paradigm shift.[19,34] One such system is next-generation risk assessment (NGRA), an "exposure-led, hypothesis-driven risk assessment approach that integrates NAMs to assure safety without the use of animal testing."[26] NGRA would allow for a more flexible, adaptive regulatory system, which could incorporate different NAMs for effective safety assessment.[35,36]

Fragmented EU safety assessment governance

Spread over 40 pieces of legislation, the existing governance of chemical safety assessment in the EU is described as slow, incoherent,[37] and inefficient in reaching its goal[38–40] of implementing the EU's main policy of striving for "a toxic-free environment."[41] The EU approaches regulatory safety assessment in a fragmented way, with a heavy burden placed on private actors to report on the safety of the chemical in question.[32]

EU policy on chemicals in the past political term (2020–2024), when the roadmap process was initiated, was governed by the Chemicals Strategy for Sustainability (CSS)[41] and set in concordance with the European Green Deal strategy.[42] It aimed to achieve a "toxic-free environment, where chemicals are produced and used in a way that maximizes their contribution to society."[41] It has since been replaced by the Chemicals Industry Action Plan, revealed in July 2025.[43]

Chemical safety assessment is chiefly governed by the Registration, Evaluation, Authorisation and Restriction of Chemicals legislation (REACH)[44] and the Classification, Labelling and Packaging (CLP)[45] regulations. Both pieces of legislation are primarily based on animal models, and there is a general lack of consensus on which alternatives should be used under which circumstances; this tends to result in the use of animal testing despite the existence of validated alternatives.[17,35,46]

The REACH legislation created a regulatory system wherein a chemical must be registered in order for it to be sold on the European market.[32,44] The legislation also led to the establishment of the European Chemicals Agency (ECHA) to monitor and regulate chemicals in the EU. Tragically, the total number of animals that would die as a result of implementing REACH was underestimated by 1.6 million, with an estimated 4.2 million killed from the start of REACH in 2007 to December 2022.[47] The CSS strategy has induced a REACH revision but is delayed,[40] with some fearing it could lead to an increase in the number of animals used.[48]

Under REACH, the private sector is tasked with ensuring that the substances they produce are safe. Based on the volume of the substance being produced and its "intrinsic harmfulness," companies are required to submit data on the substance to ECHA; a substance cannot be sold in the EU without such a dossier.[32,49] ECHA has only been able to evaluate a fraction of the data that is submitted—21% of submitted substances between 2009 and 2023[50]—and upon review, can either accept or reject the submitted information.[32]

CLP regulation[45] is meant to aid in the labeling and packaging of products. Although CLP Regulation does not require new studies to be conducted, the industry must obtain all the available relevant information and evaluate it using the CLP classification criteria, where data have traditionally been used to assign classes based on animal tests. In practice, substances are often classified on the basis of REACH registration data, but if it is considered that the data available is not sufficient to meet CLP classification criteria, they may choose to conduct further (animal) studies to complete their dossiers.[18,51,52]

The 2013 ban on using animal testing for cosmetics[53] was lauded for its vision when it was introduced, but it has since become clear that it is not effectively followed.[54,55] Though testing cosmetic products and their ingredients is banned in the EU, the European Court of Justice (ECJ) ruled that worker and environmental safety can be invoked to call for animal testing for cosmetic-exclusive ingredients.[56] This clash between the Cosmetics Regulation and REACH is, in part, what led to the ECI Save Cruelty Free Cosmetics initiative[57] and the response from the EC[2] in which it committed to developing a roadmap to animal-free chemical safety assessment.

Since the initiation of the roadmap, the commission has initiated processes to change some aspects of its chemical safety system, including a reform of ECHA,[58] and calls for a common data platform to streamline access to data on chemicals through a "One substance, One assessment" approach.[59] Though promising, how successful they are in effectively modernizing the EU chemical safety assessment system will depend on how well they are implemented.

Conceptual framework: Transition analysis

The social scientific debate on social change is contentious and heterogeneous as it tries to conceptualize fundamental societal change.[60–62] The field of sociotechnical transitions literature often uses concepts such as transformations, transformative change, and transformative governance, and refers to *transitions* as significant changes in specific sociotechnical systems.[62–64] The field also stresses the importance of identifying the underlying causes of barriers to transition to accelerate change.[64,65]

A widely used framework in the literature is the multilevel perspective, which identifies three levels of transitions: the niche (micro) level, where technical and social innovations are developed; the regime (meso) level, which, in our case, comprises the safety assessment regime; and the landscape (macro) level, which represents society at large.[25,66–68]

The focus of our analysis is the regime level of industrial chemical safety assessment in the EU, which can be described as a stable system of six interacting aspects. These aspects should not be seen as discrete categories but rather as elements of an integrated whole, with one aspect supporting the others, together constituting the regime (Table 1).

To understand mechanisms of change within regimes and in transitions more broadly, it is useful to differentiate between underlying causes and direct causes of change. These concepts have been adapted for this analysis[69] and are defined as follows: *Underlying causes* are defined as social, institutional, and governance arrangements that lead to direct causes.[63,69] *Direct causes* are defined as policies, laws, regulations, or practices that, in this case, lead to the perpetuation of or increase in animal testing. Causes can represent both barriers and opportunities to the transition to animal-free safety assessment, but the analysis in this text focuses on the barriers. Furthermore, although barriers and opportunities to the transition to animal-free safety assessment are increasingly being discussed in the literature, few distinctions are made between direct causes and underlying causes, and few analyses are deeply grounded in social science literature. Thus, in this chapter, we aim to contribute to the theoretical discussion on how to assess and categorize sociotechnical barriers while identifying the main bottlenecks for this transition.

The importance of anthropocentricity

The values and worldviews that form the basis of societies in the European and global North context are also important to understand the transition. Relationships between human and nonhuman animals are deeply ingrained in culture,[70] and many societies see nonhuman

TABLE 1 Adapted concepts from transitions literature; the regime aspects are grouped under three main themes: technological, economic, and social, with two associated aspects under each main category.

Technological aspects	1. Technology: the dominant technologies available, which require a material infrastructure and operate by technical standards	2. Science: the state-of-the-art of science based on the dominant paradigms of academia in a particular field
Economic aspects	3. Industry: established network relations of users and producers of technology interacting chiefly through an economic market	4. Market: interaction of user practices and producers through supply and demand, price mechanisms, user preferences, and routines
Social aspects	5. Culture: the widespread symbolic representation and significance of the regime and its various aspects in society	6. Policy: the environment of political power is expressed through laws, regulations, and policy networks (e.g., relations with industry)

Adapted by the authors, based on Geels FW, Schot J. Typology of sociotechnical transition pathways. Res Policy. 2007;36(3):399–417. doi:10.1016/j.respol.2007.01.003; Geels FW. The multi-level perspective on sustainability transitions: responses to seven criticisms. Environ Innov Societal Transitions. 2011;1(1):24–40; Smith A, Raven R. What is protective space? Reconsidering niches in transitions to sustainability. Res Policy (United Kingdom). 2012;41(6):1025–1036.

animals as organisms without political agency.[71] This basis of seeing nonhuman animals as inferior is a societal reality—a landscape (macro) influence that permeates the rest of the regime (meso) sphere. It is therefore an assumed foundation—an ontological basis—that allows for animal testing in the first place.

Methods

A critical literature review[72] was conducted with the aim of collating the main barriers at the regime level in the transition to industrial chemical safety assessment in the EU. The review was conducted until a saturation point, whereby no new barriers were found. The review was nonsystematic and was not meant to be exhaustive but aimed to identify the main causes at play in this transition.

The conceptual model helps in understanding the relationship between the barriers in the transition. The barriers were categorized as belonging to one of the six regime aspects outlined above. They were subsequently coded as being either an underlying or direct cause according to the definitions above (when direct causes were deemed to be the basis for another direct cause, both were coded as direct causes). Together, this constitutes a transition analysis of the barriers in this transition. They were subsequently assembled into a visual form and linked according to their causal relationship, informed by the definition of underlying and direct causes.

Empirical analysis

The barriers show a high number of causal connections between them, and these were assembled into Figs. 1–4. Fig. 1 describes the relationship between all underlying and direct causes identified. Mapping these causes generated three main themes or clusters (with a fourth, culture, as an ontological basis for the whole), which are described below. Each theme is presented along with its underlying causes and then the direct causes. The categorization of the themes is not discrete; rather, they are meant to show some of the self-reinforcing mechanisms at play between the different barriers.

Theme 1: Regulation, data, and validation

From the conceptual model, the key aspect relevant here is **policy**. This cluster of causes deals chiefly with the practices of regulators, and it shows how multiple barriers eventually point to the regulatory system itself being a key barrier to the transition to animal-free safety assessment. Notable as well is its self-reinforcing nature; the kind of data that is accepted, mediated through "tick-box thinking," points to the fact that, in practice, NAMs face challenges in being adopted into safety assessment practices and are therefore underutilized.

Underlying causes

Homogeneous experts

A rich body of literature questions the objectivity of science production,[73] and this is no less the case for regulatory sciences.[74] A crucial aspect is the composition of experts in any given regulatory setting, as they help determine what data will be considered.[75,76] In this case, a

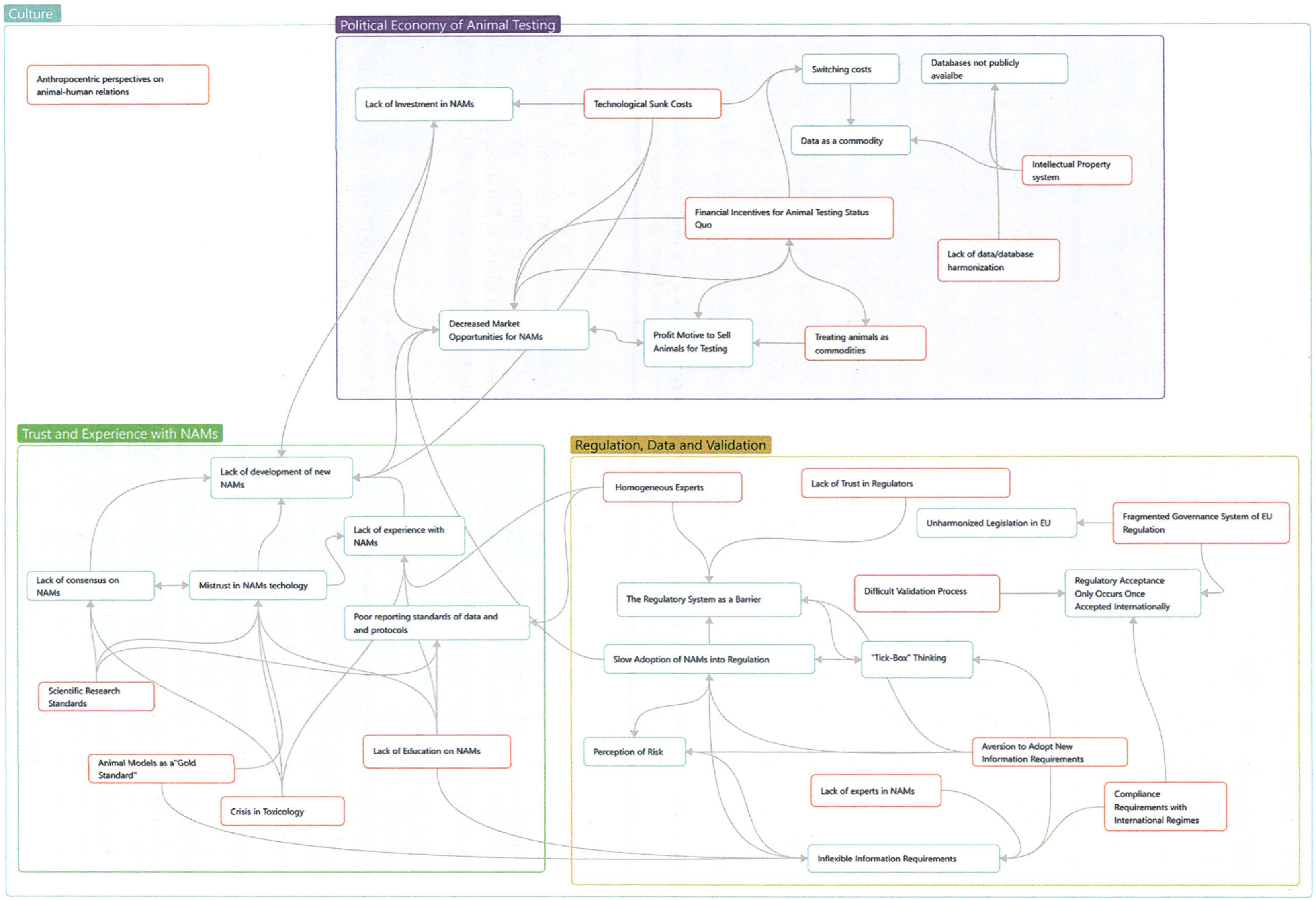

FIG. 1 Underlying causes and direct causes. The blue boxes denote direct causes, while the red boxes denote underlying causes. The relationships between causes are multiple, nonlinear, nonhierarchical, and highly intertwined. This reflects how the different aspects of the regime (see above) are highly interconnected. Culture is assumed to permeate the whole system being described, with general anthropocentric views on animals serving as an influence from the broader societal level.

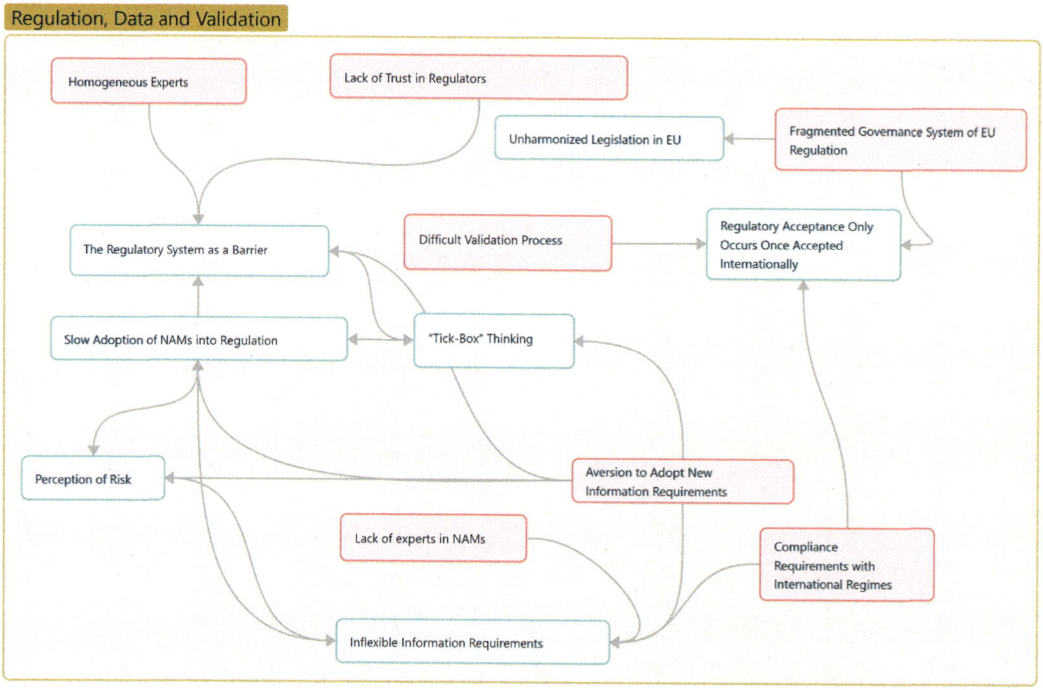

FIG. 2 Theme 1: Regulation, data, and validation. The regulatory system is a key barrier to animal-free safety assessment, with multiple self-reinforcing mechanisms. As in Figure 1, the blue boxes denote direct causes, while the red boxes denote underlying causes.

toxicological regime based on animal data is being upheld, which is, in part, predicated on dismissing the validity and data of NAMs.[77] Thus the experts can be seen as "homogeneous" vis-à-vis the "gold standard" of using animal testing. Changing the composition of expert committees would break a link in the chain that reinforces animal models. For a deeper discussion of how expertise in animal-based methods can bias research assessments, see Chapter 15.

Lack of experts in NAMs

Regulators are too often unaware of the utility of NAMs or how to use the data coming from NAMs.[78] This, as well as the lack of education on NAMs in general, means that finding experts who are well versed in how to use NAM data is more difficult and they are therefore less likely to be included in expert committees on chemical safety assessment, further reinforcing the mistrust in NAMs and the homogeneity of expert committees.

Aversion to adopting new information requirements

In part because of the gold standard paradigm, the lack of experts in the field of NAMs, and the tick-box thinking mindset (as discussed below), evidence requirements for what can be considered in dossier submissions for safety assessment are based on animal testing models and adverse effect endpoints.[55] This has created an inflexible legal system[79] that severely disenfranchises the utility of NAMs and the added quality data they can provide. If only animal data is considered acceptable data, NAMs will always be marginalized.

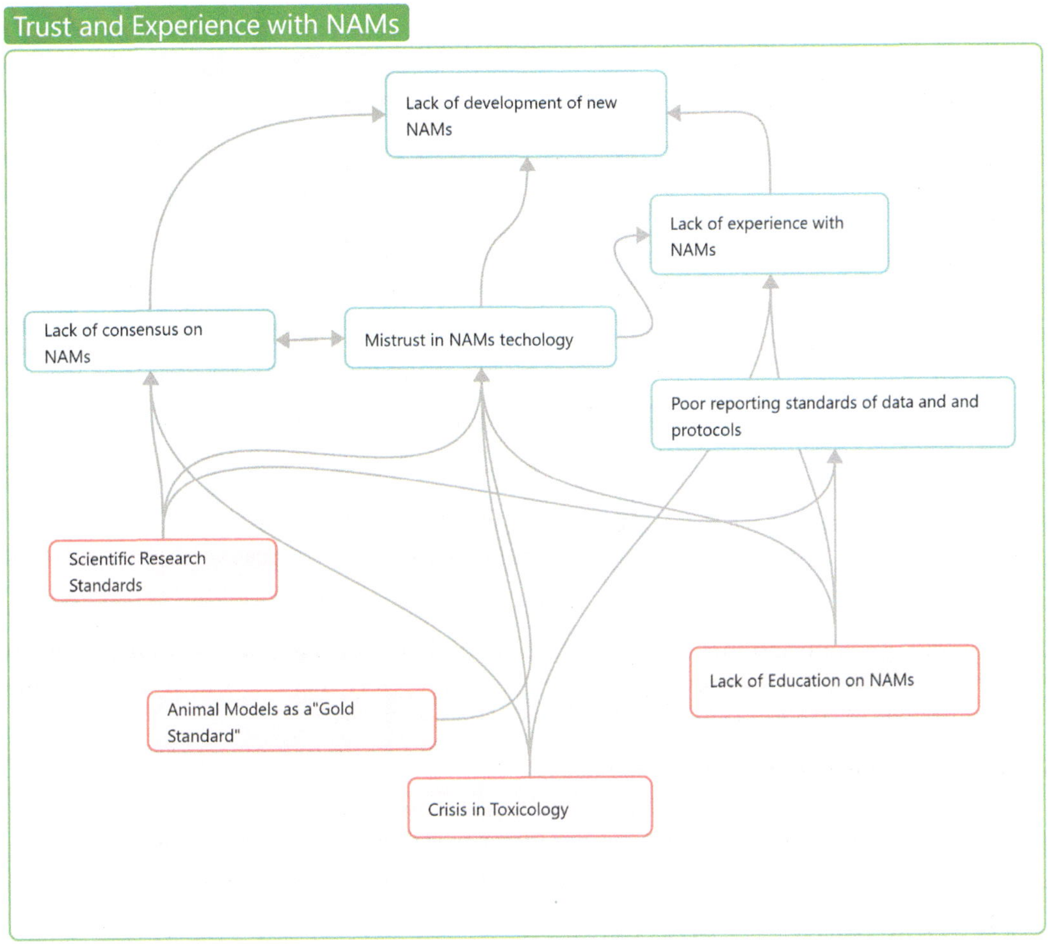

FIG. 3 Theme 2: Trust in and experience with NAMs. This figure illustrates how current standards in toxicology and scientific research contribute to the stunted growth of NAMs.

Compliance requirements with international regimes

Because the existing regulatory system is embedded in global governance systems, it is important to consider the role that the Organisation for Economic Co-operation and Development (OECD) and the Globally Harmonized System of Classification and Labelling of Chemicals (GHS) play in forming EU chemical safety assessment governance. The EU's governance system is heavily reliant on the information requirements and procedures of the OECD test guidelines, for example. This often means that regulatory acceptance of NAMs only occurs once they are accepted internationally.[25,29,67,80]

Difficult validation process

Without NAMs being validated in the regulatory safety assessment system, they will effectively not be used or implemented.[16] The existing system, which requires lengthy ring

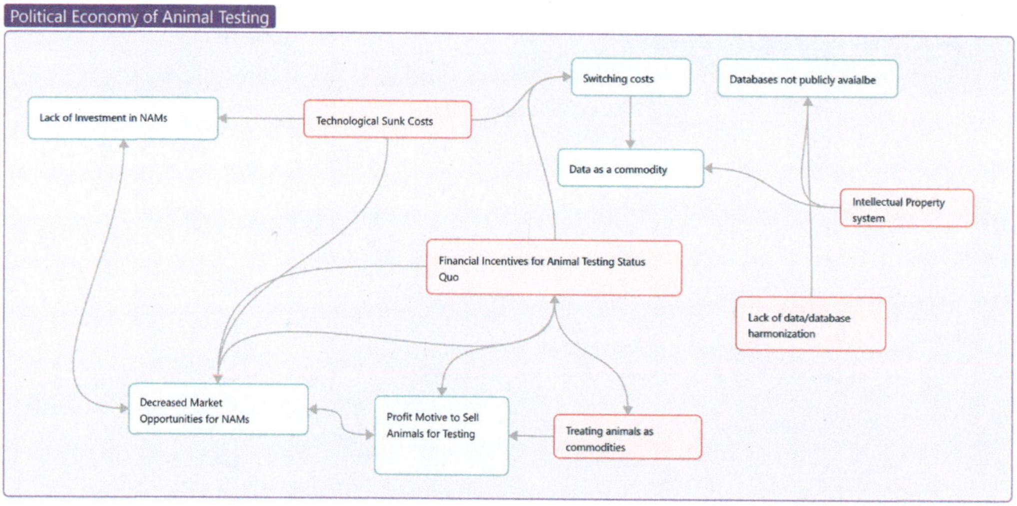

FIG. 4 Theme 3: Political Economy of Animal Testing The political economy of animal testing is strongly influenced by the decreased market opportunity of NAMs, coupled with the strong profit motive inherent in the animal testing market.

trials, inflates the time and money required.[81] Validation by the regulatory system is a crucial step to accelerating the transition; yet the path from a ready-made NAM to its acceptance into the regulatory system is excruciatingly slow and expensive.[25,29]

Fragmented governance system of EU regulation

Because regulations are spread over so many pieces of legislation[38] and multiple agencies, and because the input of multiple Member States must be considered for regulations to be implemented, the EU has systemic governance problems that hinder the adoption of NAMs.[82]

Lack of trust in regulators

Industry stakeholders do not consider regulators to be sufficiently equipped to understand data from NAMs.[78] Even the registration and evaluation practices of ECHA lack transparency and effective forms of communication to discuss dossier submissions.[46] Mistrust thus hampers innovation and reinforces the status quo.[83]

Direct causes

Unharmonized legislation in EU

The disharmonized legal-regulatory framework stands directly in contradiction with itself at times (as discussed in the introduction). The existing legal practices are too inflexible[79] and tend to perpetuate a system based on animal testing rather than precipitate the transformative change that is needed and that is called for by EU policy and law.[17,32,35,37,47,54,55,84]

The regulatory system as a barrier

The regulatory system can be seen as a barrier to the transition to animal-free safety assessment because of its complexity, fragmented governance[78] and inflexible legal requirements.[79] As can be seen in Fig. 1, this is a node of several different direct and underlying causes. As such, it speaks to the need for changes in the existing governance system of industrial chemical safety assessment in the EU.[85] Such changes must come with political support from the commission[85] and from within the regulatory system itself.

Perception of risk

The risk perception of regulators is another key component. Regulators are often seen as risk-averse due to the wide ramifications of their decisions and, therefore, less prone to adopting a fundamental shift in a new direction.[67,78,86]

Tick-box thinking

Instead of critically engaging with the evidence being presented to regulating authorities such as ECHA, safety assessment can instead become a procedure of "ticking boxes."[46,55,87] While this helps in conducting safety assessments systematically, it is not conducive to critically examining dossiers.

Inflexible information requirements

The data generated from NAMs is different from that produced by animal testing,[78] which necessarily means different kinds of competencies are needed to effectively evaluate NAMs data. Because of the unfamiliarity with such data, available information requirements are averse to data generated from NAMs.[34,38] This issue is exacerbated by the legal data requirements in place.[79]

Slow adoption of NAMs into regulation

Transitioning to NGRA and animal-free safety assessment will necessarily involve NAMs. The slow adoption of NAMs into safety assessment in the EU is, therefore, a significant barrier to the transition, since data from nonanimal methods are not valued as highly as data from animal models.[18,46,88]

Regulatory acceptance only occurs once accepted internationally

If NAMs and the data generated by them are to be considered in safety assessment, then they invariably need to be recognized and validated at an international level, such as at the OECD or GHS level first.[38] This practice constrains the ability of EU regulators to endorse NAMs and to accelerate the transition.

Theme 2: Trust and experience with NAMs

In this theme, the causes revolve around the aspect of **science**. The debate on which data to consider, the accuracy of such data, and the tools to generate them is one of the most central elements of the transition to animal-free safety assessment. There is a lack of trust in NAMs and a lack of consensus on their scientific readiness for safety assessment, which ultimately hinders the development of new NAMs and skews toward inertia and the perpetuation of animal-testing models.

Underlying causes

Scientific research standards

Some fundamental issues in scientific knowledge production are important barriers in this transition. Whether it relates to problems with the translation of animal data, the reproducibility crisis, or the overestimation of the efficacy of animal studies,[11,16,24,89–91] scientific research standards help reinforce a status quo where animal testing is considered a valid and durable option for research and safety assessment. This is further reinforced by the fact that researchers can build reputations, careers, and influence based on their expertise with animal testing.[92]

Crisis in toxicology

Toxicology is increasingly seen as a discipline in crisis, as it is struggling with large existential issues.[34,86,93] Concerns about the technological tools used, the methodological procedures, and the general problems of science production based on animal studies are calling for a revolution in how toxicology is performed.[94]

Lack of education on NAMs

Education in science is heavily biased toward animal testing rather than alternatives.[18,95–97] Without knowledge of the alternatives, greater misinformation and mistrust in NAMs arise, which further limit awareness of their effectiveness among funders, expert committees advising policymakers, and early career researchers.

Animal models as the gold standard

While the assumption that data generated from animal testing is relevant for humans is unproven,[12,98] animal testing is still considered a "gold standard" in terms of safety assessment and research.[12,27,99] Animal studies are furthermore often accompanied by poor reporting of testing protocols,[13,15] while NAMs are required to be more accurate and have their scientific validity proven. This exemplifies the social nature of the transition to animal-free safety assessment and helps explain the mistrust in NAMs and why they are not readily accepted into regulatory systems.

Direct causes

Lack of experience with NAMs

The lack of education in NAMs creates many early-career researchers who are either ignorant of or lack experience with NAMs.[25,78] It also makes both NAMs and the data generated from them unfamiliar to regulators, who are more accustomed to animal models.[38,55]

Poor reporting standards for data and protocols

Reporting standards in animal studies are poor, which, in turn, negatively affects the reproducibility of the data.[14,16,24,100] This leads to unnecessary duplication of animal studies at the cost of more accurate alternatives.

Mistrust in NAM technology

There is still a mistrust in the potential of NAMs and in their relevance for safety assessment.[12,46,101] There is also an undue burden of proof on the validity of NAMs compared to animal methods.[36] Despite some being fully validated and used, NAMs can still be considered "unready" for regulatory use by regulators,[78] which forms a major hurdle for the transition to animal-free safety assessment.

Lack of consensus on NAMs

The lack of trust in NAMs and their role in future safety assessments means that alternatives to animal testing cannot be fully embraced.[19,34,78,93,102,103] This is deeply related to the lack of experience and education with NAMs, which has shaped a safety assessment system that is skeptical of NAMs.[104,105] However, it also has to do with the lack of critical assessment of animal models.[102,105,106]

Lack of development of new NAMs

More NAMs will be needed to adequately meet the needs of a new safety assessment system.[18] However, there is great reluctance to develop new NAMs because industry stakeholders are doubtful whether they will be accepted for regulatory safety assessment, and because there are fewer opportunities for NAMs than for animal models.[78]

Theme 3: Political economy of animal testing

Further developing NAM technology is crucial to the transition to animal-free safety assessment, but regulators have generally been slow in adopting and incorporating NAMs, which disincentivizes investment in NAM technology. Industry is reluctant to develop technology that has limited application; this reinforces the economic value of the trade in animals and the data generated from animal testing. The key aspects here are markets, industry, and technology.

Underlying causes

Technological sunk costs

Sunk costs (resources and time that have been expended into a system and cannot be recovered[107]) that have already been invested in the existing safety assessment system are a significant hurdle to investing in nonanimal alternatives.[108] The demands for expertise, equipment, and technology for alternatives are more varied and vastly different from those required for animal testing. Besides the sunk costs invested in animal testing models, additional investment is still needed to switch testing regimes to accommodate NAMs.[25]

Intellectual property regime

Intellectual property (IP) laws, along with profit motives, stimulate further development of new products (chemicals in this case) and disincentivize the effective sharing of data between companies.[87] This leads to an unnecessary duplication of animal testing since data can be seen as proprietary. IP also encourages the development of more chemicals to be patented, which further necessitates more safety assessment testing.[25,32]

Lack of data and database harmonization

Easing the flow of data from previous animal testing or data on existing chemicals effectively reduces the need for additional testing and saves duplication efforts.[109,110] This is significantly hindered by the lack of easy access to datasets and databases.[21,46,111,112]

Financial incentives for animal testing status quo

Profit incentives, fueling fuel capitalist commodity production, are a deeply entrenched underlying cause of animal testing, as they incessantly create demands for more animal testing to assure the safety of new products, such as chemicals. Contract research organizations (CROs), tasked with conducting animal testing, play an important role in the transition to animal-free safety assessment, with a valuation of over US$ 10 billion for the animal testing market versus US$ 1.1 billion for the nonanimal alternatives market in 2019.[113] CROs are therefore well positioned to either continue the animal testing model or adapt and help grow the nascent nonanimal alternatives market share.

Treating animals as commodities

For the underlying cause "Treating animals as commodities" see Theme 4: Culture below.

Direct causes

Data as a commodity

Since data from animal testing has become so crucial for the development of chemicals, it has given birth to an industry whose business model is based on conducting animal testing.[33] It has even been argued that the legislation of "REACH has made test data a commodity."[33] Thus there are strong financial incentives to commodify animal testing data, reinforcing the status quo.

Lack of investment in NAMs

Since there is little regulatory demand for NAMs, there is an economic disincentive for companies to develop new NAMs[18] and a less attractive investment opportunity in companies doing so.

Lack of publicly available databases

Sharing testing data can be a powerful tool to reduce the need for safety assessments and animal testing, decrease duplication efforts, and minimize the waste of resources. The available model, where data are poorly integrated across different systems globally, not machine-readable,[21,112] or not publicly available,[46,111] makes data sharing difficult[79,114] and further drives the needless use of animal testing since it duplicates testing rather than encouraging the use of data that has already been collected.

Switching costs

"Switching costs" refers to the costs involved in changing from one product or service to another.[115] Because of the new equipment, personnel, and expertise required by NAMs, the initial cost of adopting new technologies is steep.[18] This creates a system of lock-in and path dependency in the animal testing model, which has lower upfront operating costs than a testing model that adopts new technologies.

Profit motive to sell animals for testing

It is extremely profitable to sell animals for animal testing purposes, creating a major hurdle in the transition to animal-free risk assessment.[108] The regulatory requirements that mandate animal use drive the market for animals to be used in testing.

Decreased market opportunities for NAMs

Since regulators do not always readily accept NAMs or the data generated from NAMs,[78] these methods are disincentivized in the market. This, coupled with the need for investment to develop NAMs, results in a global market for NAMs that remains vastly smaller than the market for animal testing.[33,55]

Theme 4: Culture

Underlying causes

Anthropocentric perspectives on animal-human relations

The use of animal testing is fundamentally based on the premise that nonhuman animals can be subjugated, used, and killed for human benefit. They are stripped of their agency and are not conceived of as having political rights.[71] This forms the basis for them to be seen and traded as commodities and is the foundation of a billion-dollar industry.[108] This is directly tied to the societal (macro)ontological basis, which legitimizes an anthropocentric view of nonhuman animals.[116]

Discussion and recommendations

The EU model of chemical safety assessment is in need of a paradigm shift. EU institutions and agencies can proactively contribute to accelerating the transition to assessing the safety of chemicals, increasing the quality of data being reported, and phasing out animal testing. However, the EC lacks a clear direction and specification on how it plans to fulfill its promise outlined in the European Green Deal[42] and the CSS. To be effective in this, the EC and the EU more generally must demonstrate leadership and vision by considering how they will amend the existing governing structures that steer chemical safety assessment in the EU. As a global leader in safety assessment, the EU has many avenues to effectively accelerate the transition; what is offered here is therefore a nonexhaustive list of recommendations. The above transition analysis indicates a few key areas where the EU can improve its strategy. By focusing on some of the underlying causes, such as data requirements, a number of downstream barriers will be addressed.

Recommendations

Update ECHA's dossier registration requirements to require the use of nonanimal methods if they exist and to mandate an explanation as to why they were not used. This could be complemented by updating ECHA guidance documents on NAMs[117] and ensuring easy communication between registrants and ECHA throughout the registration process[46] to avoid unnecessary testing.[88]

Provide leadership in accelerating the transition to animal-free safety assessment by setting up a steering committee consisting of a heterogeneous group of experts, including specialists in NAMs and NGRA,[88] as well as social science. This steering committee should have a clear mandate and regular checkpoints.

Fully embrace new data requirements by developing a strategy that is more flexible to multiple forms of data and is human-relevant. Multiple suggestions have been proposed: making better use of exposure science (the study of chemical, biological, and physical stressors on molecular, cellular, organic, and human receptors that potentially lead to negative effects),[118] exposomics (the study of the totality of the exposome, meaning the measure of an individual's exposure over a whole lifetime and how that relates to their health),[34,101,119,120] using human biomonitoring data (the detection of biomarkers in biological samples to detect chemical exposure),[119] integrating NAM-based toxicodynamic tests (understanding the adverse effects of chemicals and their modes of action),[38] changing information requirements to allow for the use of transcriptomics,[101,121] and transitioning to a risk-based rather than hazard-based paradigm.[122]

Provide education to scientists, regulators, and other parties involved in the development, validation, and use of safety assessment methods for regulatory needs. Assessing data from sources other than animal testing and fully utilizing new and existing competencies in other disciplines is needed to ensure a paradigm shift in safety assessment.[83]

Overhaul the existing regulatory system by creating, for example, a European Safety Sciences Institute.[82] Housing safety assessment under one European agency could coordinate the necessary measures to accelerate the transition to animal-free safety assessment and serve to increase capacity building for animal-free safety assessment in the EU. Increasing funding and the available personnel of the existing EU agencies would, otherwise, help strengthen the capacity of EU institutions to advance and accelerate the transition.

Harmonize legislation in the EU on safety assessment and animal testing to streamline and simplify it, and eliminate major hurdles and contradictions in the legislation. Particular attention should be given to the upcoming REACH revision and its alignment with the principle of using animal testing "as a last resort."

Conclusion

The transition to NGRA and an animal-free safety assessment system is desperately needed, and the EU is well poised to adopt such a system. However, to effectively accelerate the transition, efforts should focus on both the direct and underlying causes and consider the interrelatedness of barriers and opportunities.[64] The barriers to transitioning to animal-free safety assessment of chemicals in the EU are deeply entrenched in the existing regime. Some barriers are caused and influenced by others, and charting pathways to a new safety assessment system should take this into consideration. In this chapter, to help conceptualize paths to accelerate the transition to animal-free chemical safety assessment, we have categorized barriers to the transition as either underlying or direct causes. This helps to determine how an intervention or a specified policy can help address a particular barrier in the transition and whether it will be effective in doing so. As this text has argued, addressing underlying causes is more effective than focusing on direct causes in removing barriers to chemical safety assessment in the EU. Further research is needed in this field and could build on the conceptualization presented here to investigate the relationships between the barriers and thereby identify the opportunities and relationships between them.

Some underlying causes are more permeable to change than others, but all will require initiative to tackle. It is notable that many barriers lie within the regulatory system itself; this compels the European regulatory governance system to critically examine the areas where it can take concrete steps to achieve the goals it has set for itself. Addressing multiple barriers simultaneously may at times be necessary to break self-reinforcing feedback loops that impair the breakdown of the status quo. Transitioning to animal-free safety assessment is the only way to adequately face the multiple crises confronting safety assessment; human exposure to chemicals is poorly understood and relies on inadequate tools that provide poor predictions regarding the safety of chemicals. Animal testing is not fit for purpose for the task at hand and is too costly in terms of time, resources, and animals' lives. EU citizens, through the ECI, demand a change toward an ethical safety assessment system; if the EU provides such leadership, a better and more robust safety assessment system than we have today can be established, offering better protection for all and without the cost of animals' lives.

Acknowledgments

This research is part of the SAFE project (NWA.1395.20.004), supported by the Dutch Research Council.

References

[1] EU. Directive 2010/63/EU of the European Parliament and of the Council of 22 September 2010 on the Protection of Animals Used for Scientific purposesText with EEA Relevance; 2010. Available at: https://eur-lex.europa.eu/eli/dir/2010/63/oj/eng. Accessed 30 August 2025.

[2] EC. 'Communication from the Commission on the European Citizens' Initiative (ECI) 'Save Cruelty-Free Cosmetics—Commit to a Europe without Animal Testing'. Directorate-General for Internal Market, Industry. Entrepreneurship and SMEs; 2023. https://single-market-economy.ec.europa.eu/publications/communication-commission-european-citizens-initiative-eci-save-cruelty-free-cosmetics-commit-europe_en. Accessed 30 August 2025.

[3] ECI. Save cruelty free cosmetics—commit to a europe without animal testing. Text; 2024. https://citizens-initiative.europa.eu/initiatives/details/2021/000006 Accessed 30 August 2025.

[4] EP. Plans and actions to accelerate a transition to innovation without the use of animals in research, regulatory testing and education, European Parliament Resolution of 16 September 2021 on plans and actions to accelerate the transition to innovation without the use of animals in research, regulatory testing and education. P9_TA(2021)387; 2021. https://www.europarl.europa.eu/doceo/document/TA-9-2021-0387_EN.pdf.

[5] Menache A. The European citizens' stop vivisection initiative and the revision of directive 2010/63/EU. ATLA. 2016;44(4):383–390. doi:10.1177/026119291604400408.

[6] EC. Commission acts to accelerate phasing out of animal testing in response to a European Citizens' initiative. Newsroom; 2023. Available at: https://ec.europa.eu/newsroom/representations/items/797135/. Accessed 1 September 2025.

[7] EC. Summary report on the statistics on the use of animals for scientific purposes in the member states of the European Union and Norway in 2019. Commission Staff Working Document. Brussels; 2022. https://data.consilium.europa.eu/doc/document/ST-11532-2022-ADD-1/en/pdf. Accessed 30 August 2025.

[8] European Commission. Non-animal methods in science and regulation: EURL ECVAM status report 2024. Joint Research Centre. Publications Office; 2025. https://data.europa.eu/doi/10.2760/9648771. Accessed 30 August 2025.

[9] FRAME. FACT OR FICTION? Mapping perceptions of animal testing; 2020. https://replacinganimalresearch.org.uk/wp-content/uploads/2025/04/FRAME-report-v9.pdf. Accessed 30 August 2025.

[10] Alliance for Human Relevant Science. Accelerating the growth of human relevant life sciences in the United Kingdom. White Paper. UK; 2020. Available at: https://www.humanrelevantscience.org/wp-content/uploads/Accelerating-the-Growth-of-Human-Relevant-Sciences-in-the-UK_2020-final.pdf. Accessed 1 September 2025

[11] Smirnova L, Kleinstreuer N, Corvi R, et al. 3S—systematic, systemic, and systems biology and toxicology. *ALTEX*. 2018;139–162. doi:10.14573/altex.1804051.

[12] Swaters D, Van Veen A, Van Meurs W, Turner JE, Ritskes-Hoitinga M. A history of regulatory animal testing: what can we learn? *Altern Lab Anim*. 2022;50(5):322–329. doi:10.1177/02611929221118001.

[13] Henderson VC, Demko N, Hakala A, et al. A meta-analysis of threats to valid clinical inference in preclinical research of sunitinib. *eLife*. 2015;4(October):e08351. doi:10.7554/eLife.08351.

[14] Hirst JA, Howick J, Aronson JK, et al. The need for randomization in animal trials: an overview of systematic reviews. PLoS ONE. 2014;9(6):e98856. doi:10.1371/journal.pone.0098856.
[15] Kousholt BS, Præstegaard KF, Stone JC, et al. Reporting quality in preclinical animal experimental research in 2009 and 2018: a nationwide systematic investigation. PloS One. 2022;17(11):e0275962. doi:10.1371/journal.pone.0275962.
[16] Leist M, Hartung T. Inflammatory findings on species extrapolations: humans are definitely No 70 Kg mice. Arch Toxicol. 2013;87(4):563–567. doi:10.1007/s00204-013-1038-0.
[17] Fentem J. The 19th FRAME annual lecture, November 2022: safer chemicals and sustainable innovation will be achieved by regulatory use of modern safety science, not by more animal testing. Altern Lab Anim. 2023;51(2):90–101. doi:10.1177/02611929231158236.
[18] Sewell F, Alexander-White C, Brescia S, et al. New approach methodologies (NAMs): identifying and overcoming hurdles to accelerated adoption. Toxicol Res. 2024;13(2):tfae044. doi:10.1093/toxres/tfae044.
[19] Brescia S, Alexander-White C, Li H, Cayley A. Risk assessment in the 21st century: where are we heading? Toxicol Res. 2023;12(1):1–11. doi:10.1093/toxres/tfac087.
[20] Friedman KP, Gagne M, Loo LH, et al. Utility of in vitro bioactivity as a lower bound estimate of in vivo adverse effect levels and in risk-based prioritization. Toxicol Sci. 2020;173(1):202–225. doi:10.1093/toxsci/kfz201.
[21] Hartung T. Predicting toxicity of chemicals: software beats animal testing. EFSA J. 2019;17. Proceedings of the Third EFSA Scientific Conference: Science, Food and Society Guest Editors: Devos Y, Elliott KC and Hardy A. doi:10.2903/j.efsa.2019.e170710.
[22] Madorran E, Stožer A, Bevc S, Maver U. In vitro toxicity model: upgrades to bridge the gap between preclinical and clinical research. Bosnian J Basic Med Sci. 2019;20(2):157–168. doi:10.17305/bjbms.2019.4378.
[23] Cronin M, Bajot F, Enoch SJ, et al. The in chemico–in silico interface: challenges for integrating experimental and computational chemistry to identify toxicity. Altern Lab Anim. 2009;37(5):513–521. doi:10.1177/026119290903700508.
[24] Ritskes-Hoitinga M, Pound P. The role of systematic reviews in identifying the limitations of preclinical animal research, 2000–2022: part 2. J R Soc Med. 2022;115(6):231–235. doi:10.1177/01410768221100970.
[25] Abarkan FZ, Wijen AMA, van Eijden RMG, et al. Identifying key factors for accelerating the transition to animal-testing-free medical science through co-creative, interdisciplinary learning between students and teachers. Animals. 2022;12(20):2757. doi:10.3390/ani12202757.
[26] Baltazar MT, Cable S, Carmichael PL, et al. A next-generation risk assessment case study for coumarin in cosmetic products. Toxicol *Sci*. 2020;176(1):236–252. doi:10.1093/toxsci/kfaa048.
[27] Carmichael PL, Baltazar M, Cable S, et al. Ready for regulatory use: nAMs and NGRA for chemical safety assurance. ALTEX. 2022. doi:10.14573/altex.2204281.
[28] Luechtefeld T, Rowlands C, Hartung T. Big-data and machine learning to revamp computational toxicology and its use in risk assessment. Toxicol Res. 2018;7(5):732–744. doi:10.1039/c8tx00051d.
[29] Hartung T. Pyrogen testing revisited on occasion of the 25th anniversary of the whole blood monocyte activation test. ALTEX. 2021;3–19. doi:10.14573/altex.2101051.
[30] EEA. The European Environment: state and Outlook 2020: knowledge for transition to a sustainable Europe. Publications Office; 2019. https://data.europa.eu/doi/10.2800/96749. Accessed 30 August 2025.
[31] Hernández AF, Tsatsakis AM. Human exposure to chemical mixtures: challenges for the integration of toxicology with epidemiology data in risk assessment. Food Chem Toxicol. 2017;103(May):188–193. doi:10.1016/j.fct.2017.03.012.
[32] Vaughan S. EU Chemicals Regulation: New Governance, Hybridity and REACH. Edward Elgar Publishing; 2015.
[33] Meigs L. Animal testing and its alternatives—the most important omics is economics. ALTEX. 2018;275–305. doi:10.14573/altex.1807041.
[34] Hartung T, Navas-Acien A, Chiu WA Future Directions Workshop: Advancing the Next Scientific Revolution in Toxicology. Future Directions Workshop Series. Department of Defence Unites States of America; 2023. https://basicresearch.defense.gov/Portals/61/Documents/future-directions/Future%20Directions%20Workshop%20-%20Advancing%20the%20Next%20Scientific%20Revolution%20in%20Toxicology.pdf?ver=q0_CyJCAT-aj4HVv_W0a9Q%3d%3d.
[35] Fentem J, Malcomber I, Maxwell G, Westmoreland C. Upholding the EU's Commitment to 'Animal Testing as a Last Resort' under REACH requires a paradigm shift in how we assess chemical safety to close the gap between regulatory testing and modern safety science. *Altern Lab Anim.* 2021;49(4):122–132. doi:10.1177/02611929211040824.

[36] Zuang V, Dura A. Non-animal methods in science and regulation: EURL ECVAM status report 2021. Science for Policy Report by the Joint Research Centre (JRC), EUR 30960 EN. Publications Office of the European Union; 2021.
[37] Bergkamp L, Herbatschek N. Regulating chemical substances under REACH : the choice between authorization and restriction and the case of dipolar aprotic solvents. Rev Eur Comp Int Environ Law. 2014;23(2):221–245. doi:10.1111/reel.12083.
[38] Berggren E, Worth AP. Towards a future regulatory framework for chemicals in the European Union—chemicals 2.0. Reg Toxicol Pharmacol. 2023;142:105431. doi:10.1016/j.yrtph.2023.105431.
[39] EEB. The need for speed why it takes the eu a decade to control harmful chemicals and how to secure more rapid protections. EEB; 2022. Available at: https://eeb.org/library/the-need-for-speed-why-it-takes-the-eu-a-decade-to-control-harmful-chemicals-and-how-to-secure-more-rapid-protections/. Accessed 1 September 2025.
[40] EEB, and ChemTrust. Waiting for REACH: the negative impacts of delaying reform of EU chemical laws. EEB, and ChemTrust; 2023. https://eeb.org/en/library/waiting-for-reach/.
[41] EC. Chemicals strategy for sustainability towards a toxic-free environment; 2020. https://environment.ec.europa.eu/strategy/chemicals-strategy_en. Accessed 1 September 2025.
[42] EC. The European green deal. Brussels; 2019. Available at: https://eur-lex.europa.eu/legal-content/EN/TXT/HTML/?uri=LEGISSUM:4438420. Accessed 1 September 2025.
[43] EC. Communication from the commission to the European Parliament, the Council, the European Economic and Social Committee and the Committee of the Regions a European Chemicals Industry Action Plan 2025; 2025. Available at: https://eur-lex.europa.eu/legal-content/EN/TXT/PDF/?uri=COM:2025:530:FIN. Accessed 1 September 2025.
[44] EC. Regulation (EC) No 1907/2006 of the European Parliament and of the Council of 18 December 2006 Concerning the Registration, Evaluation, Authorisation and Restriction of Chemicals (REACH), Establishing a European Chemicals Agency, Amending Directive 1999/45/EC and Repealing Council Regulation (EEC) No 793/93 and Commission Regulation (EC) No 1488/94 as Well as Council Directive 76/769/EEC and Commission Directives 91/155/EEC, 93/67/EEC, 93/105/EC and 2000/21/EC. Official Journal of the European Union; 2007. Available at: https://eur-lex.europa.eu/legal-content/EN/TXT/PDF/?uri=CELEX:32006R1907. Accessed 1 September 2025.
[45] EC. Regulation (EC) No 1272/2008 of the European Parliament and of the Council of 16 December 2008 on Classification, Labelling and Packaging of Substances and Mixtures, Amending and Repealing Directives 67/548/EEC and 1999/45/EC, and Amending Regulation (EC) No 1907/2006; 2008. Available at: https://eur-lex.europa.eu/eli/reg/2008/1272/oj/eng. Accessed 1 September 2025.
[46] Macmillan DS, Bergqvist A, Burgess-Allen E, et al. The last resort requirement under REACH: from principle to practice. Reg Toxicol Pharmacol (San Diego). 2024;147(February):105557. doi:10.1016/j.yrtph.2023.105557.
[47] Knight J, Hartung T, Rovida C. 4.2 million and counting … the animal toll for REACH systemic toxicity studies. *ALTEX*. 2023. doi:10.14573/altex.2303201.
[48] Rovida C. REACH out-numbered! The future of REACH and animal numbers. ALTEX. 2023. doi:10.14573/altex.2307121.
[49] Hartung T. Food for thought … on alternative methods for chemical safety testing. ALTEX. 2010;3–14. doi:10.14573/altex.2010.1.3.
[50] ECHA. ECHA checked over 20% of REACH registration dossier; 2024. https://echa.europa.eu/-/echa-checked-over-20-of-reach-registration-dossiers-for-compliance-1.
[51] Desel H, Begeman K, de Groot R, Brekelmans P. Health hazard classification and labelling. In: Reichl F-X, Schwenk M eds. Regulatory Toxicology. Springer International Publishing; 2021:943–951. doi:10.1007/978-3-030-57499-4.
[52] ECHA. Factsheet; role of animal testing in ensuring the safe use of chemical substances; n.d. https://echa.europa.eu/documents/10162/13630/reach_factsheet_animal_testing_en.pdf/fb82d34b-7e58-4446-a3fb-f87b242d6829. Accessed 25 July 2024.
[53] EC. Regulation (EC) No 1223/2009 of the European Parliament and of the Council of 30 November 2009 on Cosmetic Products; 2009. Available at: https://eur-lex.europa.eu/eli/reg/2009/1223/oj/eng. Accessed 1 September 2025.
[54] Knight J, Rovida C, Kreiling R, et al. Continuing animal tests on cosmetic ingredients for REACH in the EU. ALTEX. 2021;38(4):653–668. doi:10.14573/altex.2104221.

[55] Pistollato F, Carpi D, Mendoza-de Gyves E, et al. Combining in vitro assays and mathematical modelling to study developmental neurotoxicity induced by chemical mixtures. Reproduct Toxicol. 2021;105(October):101–119. doi:10.1016/j.reprotox.2021.08.007.

[56] Court of Justice of the EU. Case T-655/20; 2023. https://curia.europa.eu/jcms/jcms/j_6/en/.

[57] EC. Save cruelty free-cosmetics: commit to a Europe without Animal Testing. Publications Office; 2023. https://op.europa.eu/en/publication-detail/-/publication/e7dd1cfd-3a4b-11ee-bbad-01aa75ed71a1/language-en. Accessed 1 September 2025.

[58] EC. Regulation of the European Parliament and of the Council on the European Chemicals Agency and Amending Regulations (EC) No 1907/2006, (EU) No 528/2012, (EU) No 649/2012 and (EU) 2019/1021; 2025. https://eur-lex.europa.eu/legal-content/EN/TXT/?uri=celex:52025PC0386. Accessed 1 September 2025.

[59] Eurogroup for Animals. A turning point for animal testing phase-out in Europe: EU reaches final deal on OSOA legislation. Eurogroup for Animals; n.d. https://www.eurogroupforanimals.org/news/turning-point-animal-testing-phase-out-europe-eu-reaches-final-deal-osoa-legislation. Accessed 7 August 2025.

[60] Feola G. Societal transformation in response to global environmental change: a review of emerging concepts. Ambio. 2015;44(5):376–390. doi:10.1007/s13280-014-0582-z.

[61] Linner BO, Wibeck V. Conceptualising variations in societal transformations towards sustainability. Environ Sci Policy. 2020;106:221–227. WOS:000521508500025. doi:10.1016/j.envsci.2020.01.007.

[62] Linnér BO, Wibeck V. *Sustainability Transformations: Agents and Drivers across Societies*. Cambridge University Press; 2019. doi:10.1017/9781108766975. Available at: https://files.ipbes.net/ipbes-web-prod-public-files/inline/files/ipbes_global_assessment_report_summary_for_policymakers.pdf. Accessed Aug 15, 2026.

[63] IPBES. Global assessment report on biodiversity and ecosystem services of the intergovernmental science-policy platform on biodiversity and ecosystem services. Zenodo; 2019. doi:10.5281/ZENODO.3831673.

[64] Visseren-Hamakers IJ, Cashore B, Loorbach D, et al. How to Save a Million Species? Transformative Governance through Prioritization. In: Visseren-Hamakers IJ, Kok MTJ eds.. Transforming Biodiversity Governance. 1st ed. Cambridge University Press; 2022:67–89. doi:10.1017/9781108856348.005.

[65] IPBES. Summary for Policymakers of the Global Assessment Report on Biodiversity and Ecosystem Services of the Intergovernmental Science-Policy Platform on Biodiversity and Ecosystem Services. IPBES Secretariat; 2019. doi:10.5281/zenodo.3553579.

[66] Geels FW, Schot J. Typology of sociotechnical transition pathways. Res Policy. 2007;36(3):399–417. doi:10.1016/j.respol.2007.01.003.

[67] NCad. Transition to non-animal research on opportunities for the phasing out of animal procedures and the stimulation of innovation without laboratory animals. No. 201609EN. Netherlands National Committee for the Protection of Animals Used for Scientific Purposes. The Hague; 2016. https://english.ncadierproevenbeleid.nl/.

[68] RIVM. Roadmap for Animal-Free Innovations in Regulatory Safety Assessment. National Institute For Public Health and the Environment; 2022. https://www.rivm.nl/sites/default/files/2022-11/Roadmap%20animal-free%20innovations%20in%20regulatory%20safety%20assessment.pdf.

[69] Díaz S, Demissew S, Carabias J, et al. The IPBES conceptual framework—connecting nature and people. Curr Opin Environ Sustain. 2015;14:1–16. doi:10.1016/j.cosust.2014.11.002.

[70] Haraway D. When Species Meet. Nachdr. Posthumanities 3. University of Minnesota Press; 2008.

[71] Kymlicka W, Donaldson S. Locating animals in political philosophy. Philos Compass. 2016;11(11):692–701. doi:10.1111/phc3.12365.

[72] Booth A, Papaioannou D. Systematic Approaches to a Successful Literature Review. SAGE; 2012.

[73] Latour B, Woolgar S. Laboratory Life: The Construction of Scientific Facts. 1. Princeton Paperback Print. Princeton Paperbacks Science, Sociology. Princeton University Press; 1986.

[74] Jasanoff S. The Fifth Branch_Science Advisers as Policymakers. Harvard University Press; 1998.

[75] Demortain D. Experts and the production of international policy knowledge do epistemic communities do the job? In: Littoz-Monnet A ed . *Politics of Expertise in International Organizations: How International Bureaucracies Produce and Mobilize Knowledge*. Routledge; 2017:76–92. WOS:000449211900005. Available at: https://hal.science/hal-03124858/document. Accessed 1 September 2025.

[76] Fischer F, Torgerson D, Durnová A, Orsini M eds. Handbook of Critical Policy Studies. Edward Elgar Publishing; 2015. doi:10.4337/9781783472352.

[77] Herrmann K, Jayne K, eds. Animal Experimentation: Working Towards a Paradigm Change. BRILL; 2019. doi:10.1163/9789004391192.

[78] Čavoški A, Holden L, Lee R 26/01 D6.1 report on socio-technical barriers. PrecisionTox; 2023. Available at: https://precisiontox.org/wp-content/uploads/2024/02/D6.1-Report-on-Socio-Technical-Barriers-26Jan.pdf. Accessed 1 September 2025

[79] Stoykova K. Towards non-animal testing in European regulatory toxicology: an introduction to the REACH framework and challenges in implementing the 3Rs. Eur J Risk Regulation 2025;16:985–1016. doi:10.1017/err.2025.1.

[80] van der Zalm AJ, Barroso J, Browne P, et al. A framework for establishing scientific confidence in new approach methodologies. Arch Toxicol. 2022;96(11):2865–2879. doi:10.1007/s00204-022-03365-4.

[81] Burden N, Clift MJD, Jenkins GJS, Labram B, Sewell F. Opportunities and challenges for integrating new in vitro methodologies in hazard testing and risk assessment. Small. 2021;17(15):2006298. doi:10.1002/smll.202006298.

[82] Busquet F, Hartung T. The need for strategic development of safety sciences. ALTEX. 2017;34(1):3–21. doi:10.14573/altex.1701031.

[83] European Commission. Addressing evidence needs in chemicals policy and regulation. Joint Research Centre. Publications Office; 2022. https://data.europa.eu/doi/10.2760/9130.

[84] Pereira L, Asrar G, Fisher LH, et al. Bottom-up initiatives and participatory approaches for outlooks. In: Ekins P, Gupta J, Boileau P eds . Global Environment Outlook (GEO-6): Healthy Planet, Healthy People. Cambridge University Press; 2024:544–579. doi:10.1017/9781108627146.

[85] Hoogstraaten MJ, Vriend J, de Leeuw VC, et al. Animal-free safety assessment of chemicals: an innovation system perspective. Arch Toxicol. 2015;99(1):43–56. doi:10.1007/s00204-024-03878-0.

[86] Demortain D. Regulatory toxicology in controversy. Sci Technol Hum Values. 2013;38(6):727–748. doi:10.1177/0162243913490201.

[87] Schiffelers MJ. Animal Testing, 3R Models and Regulatory Acceptance: Technology Transition in a Risk-Averse Context. Utrecht University; 2016. https://dspace.library.uu.nl/handle/1874/334103.

[88] Pereira M, Macmillan DS, Willett C, Seidle T. REACHing for solutions: essential revisions to the EU chemicals regulation to modernise safety assessment. Reg Toxicol Pharmacol. 2022;136(December):105278. doi:10.1016/j.yrtph.2022.105278.

[89] Begley CG, Ioannidis JPA. Reproducibility in science: improving the standard for basic and preclinical research. Circ Res. 2015;116(1):116–126. doi:10.1161/CIRCRESAHA.114.303819.

[90] Sena ES, van der Worp HB, Bath PMW, Howells DW, Macleod MR. Publication bias in reports of animal stroke studies leads to major overstatement of efficacy. PLoS Biol. 2010;8(3):e1000344. doi:10.1371/journal.pbio.1000344.

[91] Tsilidis KK, Panagiotou OA, Sena ES, et al. Evaluation of excess significance bias in animal studies of neurological diseases. PLoS Biol. 2013;11(7):e1001609. doi:10.1371/journal.pbio.1001609.

[92] Pound P. Rat Trap: The Capture of Medicine by Animal Research—and How to Break Free. MATADOR; 2023.

[93] Hartung T, Tsatsakis AM. The state of the scientific revolution in toxicology. ALTEX. 2021. doi:10.14573/altex.2106101.

[94] Hartung T. Evolution of toxicological science: the need for change. Int J Risk Assess Manag. 2017;20(1/2/3):21. doi:10.1504/IJRAM.2017.082570.

[95] Daneshian M. A framework program for the teaching of alternative methods (replacement, reduction, refinement) to animal experimentation. ALTEX. 2011;28(4):341–352. doi:10.14573/altex.2011.4.341.

[96] Hartung T, Blaauboer BJ, Leist M. Food for thought … on education in alternative methods in toxicology. ALTEX. 2009;255–263. doi:10.14573/altex.2009.4.255.

[97] Krul C, De Moor A, Stegmeijer K, et al. Beyond animal testing index: benchmarking tool for a world beyond animal testing. *ALTEX*. 2023. doi:10.14573/altex.2304161.

[98] Greek R, Pippus A, Hansen LA. The Nuremberg Code subverts human health and safety by requiring animal modeling. BMC Med Ethics. 2012;13(1):16. doi:10.1186/1472-6939-13-16.

[99] Ritskes-Hoitinga M, Leenaars C, Beumer W, et al. Improving translation by identifying evidence for more human-relevant preclinical strategies. Animals. 2020;10(7):1170. doi:10.3390/ani10071170.

[100] Bracken MB. Why animal studies are often poor predictors of human reactions to exposure. J R Soc Med. 2009;102(3):120–122. doi:10.1258/jrsm.2008.08k033.

[101] Dent MP, Vaillancourt C, Thomas RS, et al. Paving the way for application of next generation risk assessment to safety decision-making for cosmetic ingredients. Reg Toxicol Pharmacol. 2021;125:105026. doi:10.1016/j.yrtph.2021.105026.

[102] Hartung T. Opinion versus evidence for the need to move away from animal testing. ALTEX. 2017;193–200. doi:10.14573/altex.1703291.

[103] Krewski D, Acosta D, Andersen M, et al. Toxicity testing in the 21st century: a vision and a strategy. J Toxicol Environ Health B. 2010;13(2–4):51–138. doi:10.1080/10937404.2010.483176.

[104] Von Aulock S, Busquet F, Locke P, Herrmann K, Hartung T. Engagement of scientist with the public and policymakers to promote alternative methods. ALTEX. 2022;543–559. doi:10.14573/altex.2209261.

[105] Hartung T. Food for thought … on animal tests. ALTEX. 2008;3–9. doi:10.14573/altex.2008.1.3.

[106] Hartung T. Look back in anger—what clinical studies tell us about preclinical work. ALTEX. 2013;30(3):275–291. doi:10.14573/altex.2013.3.275.

[107] Rogers A, Castree N, Kitchin R. . In: A Dictionary of Human Geography. Oxford University Press; 2013. https://www.oxfordreference.com/display/10.1093/acref/9780199599868.001.0001/acref-9780199599868-e-1820. Accessed 07 May 2026.

[108] Bottini A, Hartung T. Food for thought … on the economics of animal testing. ALTEX. 2009;26:3–16. doi:10.14573/altex.2009.1.3.

[109] Hartung T. Making big sense from big data in toxicology by read-across. ALTEX. 2016;83–93. doi:10.14573/altex.1603091.

[110] Luechtefeld T, Hartung T. Computational approaches to chemical hazard assessment. ALTEX. 2017;459–478. doi:10.14573/altex.1710141.

[111] Luechtefeld T, Maertens A, Russo DP, et al. Global analysis of publicly available safety data for 9,801 substances registered under REACH from 2008-2014. ALTEX. 2016;33(2):95–109. doi:10.14573/altex.1510052.

[112] Luechtefeld T, Marsh D, Rowlands C, Hartung T. Machine learning of toxicological big data enables read-across structure activity relationships (RASAR) outperforming animal test reproducibility. Toxicol Sci. 2018;165(1):198–212. doi:10.1093/toxsci/kfy152.

[113] The Business Research Company. Non-animal alternative testing market size and share 2024–2032. 2023. https://www.gminsights.com/industry-analysis/non-animal-alternative-testing-market.

[114] Rectanus M, Peters D. Global data sharing: requirements from chemical regulation. Int Chem Regulatory Law Rev. 2019;(2):61–64. doi:10.21552/icrl/2019/2/6.

[115] Law J, Switching cost. In: Law J ed. A Dictionary of Business and Management. Oxford University Press; 2016. https://www.oxfordreference.com/display/10.1093/acref/9780199684984.001.0001/acref-9780199684984-e-6258.

[116] Lopez PJ, Gillespie K eds.. Economies of death: economic logics of killable life and grievable death. In: Routledge Frontiers of Political Economy 199. Routledge, Taylor & Francis Group; 2015.

[117] Westmoreland C, Bender HJ, Doe JE, et al. Use of new approach methodologies (NAMs) in regulatory decisions for chemical safety: report from an EPAA deep dive workshop. Regul Toxicol Pharmacol. 2022;135:105261. doi:10.1016/j.yrtph.2022.105261.

[118] de Bruin Y B, Bridges J, Coggins M, et al. Assessment of Needs for a European Strategy on Exposure Science: summary Report of the First European Exposure Science Strategy Workshop: building a Roadmap 2020–2030. ISES Europe; 2018. https://www.academia.edu/87740322/Assessment_of_needs_for_a_European_strategy_on_exposure_science_summary_report_of_the_first_European_Exposure_Science_Strategy_Workshop_Building_a_roadmap_2020_2030_. Accessed 1 September 2025.

[119] Jeddi MZ, Hopf NB, Louro H, et al. Developing human biomonitoring as a 21st century toolbox within the European exposure science strategy 2020–2030. Environ Int. 2022;168:107476. doi:10.1016/j.envint.2022.107476.

[120] Wan M, Simonin EM, Johnson MM, et al. Exposomics: a review of methodologies, applications, and future directions in molecular medicine. EMBO Molecular Med. 2025;17(4):599–608. doi:10.1038/s44321-025-00191-w.

[121] Transcriptomics—an Overview. ScienceDirect Topics; n.d. https://www.sciencedirect.com/topics/biochemistry-genetics-and-molecular-biology/transcriptomics. Accessed 25 April 2025.

[122] Alexander-White C. In NAMs we trust—an innovative paradigm-shift in risk-based chemicals management for globally harmonized protection goals. Expert Opin Drug Metab Toxicol. 2024;20(7):545–547. doi:10.1080/17425255.2024.2353765.

Index

Note: Page numbers followed by "*f*" indicate figures, "*t*" indicate tables.

N

O

P

9780443265143